World Regional Geography Book Series

Series Editor

E. F. J. De Mulder, DANS, NARCIS, Utrecht, Noord-Holland, The Netherlands

What does Finland mean to a Finn, Sichuan to a Sichuanian, and California to a Californian? How are physical and human geographical factors reflected in their present-day inhabitants? And how are these factors interrelated? How does history, culture, socio-economy, language and demography impact and characterize and identify an average person in such regions today? How does that determine her or his well-being, behaviour, ambitions and perspectives for the future? These are the type of questions that are central to The World Regional Geography Book Series, where physically and socially coherent regions are being characterized by their roots and future perspectives described through a wide variety of scientific disciplines. The Book Series presents a dynamic overall and in-depth picture of specific regions and their people. In times of globalization renewed interest emerges for the region as an entity, its people, its landscapes and their roots. Books in this Series will also provide insight in how people from different regions in the world will anticipate on and adapt to global challenges as climate change and to supra-regional mitigation measures. This, in turn, will contribute to the ambitions of the International Year of Global Understanding to link the local with the global, to be proclaimed by the United Nations as a UN-Year for 2016, as initiated by the International Geographical Union. Submissions to the Book Series are also invited on the theme 'The Geography of…', with a relevant subtitle of the authors/editors choice. Proposals for the series will be considered by the Series Editor and International Editorial Board. An author/editor questionnaire and instructions for authors can be obtained from the Publisher.

This book series is published in cooperation with the International Geographical Union (IGU). The IGU is an international, non-governmental, professional organization devoted to the development of the discipline of Geography. The purposes of the IGU are primarily to promote Geography through initiating and coordinating geographical research and teaching in all countries of the world.

Lola Gulyamova

The Geography of Uzbekistan

At the Crossroads of the Silk Road

Springer

Lola Gulyamova
Geological Prospecting and Mining Metallurgy Faculty
Tashkent State Technical University (named after Islam Karimov)
Tashkent, Uzbekistan

ISSN 2363-9083 ISSN 2363-9091 (electronic)
World Regional Geography Book Series
ISBN 978-3-031-07872-9 ISBN 978-3-031-07873-6 (eBook)
https://doi.org/10.1007/978-3-031-07873-6

Picture Credit: Anton Balazh - Fotolia.com

This Springer imprint is published by the registered company Springer Nature Switzerland AG
The registered company address is: Gewerbestrasse 11, 6330 Cham, Switzerland

To my parents, Rabija Axmedova and Khadji Akbar Gulyamov,
who were devoted to this country.

Preface

In 1996–1997, during my Fellowship under the Fulbright Program in the USA, my colleagues called me "the Russian professor." Only a few knew about Uzbekistan, and usually geographers asked me to show its location on a map relative to China and Pakistan. The country was locked not just geographically but also informationally. Today, one may find information about this "terra incognita" in the news, on the internet, and in academic journals.

The main purpose of writing this book is to bring together different views, opinions, and insights into the current state of the country and its prospects for the future.

It is not easy to describe in one book the historical roots and geographical features of Uzbekistan. It is a unique country in the Central Asian region because of its central location at the crossroads of the Great Silk Route. Today, this route is vital for strengthening political, economic, and cultural ties between East and West through the lands of Uzbekistan.

This book, which is based on a literature review of publications by local and foreign authors, covers just part of a big issue in a rapidly changing world and discusses Uzbekistan from geographic, cultural, and historical perspectives.

Tashkent, Uzbekistan Lola Gulyamova

Acknowledgments

I thank Dr. Arthur Guljamow, Mrs. Dilyora Rahimova, and Mr. Behruz Habibullaev for their technical assistance during the preparation of this book. My deep appreciation to late Mr. Shavkat Boltaev from Bukhara and Mrs. Magfirat Mumnova for their permission to use images, the State Committee for Land Resources, Geodesy, Cartography and State Cadastre of the Republic of Uzbekistan for the permission to publish maps from the Atlas Physical Geography of Central Asia and Uzbekistan, the Geological Survey of America, the Society of Economic Geologists, and NASA Earth Observatory for the permission to share their materials.

Contents

Introduction

1

Abstract

This part of the book discusses the centrality of the location of Uzbekistan and its significant geographic features. This chapter provides an understanding of Uzbekistan's importance in the world and its regional history and development. It also shows how Uzbekistan's geographic location in the center of the Eurasian continent determines the natural conditions of the country. The history of the country is briefly reviewed from the political, economic, social, cultural, and geographic points of view. Various patterns and processes in the region and recent geopolitical developments are also discussed.

Keywords

Uzbekistan · Location · Nature · History

1.1 The Gem of Central Asia

Why is Uzbekistan considered the gem of Central Asia? What makes this country unique? What are the influences on the development of this country? How has its location on the ancient Great Silk Route (also known as the Great Silk Road) helped people to share knowledge and ideas?

The unique combination of its natural resources, the heritage of its scientific achievements, its rich culture, and its history are the peculiarities of the geography of this country.

To visit Uzbekistan is to travel in time when you observe ancient monuments in Bukhara (Buxoro in Uzbek), Khiva (Xiva), Samarkand (Samarqand in Uzbek), or Shakhrisabz; or to travel in space while hiking in the mountains of the Western Tien Shan or Gissar (also known as Hissor), or while wandering in the Kyzylkum Desert or on the dried seabed of the Aral Sea. All of these are in one single place with its vast natural resources, traditions, and beliefs that combine pre-Islamic and Islamic ones, as well as Soviet ones. Today, this country is in economic, social, and cultural transition. This transition is just one experiment in a range of changes that the country has faced over time.

1.1.1 Contributions to Science

People who have lived in this land have developed many innovations in science and engineering, which were spread further via the Great Silk Route. The Silk Route served as a significant medium for sharing knowledge. Merchants brought not only silk, incense, and jewels but also books. Scientists in the West and the East welcomed this trade. Many scientists in Central Asia used the results of the scientific works of their Eastern and Western colleagues, and their contributions to science were based on ideas and knowledge from their Greek, Roman, Chinese, and Indian scientific ancestors.

The well-known concept of the "algorithm"[1]—a procedure for solving a mathematical problem in a finite number of steps that frequently involves repetition of an operation—was developed by Muhammad Ibn Musa al-Khwarizmi (Fig. 1.1).

His contributions to science include a long list of innovations. The Arabic numerals were introduced to the Western world in his works on arithmetic; he suggested the algorism technique for performing arithmetic. The term *algorithm* is derived from the Latinized form of his name, al-Khwarizmi. He was born in 780 in Khorezm (also known as Xorazm, Khorezm, Khwarazmia,[2] and Chorasmia) and spent many years as a scholar at the Mamun Academy in Baghdad, Iraq. He wrote a mathematical book with examples and applications in a wide range of disciplines. The term *algebra* is also

[1]Definition listed in the Merriam Webster Dictionary, 2017 (https://www.merriam-webster.com/dictionary/algorithm).

[2]Xorazm is a large oasis region at the Amu Darya delta, bordered in the north by the Aral Sea, in the east by the Kyzylkum Desert, in the south by the Karakum Desert, and in the west by the Ustyurt Plateau.

L. Gulyamova, *The Geography of Uzbekistan*, World Regional Geography Book Series,
https://doi.org/10.1007/978-3-031-07873-6_1

Fig. 1.1 The statue of al-Khwarzimi in Urgench, Khorezm (Xorazm) region

Fig. 1.2 The statue of Abu Rayhan Muhammad Ibn Ahmad Biruni near the metro station of the same name in Tashkent

derived from the name of his book (*al-jabr* means one of the operations to solve a quadratic equation), which was used as the principal mathematical textbook in European universities until the sixteenth century. The spread of his works in Europe had a profound impact on the advance of mathematics there.

Alfaraganus (Ahmad Ibn Muhammad Ibn Kathir al-Farghani) was a prominent astronomer, and a lunar crater is named after him. He was born in 800 and died in 870. His work titled "Elements of Astronomy on the Celestial Motions" was translated into Latin in the twelfth century and widely used in Europe. His estimates of the Earth's circumference were used by Christopher Columbus for his voyages to America. His idea of the nilometer[3] on the island of al-Rawda in Old Cairo is still implemented in hydrology.

Alpharabius (Abu Nasr Muhammad Ibn Muhammad Farabi) was born in 872 in Farab and died in 950 in Damascus. The dearth of information about his life and family have created speculation regarding his ethnicity and place of birth. He was honored with the title "the Second Master" after Aristotle in the Arabic philosophical tradition. Through his commentaries and treaties, he influenced many prominent philosophers such as Ibn Sina. Farabi's synthesis of philosophy and Sufism paved the way for Avicennism.[4] His works were well known in the East and the West. His contributions to logic, music, philosophy, psychology, and education influenced the advance of knowledge. His ideas are still subjects of discussion with regard to the philosophical principles of music and its cosmic qualities, as well as the therapeutic effects of music on the soul.

Another famous son of the Great Khorezm[5] (Xorazm) was Abu Rayhan Muhammad Ibn Ahmad Biruni; he was born in 973 in Kath (a city in modern Karakalpakstan) and died in 1048 (Fig. 1.2).

He is considered the father of anthropology, geodesy, and experimental psychology. His list of contributions to science is also long and includes innovations in many fields. The technique used to measure the circumference of the Earth on the basis of trigonometry is one of them. He made significant contributions in the fields of cartography, geodesy, geography, geology, and mineralogy. His estimation of the Earth's radius was only 16.8 km less than the modern measurement. In recognition of his immense contributions to geography, Scientist from Uzbekistan H. Khasanov called the eleventh and twelfth centuries the "Central Asian geography of Biruni" (Khasanov 1981).

Avicenna (Abu Ali Ibn Sina), was also a polymath. He is regarded as one of the most significant physicians, astronomers, and great thinkers, and as the father of early modern medicine (Fig. 1.3).

[3]A structure for measuring the Nile River's clarity and water level during the flood season.

[4]Islamic Avicennism (established by Ibn Sina)—and, later, Latin Avicennism—is a school of philosophy attempting to describe the structure of reality with a rational system of thinking.

[5]All geographic names are given in the Uzbek language, and the corresponding names in other languages are shown in parentheses.

Fig. 1.3 The statue of Avicenna (Abu Ali Ibn Sina) in front of the Republican Specialized Center of Surgery in Tashkent

His famous book *The Canon of Medicine* was used as a standard medical text at many medieval universities until the end of the seventeenth century. In addition, his works on psychology and the theory of knowledge influenced William of Auvergne and Albertus Magnus, and his metaphysics influenced the thought of Thomas Aquinas. Ibn Sina was born in 980 in Afshana Village near Bukhara and died in 1037 in Hamadan, Iran. Dante Alighieri illustrated him in his "Divine Comedy" in Limbo alongside Virgil, Homer, Horace, Socrates, and Plato. This demonstrates that Ibn Sina was recognized by both East and West. Many researchers at different times have described Ibn Sina as one of the most famous people among all races, places, and times, and he has been referred to as the "prince of physicians" (*princeps medicorum* in Latin).

Mir Alisher Navoyi (Ali-Shir Navai) was born in 1441 in Herat (which is now a city in modern Afghanistan) and buried there in 1501 (Figs. 1.4 and 1.5). He was the first Turkic poet, writer, and linguist, and the greatest contributor to the Chagatai literature. As a musician, composer, painter, and sculptor, he was respected in both the East and the West. Many places and institutions in Uzbekistan, as well as in other parts of Central Asia, are named after him.

In his works, he suggested that Turkic languages had greater richness, precision, and malleability of vocabulary than Persian. This was an uncommon view at the time of the prevailing Persian traditions in art and literature. Navoyi is considered the founder of early Turkic literature. He was one of the key writers who revolutionized literary use of Turkic languages. His four poetry collections (diwans) include total roughly 50,000 verses and were created during different periods of his personal life. His collection of over 450 biographical sketches of contemporary poets is a good source of information about the culture of that period. His Quintuple (*Khamsa*) is composed of five epic poems inspired by Nizami Ganjavi's *Khamsa*, with fine mastership.

Fig. 1.4 The statue of Alisher Navoyi in front of the Alisher Navoyi Museum on Navoyi Street near Alisher Navoyi Metro Station in Tashkent

Fig. 1.5 The National Library of Uzbekistan, named after Alisher Navoyi, in Tashkent

As a public administrator, he contributed to the architecture of Herat, including construction of madrassas,[6] libraries, hospitals, caravanserais,[7] and educational, spiritual, and charitable institutions in Khorasan.

[6]Any type of educational institution (in Arabic).

[7]A roadside inn where travelers (caravaners) rested and recovered. They were an important element of the network of transportation routes for commerce, information, and people along the Silk Route.

Fig. 1.6 The statue of Mirza Ulugh Beg (Ulugh Beigh) in the Mirza Ulugh Beg district, Tashkent

Mirza Muhammad Taraghay Bin Shahrukh (Fig. 1.6), widely known as Ulugh Beg, was born in 1394 in Soltaniyeh, Persia. He was beheaded by the order of his own eldest son in 1449 on his way to Mecca and was buried in the mausoleum of Timur in Samarkand.

His measurement of the tropical year as consisting of 365 days, 5 h, 49 min, and 15 s has an error of just +25 s. It is more accurate than the estimate of Copernicus, which has an error of +30 s. His determinations of the length of the sidereal year, the Earth's axial tilt, and the magnitude of 1° of longitude—and, on this basis, the length of the Earth's equator—remained the most accurate such measurements for 100 years. In 1428, he built an observatory in Samarkand, which is considered by scholars to have been one of the finest and largest in Central Asia. The sextant there (Fig. 1.7) had a radius of about 36 m. He compiled the most original star catalogue in the Middle East. His name is listed among those of the great medieval astronomers of the West and the East. The Moon crater Ulugh Beigh and the main-belt asteroid 2439 Ulugbek are named after him. The Ulugh Beg Madrassas in Samarkand and Bukhara were built by him, and he transformed these cities into cultural centers of learning in Central Asia by establishing an academy of sciences in Samarkand. His books *Tarihiarba'ulus* and *ZijiKo'rag'oniy* describe the political life, culture, and economy of Genghis Khan's empire, as well as the results of his own astronomical observations.

Fig. 1.7 The sextant in the Ulugh Beg Observatory in Samarkand

1.2 Central Asia and Uzbekistan in the Works of Ancient Authors

The first information about the natural world, the population, and the culture of Central Asia is contained in an ancient book, *Zend Avesta*, written between the twelfth and eighth centuries BCE[8] by authors living in Central Asia (presumably in Bactria[9]) (Fig. 1.8). This book contained the most important texts of Zoroastrianism, the world's oldest religion, which spread widely in Central Asia, Iran, and the Caucasus. It emphasizes protection of nature and is frequently called an ecological religion. In this book, the authors wrote about Sogdiana[10], Bactria (Balkh), Margiana[11] (Merv), Herat (Haravia), Khorasan,[12] and the city of Samarkand.

In many other historical sources, ancient Khorezm (Xorazm), Sogdiana, and Bactria are described as rich states with diverse natural environments, economies, and cultures.

[8]In dates, "BCE" means "before the Common Era" and "CE" means "of the Common Era", based on the year in which Jesus is believed to have been born.

[9]Bactria, or Bactriana, was the name of a historical region in the north of the Hindu Kush mountain range and south of the Amu Darya river, west of the Pamirs and south of the Tian Shan, with the Amu Darya flowing westward through the center.

[10]Sogdiana, or Sogdia, was an ancient Iranian civilization between the Amu Darya and Syr Darya rivers, which, at different times, inhabited territory located in present-day Uzbekistan (Samarkand and Bukhara) and Tajikistan.

[11]Margiana is a historical region, bordered by Sogdia in the north and Bactria in the east.

[12]Khorasan is a historical region that lies mostly in modern-day Uzbekistan and northern/northwestern Afghanistan, northeastern Iran, and southern Turkmenistan.

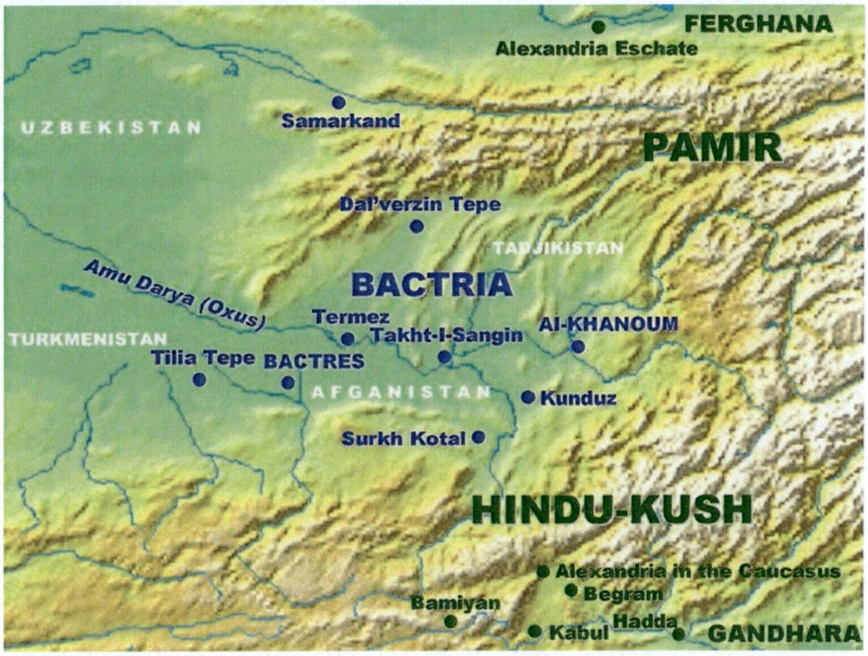

Fig. 1.8 A map of the Bactria region. [From World Imaging (public domain) via Wikimedia Commons]

In their books, the ancient Greek historians Herodotus (484–425 BCE), who wrote *The Histories*, and Claudius Ptolemy (100–170 CE) stressed the importance of nature and described the populations of the plain parts of the region, as well as the caravan routes and major cities.

In the ninth and eighth centuries BCE, human settlements were widely spread in the Khorezm (Xorazm) region and in the Zarafshan and Surkhandarya (Surxondaryo) Valleys. In the middle of the sixth century BCE, by the time of the emergence of the Achaemenid, Bactria (Balkh) and Samarkand were well developed as large cities of the state. Agriculture on arable lands was stimulated by rich natural resources in the forms of water resources, good soils, and valley plains, which were convenient for farming.

In 138–126 BCE, the Chinese historian Chjan-Zyan traveled in Central Asia and intensively studied the natural environments, populations, and economies in places around Issyk Kul Lake, the Fergana Valley, and Khorezm (Xorazm). All ancient authors noted that the well-developed cities and states were economically strong and culturally diverse with good governance.

Knowledge of these lands was increased by the scientists Alfraganus (Ahmad Farg'oniy) in the ninth century; Abu Zaid Balhiy, Jayhoniy, and Abu Abdulla Khorazmiy in the tenth century; Mahmud Kashgariy (Qoshg'ariy) in the eleventh century; Hofizi Abru and Abdurazzak Samarkandiy in the fifteenth century; and Zahir ud-Din Muhammad Babur (Zahriddin Muhammad Bobur), Mirza Muhammad Haidar Dughlat beg, Hafiz Muhammad Bukhari (Hofiz Buxoriy), and Sultan Balkhi (Sulton Balxiy) in the sixteenth century. In their books, they wrote about economics, trade, agriculture, and struggles for natural resources among different tribes and states.

In 629 CE, the Chinese traveler Suan Tzan wrote about the social and economic lives of citizens in the cities of Tashkent, Samarkand, and Bukhara. Many scientists of that time devoted their work to refining their mapping tools, and depicted and defined the coordinates of states, lands, cities, mountains, seas, oceans, rivers, and lakes.

One of the famous resources concerning the geography of these lands is *Devonu-lug'at'it*, written by Mahmud Qoshg'ariy. He wrote about the geography and population of Central Asia and neighboring countries, and the origin of their geographic names. He also provided a map of the known lands. The Uzbekistani geographers R. Rahimbekov and Z. Dontsova (1982a, b) noted that the location of Central Asia in Eurasia, the main geographic objects, the mountains and ridges, and the valleys were shown in good detail on that early map.

Besides local scientists and researchers, various Arabic and Iranian travelers—Ibn Hurdodbeh, Ibn Rusta, Al Ma'sudiy, Ibn Fazlon, and Ibn Battuta—and the Europeans Plano Karpini, Wilhelm Rubruks, and Marco Polo wrote interesting stories about the natural features, people, economies, and cultures of sites they visited. During his travel to India, Abdurazzoq Samarkandiy (1413–1482) wrote about the Zarafshan Valley and the Kyzylkum Desert. Another great governor and scientist, Zahir ud-Din Muhammad Babur (1488–1530), was the author of such remarkable geographic sketches as Babur-Nama (*Boburnoma*). Khasanov (1983) stressed the importance of this 3-part book, each of which gives a very detailed review of the natural environments, cultures, economies, populations, and governance of different states of the Timurid Empire.

1.3 Location

The location of Uzbekistan is unique because of its central position in the center of Eurasia at the crossroads of all routes from the west to the east and from the north to the south. Uzbekistan is situated in the northern hemisphere between the Amu Darya (Amudaryo) and Syr Darya (Sirdaryo) rivers, surrounded by mountains in the east and southeast (Fig. 1.9).

Uzbekistan covers an area of 448,960 km^2. The northernmost point is at 45° 36′ N, the southernmost point at 37° 11′ N, the westernmost point at 56° 00′ E, and the easternmost point at 73° 10′ E. The country stretches 930 km from north to south and 1425 km from east to west. Since the territory is broader in the meridian direction, this determines the relative uniformity of its landscapes. The surface of Uzbekistan is mainly plains, which are convenient for caravan routes. The sixteenth largest desert in the world, the Kyzylkum Desert, is located between the Amu Darya and Syr Darya rivers in the central and western parts of Uzbekistan.

Uzbekistan borders five countries of Central Asia and shares with them not just a natural environment but also a history, culture, and future. The length of Uzbekistan's borders is more than 5300 km, including a 2100-km border with Kazakhstan, situated to the north. These two countries share the Kyzylkum Desert, the Aral Sea, and the Tien Shan mountain system. Uzbekistan shares the Fergana Valley, Piskom Mountains, Chatkal Mountains, Fergana Mountains, and Alai Mountains with Kyrgyzstan (situated to the east) and shares the Kurama Mountains, Turkestan Mountains, Zarafshan Mountains, and Gissar Mountains with Tajikistan. All of these ranges are part of the Tien Shan mountain system. Uzbekistan shares a border with Afghanistan in the south. The Amu Darya river serves as the natural barrier that separates Uzbekistan from Afghanistan and Turkmenistan in the south (Figs. 1.10 and 1.11).

Another peculiarity of Uzbekistan is its double-landlocked location, as neither Uzbekistan nor any of its neighbors have direct access to any ocean or sea. Liechtenstein is the only other double-locked country in the world. This geography strongly influences the nature, climate, culture, and history of Uzbekistan.

The Tien Shan mountain system—the longest in Central Asia—stretches for about 2800 km from Tashkent toward Kyrgyzstan and China in the east. This system is a part of the highest mountain range on Earth, the Himalayan orogenic belt. Several ranges in the Tien Shan cross eastern Uzbekistan (Fig. 1.12).

In the south, the Pamir Mountains extend into part of the Tien Shan, serving as a junction of the Himalayas with the Tien Shan. These mountains play an important role as a source feeding the major rivers of Central Asia. In eastern Uzbekistan, the relief is a combination of mountain ranges and intervening valleys oriented from east to west.

The tallest mountain in the Central Asian Tien Shan, Victory Peak (7439 m high), is in neighboring Kyrgyzstan. The mountain Khazret (Hazreti) Sultan (4643 m high) is the highest point in Uzbekistan and is part of the Gissar Range of the Pamir–Alai Mountains in Surkhandarya (Surxandaryo) region in Uzbekistan. The tallest peak in the Western Tien Shan is Mount Adelung (4301 m high) in the Piskom Ridge in Tashkent region (Fig. 1.13). The major mountain ranges in Uzbekistan and their highest peaks are listed in Table 1.1.

Fig. 1.9 The location of Uzbekistan between the Syr Darya and Amu Darya rivers. [From Poulpy, based on work by NASA World Wind (public domain) via Wikimedia Commons]

1.4 Nature

The relief of Uzbekistan is quite complex and consists of combinations of mountains, valleys, plateaus, and sinks. The mountains in the southern and southeastern parts of the country constitute about 21% of its total area, and the plains and plateaus in the western and northern parts make up the remaining 79% of the country. These physical conditions determine the biodiversity and the wide range of various desert, steppe, and mountainous landscapes. Such surface construction determines the formation of horizontal nature zones and vertical belts. Thus, desert and semidesert landscapes are widely spread in the plains and steppes, whereas glacial and alp landscapes are predominantly found in the high mountains. The variety of flora and fauna is another unique natural feature of this country.

Fig. 1.10 The countries neighboring Uzbekistan. [Map and software from Ian Macky (public domain)]

Fig. 1.11 Central Asia. [Map and software from Ian Macky (public domain)]

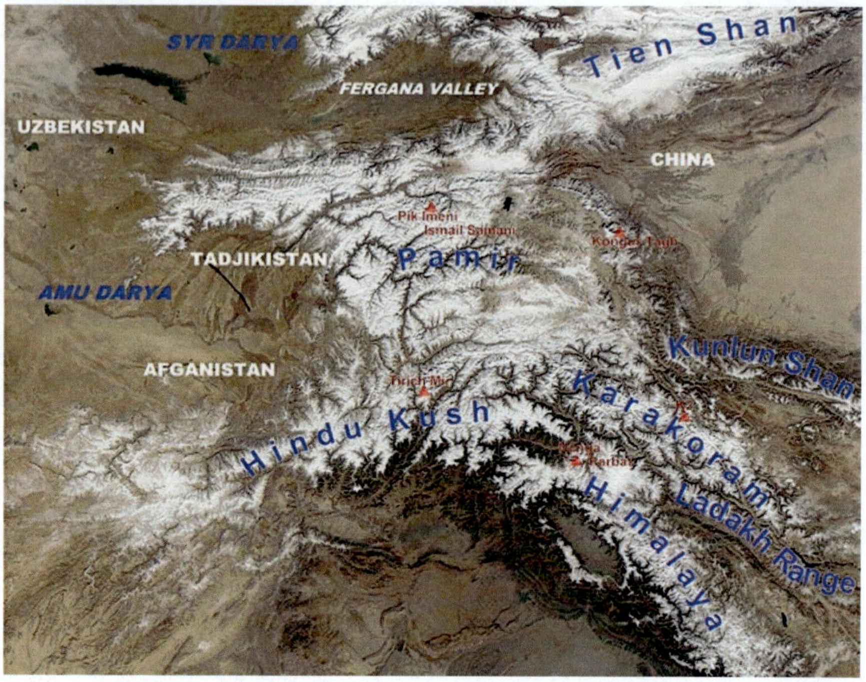

Fig. 1.12 The mountains in the eastern part of Uzbekistan. [Modified by Jeff Schmaltz, MODIS Land Rapid Response Team at NASA GSFC; derivative work by Rupert Pupkin (public domain), via Wikimedia Commons]

Fig. 1.13 Mount Adelung (4301 m high), Piskom Ridge, Western Tien Shan. [Public domain via Wikimedia Commons]

1.4.1 Climate

The climate of Uzbekistan is a result of such factors as (1) its location, (2) solar radiation, (3) the dominant flow of air masses from the north and the west, (4) the relief, (5) the landscape, and (6) human activity (Khasanov and Gulomov 2006).

The geographic location of Uzbekistan in the inner part of Eurasia and its remoteness from oceans and seas determine the continental character of its climate. This is manifested by high temperatures in summer, small amounts of precipitation and high evaporation of moisture, a long and hot summer season, large daily and annual amplitudes of air temperature, and a winter season that is relatively cold (for these latitudes).

Because of its location between 37.2° N and 45.5° N, the climates in the western part of the country are of the desert and semidesert types, and the climates in the central and eastern parts of the country are of the arid and semiarid steppe types. Thus, the amounts of precipitation vary between different locations. In the Kyzylkum Desert, there is either too little precipitation to sustain any vegetation at all, or only very sparse shrub vegetation is found.

Table 1.1 Major mountain ranges in Uzbekistan and their highest peaks

Mountain range	Highest peak	Location	Altitude (m)
Gissar[a] (Hissor)	Khazreti (Hazrati) Sultan (former name: Peak of the 22nd Congress of the Communist Party)	Alai mountain system	4643
Alas Ala-Too	Manas	Western Tien Shan	4484
Baysuntau	Hodjapiryah	Alai mountain system	4425
Piskom	Mount Adelung (Adelunga Tog`i)	Western Tien Shan	4301
	Beshtar (Beshtor)		4299
Ugam	Sayram	Western Tien Shan	4236
Turkestan	Shavkartau	Alai mountain system	4029
Chatkal	Greater Chimgan	Western Tien Shan	3309
Kyzylnura (Qizilnura)	Kyzylnura (Qizilnura)	Western Tien Shan	3267
Karjantau	Minbulak	Western Tien Shan	2824
Babaytag	Babaytag	Western Tien Shan	3555
Nuratau	Hayatbashi	Alai mountain system	2169
Tamditau	Aktau	Alai mountain system	974
Kuljuktau	Kuljuktau	Alai mountain system	785

[a]Russian name

According to the Köppen climate classification, Uzbekistan has a cold semiarid climate with hot and dry summers, cold winters, and major temperature swings between day and night (Fig. 1.14).

In Uzbekistan, the dry continental climate determines the prevalence of the desert, semidesert, near-desert, and steppe types. The climate of this double-landlocked country is formed by cold, dry air easily entering the region from the Arctic and Siberian regions to the north. The most severe weather is found in the western plain part of the country on the Ustyurt Plateau, on the dried seabed of the Aral Sea, in the lower Amu Darya river region, and in the Kyzylkum Desert. It is only at selected sites in the southern Kashkadarya and Surkhandarya (Surxondaryo) Valleys that the climate is relatively mild.

The main parts of the plain deserts and steppes consist of sand, clay, and stone. The uniqueness of the natural environment is the crucial difference in natural conditions between the northern and southern parts of the desert and steppe regions. This is demonstrated by the aridity of the climate and greater temperature extremes in the summer and winter seasons.

In a more general sense, the following subtropical and temperate climate zones are found in Uzbekistan: the Ustyurt Plateau and the Amu Darya delta are in the southern part of the temperate zone, and the Bukhara, Kashkadarya, and Surkhandarya regions are in the northern part of the subtropical zone (Fig. 1.14).

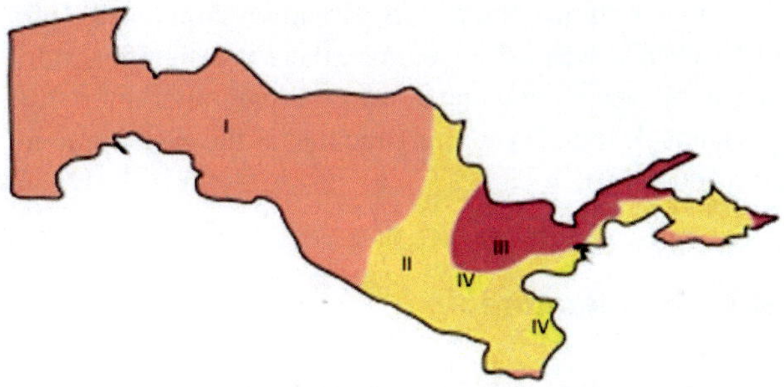

Fig. 1.14 The Köppen climate classification of Uzbekistan. *I* indicates a cold desert climate; *II* indicates a cold, semiarid climate; *III* indicates a warm continental climate/Mediterranean climate; and *IV* indicates a warm Mediterranean climate. [Map by Ali Zifan (enhanced, modified, and vectorized); CC BY-SA 4.0, via Wikimedia Commons]

Because of the influence of the Central Asian anticyclone, in the northern regions the winter season is longer and severe, with little precipitation. In the southern regions, the climate is mainly influenced by cyclones, and this results in more winter precipitation and thinner snow cover because of the warmer, shorter winter seasons. Arctic air masses also influence the climate in winter by freezing the whole country and lowering the air temperature to between −20 and −30 °C. In this region, the climate is influenced by solar radiation, particularly in summer. In summertime, the air circulation is dependent on thermal depression on the plains, which results in creation of dry, hot, local air masses named "Turan" (after the Turan lowland), which cover the whole country. This leads to formation of an area of low pressure, which facilitates penetration of warm and more humid air from the northwest and west. Because of the high temperature, the air heats up quickly and no precipitation falls.

This causes considerable differences in average temperatures and precipitation between the southern and northern regions, and between the mountains and plains. The dry and arid continental climate and the different climatic zones (including those of vertical types in the mountains) are the main characteristics of Uzbekistan's natural weather conditions.

In the mountains, depending on the elevation, the climate changes from a continental, arid, dry, hot climate to a milder one. The relief factor and the mountain valley air circulation result in shorter spring and summer seasons.

Natural barriers prevent entry of warm, moist air from the Indian Ocean. The openness of the northwestern and central–eastern parts of the country makes them vulnerable to cold air masses from the north. The wind direction changes from northeast in winter (as a result of the Siberian anticyclone) to south and southeast in summer. Clashes of cold and warm air masses frequently lead to strong winds and dust storms, which can become disastrous, damaging agriculture and settlements.

The major climatic indicators vary throughout the country and over the seasons. For example, there is a significant difference between the annual average precipitation in the Kyzylkum Desert (70–80 mm) and that in the eastern mountains (1000 mm).

1.4.2 Temperature

In the northernmost area (i.e., the Aral Sea and the northern part of the Kyzylkum Desert), which is open to cold air masses from Siberia, the winter season is usually longer and the average temperature in the coldest month of the year (January) is lower than those in other parts of the country.

The absolute minimum temperature in the northern part of the country is between −35 and −40 °C in January, and the absolute maximal temperature in the south is +50 °C in July. According to Khasanov and Gulomov (2006), the average temperature in January is about −6 °C to −7 °C in the Khorezm (Xorazm) region and −8 °C to −11 °C on the Ustyurt Plateau. Sometimes it falls to −35 °C. The temperature in January in the central part of the country is −2 °C to −4 °C. The longest season is summer, which lasts for 106 days in the south and 105 days in the north. The summer is hotter in the southern part of the plain; in July, the temperatures are 26 °C to 27 °C in the north but can exceed 35 °C in the southern part of the plain.

Figures 1.15, 1.16, and 1.17 were compiled by the author on the basis of data obtained from 8 years of historical weather readings (Climate-Zone.com 2018). Table 1.2 provides information about the locations of these selected sites.

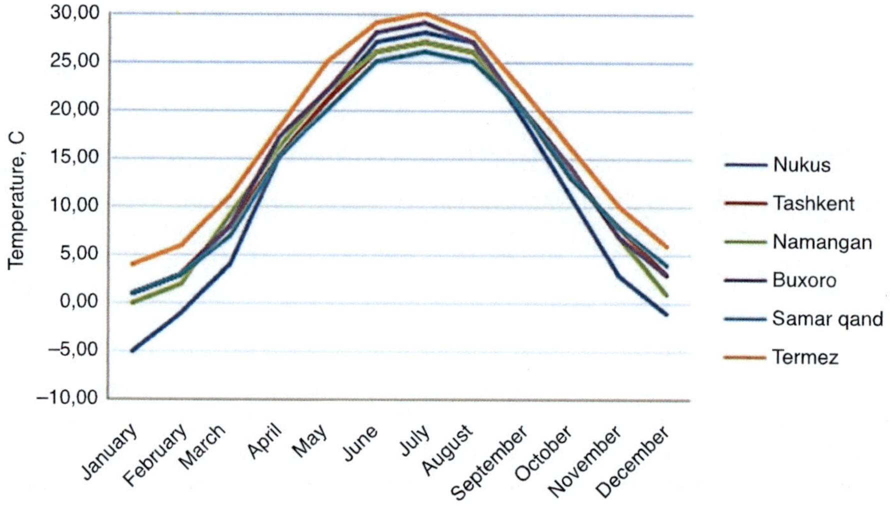

Fig. 1.15 Average monthly temperatures at selected sites in Uzbekistan. [Compiled by the author using data from Climate-Zone.com (2018), based on *The World Factbook*, a public-domain work published by the US Central Intelligence Agency]

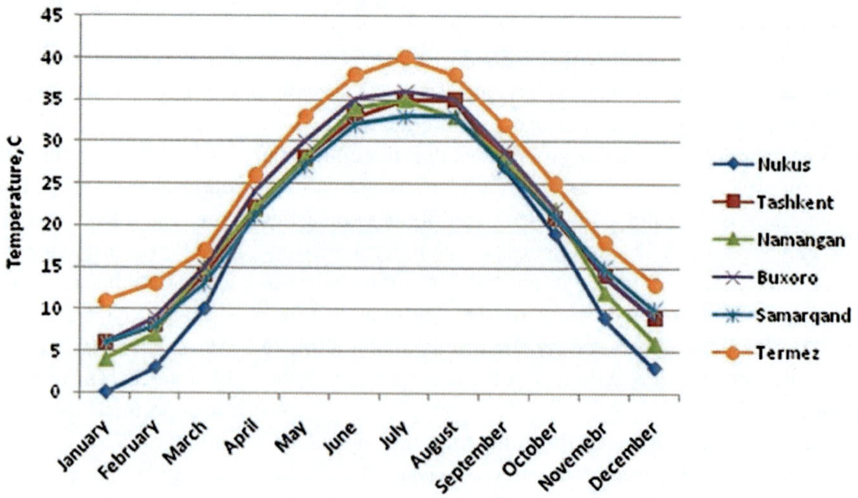

Fig. 1.16 An average maximal temperature in selected points of Uzbekistan (Compiled by author according to Climate Zone 2018. Portions of this site are based on the CIA World Fact Book, a public-domain work)

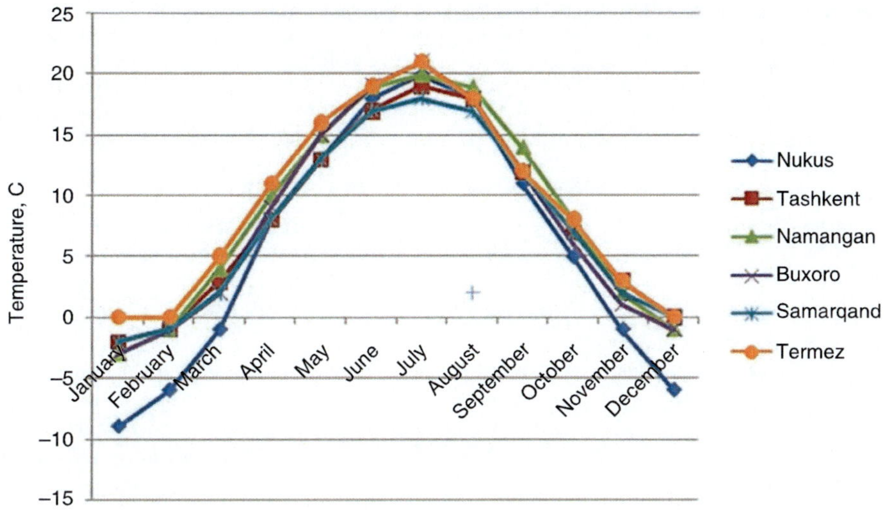

Fig. 1.17 An average minimal temperature in selected points of Uzbekistan (Compiled by author according to Climate Zone 2018. Portions of this site are based on the CIA World Fact Book, a public-domain work)

Table 1.2 Climate zones in Uzbekistan

Place	Climate zone (according to the Köppen climate classification)	Location		
		Latitude	Longitude	Elevation above sea level (m)
Nukus	I: cold desert climate	42.45 N	59.61 E	77
Tashkent	III: warm continental climate/Mediterranean climate	41.26 N	69.26 E	489
Namangan	II: cold semiarid climate	40.98 N	71.58 E	476
Bukhara	II: cold semiarid climate	39.71 N	64.61 E	230
Samarkand	III: warm continental climate/Mediterranean climate	39.70 N	67.00 E	724
Termez	III: warm continental climate/Mediterranean climate	37.23 N	67.26 E	302

Source: Climate-Zone.com (2018)

1.4.3 Solar Radiation

The 925-km length of the territory of Uzbekistan from north to south determines differences in the duration and angle of sunshine between the northern and southern parts of the country. Khasanov and Gulomov (2006) note that on June 22, the sun's rays hit the ground at a 68° angle in Ustyurt (the northernmost point), at a 72° angle in Tashkent, and at a 76° angle in Termez (the southernmost point). In winter, the sun's rays hit the ground at a 21° angle in Ustyurt and at a 29° angle in Termez. In general, the daytime lasts for 15 h in summer and for no less than 9 h in winter. The significant height of the sun above the horizon and the mostly cloudless days are among the reasons for the long sunshine duration. The potential sunshine duration is 4447 h per year at a latitude of 35° and 4455 h per year at a latitude of 40°.

According to Abdullaev and Isaev (2005), the annual sunshine durations are about 3095 h in Termez (the southernmost point), 2999 h in Nukus (a northern point), 2000 h in Ustyurt (the northernmost point), and 2870 h in Tashkent (Table 1.3). Depending on the sun exposure of the slopes in the mountains, the average annual sunshine duration varies from 2124 h to 2692 h.

However, the complexity of the ground surfaces and the numbers of cloudy days lead to differences in the sunshine duration between the seasons. Rhodes (2010) notes that about 70% of the incomeing solar energy reaches the Earth's surface. The difference between the average monthly sunshine durations in winter and summer is significant.

Ayzenshtat et al. (1986) note that the average monthly sunshine durations in Nukus are 109 h in December and 384 h in July. According to Ayzenshtat et al. (1983), the average monthly sunshine durations in Fergana are 97 h in December and 358 h in July.

The total amount of heat from solar radiation varies depending on the location (Table 1.3). Figure 1.18 was compiled by the author on the basis of data obtained from Khasanov and Gulomov (2006), Worldwide weather forecasts and climate information (n.d.), and Ayzenshtat et al. (1983, 1986) (Table 1.4).

According to estimates (Asian Development Bank 2011), the solar irradiance[13] (annual total global horizontal irradiance) is 1400–1800 kWh/m^2.

[13] Irradiance is a measurement of solar power and is defined as the rate at which solar energy falls onto a surface.

Table 1.3 Total amount of heat from solar radiation in different locations in Uzbekistan

Place	Total amount of heat from solar radiation (kcal/cm^2)[a]	Sunshine duration per year (hours)[b]	Location			Köppen climate classification
			Latitude	Longitude	Elevation above sea level (m)	
Tashkent	139.4	2870	41.2995 N	69.2401 E	477	Warm continental climate
Samarkand	143.9	2750	39.6270 N	66.9750 E	726	Warm continental climate
Fergana	144.7	2673	40.3842 N	71.7843 E	578	Cold semiarid climate
Takhiatash	147.1	2000	42.3367 N	59.5765 E	76	Desert climate
Tomdibuloq	153.1	2500–3000	41.7565 N	64.6319 E	236	Desert climate
Termez	159.8	3095	37.2611 N	67.3086 E	308	Desert climate
Qizilcha	160.8	2692	41.010 N	69.7500 E	2075	Warm continental climate

[a]Source: Khasanov and Gulomov (2006)
[b]Sources: Khasanov and Gulomov (2006), World Weather and Climate Information (2019), Ayzenshtat et al. (1983, 1986), and Abdullaev and Isaev (2005)

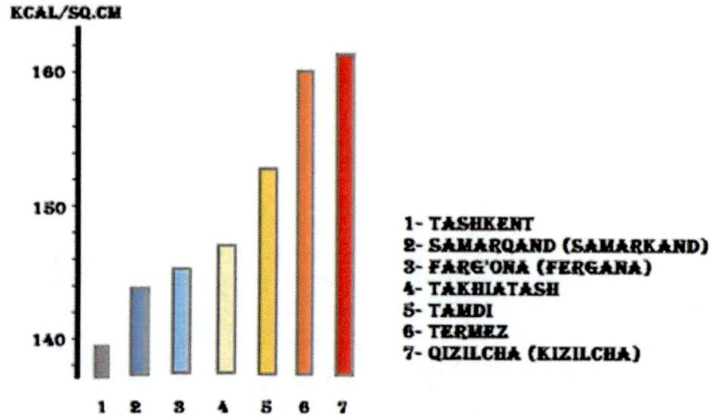

Fig. 1.18 Total amounts of average annual heat received from solar radiation at selected sites in Uzbekistan. [Compiled by the author using data from Khasanov and Gulomov (2006), World Weather and Climate Information (2019), and Ayzenshtat et al. (1983, 1986)]

At the same time, Khasanov and Gulomov (2006) argue that the annual radiation budget[14] is not too high. In Tamdibulak (Tomdibuloq in the central Kyzylkum Desert), it is 43.0 kcal/cm^2; in Fergana, it is 47.8 kcal/cm^2; in Tashkent, it is 53.1 kcal/cm^2; and in Termez, it is 67.1 kcal/cm^2. During the summer, the main part of the incoming solar radiation is about four or five times that during the winter. In the mountainous regions, depending on the altitudes and slopes of the hills and mountains, less solar radiation is absorbed; on Kizilcha (Qizilcha) mountain, 2075 meters above sea level), the radiation budget is 32.5 kcal/cm^2. Since the albedo is high (80–90%) in the high mountains, the budget may even be negative in winter.

Eshkulov (2012) points out that solar radiation is attenuated in the desert zones of Uzbekistan because of a reduction in the transparency of the atmosphere as a result of an increase in its aerosol component. In the western and central parts of Uzbekistan, the reduction in the transparency of the atmosphere is caused by advective aerosol and aerosol lifted from the bottom of the drying Aral Sea, whereas in the eastern areas, the main factor increasing atmospheric aerosol is industrial emissions. In the Fergana Valley, especially in winter, accumulation of aerosol is favored by a concave relief. The coefficient of the transparency of the atmosphere varies over the seasons, reaching its highest value in summer.

Uzbekistan is considered one of the sunnier countries among those located at the same latitude. For comparison, according to research (Osborn 2019), the sunniest city in the world is Yuma in Arizona, USA, with an average of 4015 h of sunshine per year.

1.4.4 Precipitation

Precipitation is the main source of water resources in Central Asia. It is formed from moist air masses coming from the north. Depending on the latitude, the orientation and exposure of the terrain, the absolute altitude, and the physiographic conditions, precipitation is uneven between the plains and the mountains. The annual precipitation is less than 80–100 mm on the northwestern plain (the Kyzylkum Desert and the Lower Amu Darya river) and in the western part of the Fergana Valley. Khasanov and Gulomov (2006) show that the precipitation gradually increases from the northwest to the southeast; on the plains, it may reach 100–200 mm/year, in the foothills 200–400 mm, and in the mountains 800–900 mm. The greatest amounts of precipitation occur in places facing toward moist air flows on the slopes of the Western Tien Shan and the Gissar–Zarafshan Mountains. The slopes of the Ugam, Korjantog, Piskom, and Fergana mountain ranges, which are oriented to the west and the southwest, are in more favorable positions (Khasanov and Gulomov 2006). About 90% of the annual precipitation falls in the autumn, winter, and spring seasons—30% in winter and 40% in spring. Precipitation is observed only on 35–40 days per year in the lowlands and on 70–90 days per year in the mountains (Fig. 1.19).

The snow cover is unstable, lasting for 40–50 days in the northwestern part of the country and for 90–100 days in the mountains. The thickness of the snow is 1–8 cm (reaching

[14]The radiation budget represents the accounting of the balance between incoming radiation, which is almost entirely solar radiation, and outgoing radiation, which is partly reflected solar radiation and partly radiation emitted from the Earth system (NASA Science Mission Directorate 2010).

Table 1.4 Solar energy characteristics

No	Characteristics	Winter solstice	Summer solstice
1	Solar elevation		
	Northernmost point 45°35'	21°	68°
	Southernmost point 37°10'	29°	76°
2	Daily sunshine duration, hours/day		
	Northernmost point 45°35'	3–5 (in average)	10–13
	Southernmost point 37°10'	3–5 (in average)	8–10
3	Days without sun, days/year		
	Northernmost point 45°35'	45–50	10–15
	Southernmost point 37°10'	22–25	1–4
4	Average sunshine duration, hour/year		
	Northernmost point 45°35'	2,800	
	Southernmost point 37°10'	3,050	
5	Direct solar radiation intensity (S), kW/m²		
	On plains	0.80–0.94	
	On high-altitude stations	0.94–1.06	
	On Kyzilcha station	1.21	
6	Duration of minimal sunshine, hours/month	80–100	

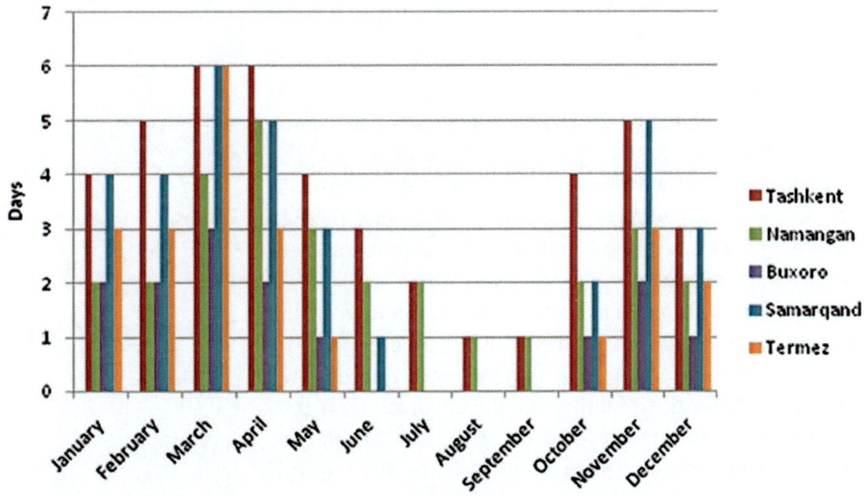

Fig. 1.19 Average numbers of rain days at selected sites in Uzbekistan. [Compiled by the author using data from Climate-Zone.com (2018), based on *The World Factbook*, a public-domain work published by the US Central Intelligence Agency]

30 cm), 10–20 cm (but can reach as much as 60 cm) in the foothills, and 60 cm (up to 1.5–2.0 m) in the mountains (Khasanov and Gulomov 2006).

The evaporation rate is several times the precipitation rate. For example, it is 3.5 times the precipitation rate in Tashkent and 27 times that in Nukus (Khasanov and Gulomov 2006). This leads to low humidity of the air on the plains. In the southern parts of Uzbekistan, the relative humidity of the air is 20–30% in summer and 65–80% in winter. The humidity of the air is also changeable over the seasons and varies between locations (Fig. 1.20).

1.4.5 Climate Change

Climate change poses a serious threat to the entire natural–economic system of Central Asia, including the water and land resources of the region.

Global warming forecasts are concerning in this respect. Glaciers presently contribute up to 70% of the water flow in the major river systems of the region during the hot and dry summers (Makhmadaliev et al. 2008).

According to researchers (Borisova 2012; CAWater-Info 2019; Ibatullin 2011; Makhmadaliev et al. 2008), the water reserves in the glaciers decreased by more than 25% between 1957 and 2000. It has been predicted (Borisova 2012; CAWater-Info 2019; Ibatullin 2011; UNDP 2011) that by 2025, thousands of small glaciers will disappear, the total area covered by glaciers will shrink by 20%, and ice reserves will decrease by 25%. As a result (Borisova 2012; CAWater-Info 2019; Ibatullin 2011; UNDP 2011), the mountainous areas may lose major parts of their glaciers, and this will have a significant impact on surface runoff: by 2050, the runoff of the Amu Darya river will decline by 10–15% and that of the Syr Darya river by 6–10%.

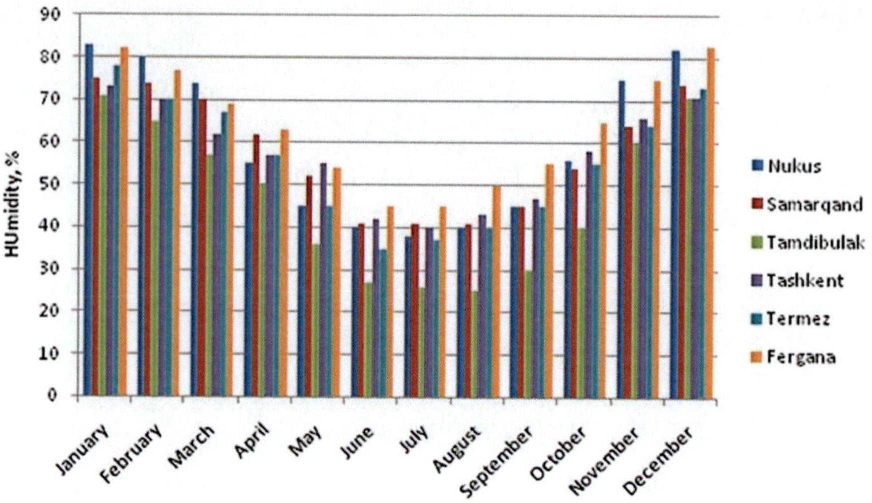

Fig. 1.20 Annual variability in air humidity at selected sites in Uzbekistan. Note: Tamdibulak is situated in central Kyzylkum. [Compiled by the author using data from Physical Geography of Uzbekistan (2017)]

The reductions in water flows from such changes in glacial areas is at present poorly understood but is likely to be dramatic, particularly in hot and dry summers. Large reductions in water flow will have severe consequences for the ecological functioning of the rivers as well as for the water, energy, and food security of all Central Asian countries. The water shortages in Central Asia have already triggered conflicts between communities. The effects of glacial melting will exacerbate such conflicts unless timely adaptation measures are implemented.

According to data from the executive committee of the International Fund for Saving the Aral Sea (Ibatullin 2011), in most areas of Central Asia, winter temperatures have increased, as have the variability and the intensity of the precipitation; rain showers are followed by periods of drought, increasing soil erosion. The risks of flooding have grown significantly.

According to the World Bank Group's Climate Change Knowledge Portal, from 1901 to 2015 the average temperature in Uzbekistan in July (the hot summer period) increased from 27.0 °C to 27.8 °C and that in January (the coldest winter month) decreased from −3.5 °C to −2.3 °C (Table 1.5).

The average rainfall in March–April increased from 29.4 mm in 1901–1930 to 33.6 mm in 1991–2015, and that in July–August increased from 1.6 mm in 1901–1930 to 1.9 mm in 1991–2015 (Table 1.6).

According to data from the Tashkent meteorological station, the coolest summers were recorded in 1901–1910 and in 1951–1960; the hottest ones were recorded in 1971–1980. The coldest winter seasons occurred in 1891–1900, 1931–1940, and 1970–1980. From 1981 to 1990, the average temperature in January was 2 °C warmer than those in previous years. During last 100 years, the average July temperatures rose by 0.5 °C and the average January temperatures rose by 2.4 °C (Khasanov and Gulomov 2006). The shrinking of the Aral Sea and the Lower Amu Darya delta resulted in colder winters and hotter summers in places located up to several hundred kilometers away (World Bank 2019).

Table 1.5 Average monthly temperatures in Uzbekistan

Years	Average monthly temperature (°C)	
	January	July
1901–1930	−3.5	27.0
1931–1960	−3.6	26.6
1961–1990	−3.3	27.5
1991–2015	−2.3	27.8

License (CC-BY-4.0); data source: World Bank Group Climate Change Knowledge Portal

Table 1.6 Average monthly rainfall

Years	Average monthly rainfall (mm)			
	March	April	July	August
1901–1930	29.4			1.6
1931–1960	33.9			1.5
1961–1990		33.8	1.6	
1991–2015		33.6	1.9	

License (CC-BY-4.0); data source: World Bank Group Climate Change Knowledge Portal

1.4.6 Flora and Fauna

The variety of flora and fauna is another unique property of Uzbekistan's natural environment. The country's central geographic location at the junction of several biogeographic provinces within the Central Asian region determines the biodiversity and richness of the flora and fauna.

1.4.7 Flora

Uzbekistan has rich biodiversity; according to some estimates (Gintzburger et al. 2003; Mabberley 2017; Sennikov et al. 2016; Zakirov 1989) it has 3110–3700 species of vascular plants, 366 (20%) of which are endemic. A more recent study (Tojibaev et al. 2017) reported that the flora is made up of more than 4250 vascular plant species, including a large number of endemic, endangered, and globally important species. This biodiversity is at risk because of global climate change and human activities, and 313 species are listed in *The Red Data Book of the Republic of Uzbekistan* (Khassanov 2016). The most vulnerable sites are in desert and semidesert regions, where degradation of pastures and ranges affects more than 30% of the total area (Gintzburger et al. 2003). The flora of the steppes and deserts consists of unique bush species. Wooden, bushy, and grassy plants are well developed on the low plains. Tugays feature reeds and kendyrs.

The majority of endemic plants grow in the mountains. The unique climate, weather conditions, and soils of each of the high-altitude belts of the relief determine the wealth and diversity of the plants. Distribution of various species is significantly related to features of the relief of the mountains, such as the exposure of the slopes, their orientation, and the exposure to moister and solar radiation, as well as the steepness of the slopes and the altitudinal zoning composition of the soils.

Muhamedjanova (2004) noted that the Western Tien Shan region contains a high level of endemism of local character. Egamberdieva et al. (2013) pointed out that the flora of the Western Tien Shan is central to formation and conservation of species because of the abundance of endemic, rare, and relict plants, the absolute number of endemic species, and the percentage of endemic flora. In this sense, the Western Tien Shan significantly surpasses other regions. Mukhamejanova estimated that more than 2500 species of vascular plants grow in this area. Chatkal Biosphere Reserve is located in Ugam–Chatkal National Park and is the most interesting site for the study of biodiversity. Mukhamejanova (2004) and Egamberdieva et al. (2013) noted that almost all types of vegetation in the Central Asian mountains grow within this region. Figures 1.21, 1.22, 1.23, 1.24, 1.25, 1.26, 1.27, 1.28, 1.29, 1.30, 1.31, 1.32, 1.33, 1.34, and 1.35 give an overview of the endemic species found in the Greater Chimgan (3309 m) in the Chatkal Range of the Western Tien Shan, which is located in Ugam–Chatkal National Park.

The landscape of the submountainous plains is characterized by grass; there are no trees, but bushes may be found alongside waterways. Various species of onion, tulip, rhubarb, and iris grow here. High foothills feature dry steppes with motley grass on gray earth. Bushes such as almond, kurchava, and cherry grow on stony sites.

The most valuable wood species, the *Zarafshani archa (mountain juniper)*, grows at moderate altitudes in mountains (Fig. 1.36). Deciduous species such as maple, hawthorn, various sorts of wild apple, pistachio, walnut, birch, willow, poplar, and cherry trees are also widespread in the middle mountain zone (Fig. 1.37).

The lower mountains are rich in bushes: honeysuckle, barberry, dog rose, meadowsweet, and wild grape (Fig. 1.38). The grasses are also very diverse and include muscat sage, rhubarb, tulip, and Piskom onion (a precious herb). The middle mountains feature dog rose and other bushes. Only 30% of the high mountains are covered with plants, mostly tipchak. Figures 1.39, 1.40, 1.41, 1.42, 1.43, 1.44, 1.45, 1.46, 1.47, and 1.48 show the diversity of landscapes in different locations.

According to a botanical and geographical provinces classification (Tojibaev et al. 2017)—based on taxonomy, mor-

Fig. 1.21 *Allium pskemense*, an endemic species of the Western Tien Shan, found on the Piskom Ridge, Tashkent region

Fig. 1.22 *Astragalus baranovii Popov*, an endemic species of the Chatkal Range, found in the subalpine mountain belt on Greater Chimgan, Tashkent region

Fig. 1.23 *Nanophyton botschantzevii* U.P.Pratov, a relict endemic species of the Western Tien Shan, found at an altitude of 1600–2000 m at Chimgan on the Chatkal Ridge, Tashkent region

Fig. 1.24 *Salvia tianschanica*, an endemic species of the Western Tien Shan, found on Chatkal Ridge, Chimgan, Tashkent region

Fig. 1.25 *Cousinia sewerzowii*, an endemic species of the Western Tien Shan, found on Chatkal Ridge, Chimgan, Tashkent region

Fig. 1.26 *Cousinia*, an endemic species of the Western Tien Shan, found on Chatkal Ridge, Chimgan, Tashkent region

Fig. 1.27 A wild apple tree found on the Chatkal Ridge, Chimgan, Tashkent region

Fig. 1.28 A greek walnut found on the Chatkal Ridge, Chimgan, Tashkent region

Fig. 1.29 *Prunus cerasifera* found on the Chatkal Ridge, Chimgan, Tashkent region

Fig. 1.30 *Prunus cerasifera* found on the Chatkal Ridge, Chimgan, Tashkent region

Fig. 1.31 *Hypericum scabrum* found on the Chatkal Ridge, Chimgan, Tashkent region

Fig. 1.32 *Berberis nummularia* Bunge found on the Chatkal Ridge, Chimgan, Tashkent region

Fig. 1.33 *Ziziphora*, a species of the Western Tien Shan, found on the Chatkal Ridge, Chimgan, Tashkent region

Fig. 1.34 *Achillea millefolium*, a species of the Western Tien Shan, found on the Chatkal Ridge, Chimgan, Tashkent region

Fig. 1.35 *Achillea millefolium*, a species of the Western Tien Shan, found on the Chatkal Ridge, Chimgan, Tashkent region

Fig. 1.36 *Zarafshani archa*, found on the Chatkal Ridge, Tashkent region

Fig. 1.37 The Ugam–Chatkal National Park, Tashkent region

Fig. 1.38 Lower mountains, Galavasay, Tashkent region

Fig. 1.39 These slopes with a southern orientation in Tashkent region do not have enough vegetation to prevent erosion

Fig. 1.40 Gallery poplar (tugai) forests along riverbeds on the plains and in the deltas of major rivers, such as the Chirchiq River in Tashkent region

Fig. 1.41 *Elaeagnus orientalis* found in the foothills of the Korjantog Mountains, Tashkent region

Fig. 1.42 The landscape of the Mirzachul Steppe, Jizzakh region

Fig. 1.43 Semidesert in Bukhara region

Fig. 1.44 Tugai in the Lower Amu Darya river, Khorezm region

Fig. 1.45 These species in the Amu Darya delta, Khorezm region, are threatened with extinction

Fig. 1.46 The Kyzylkum Desert, Khorezm region

Fig. 1.47 The landscape of the Kyzylkum Desert, Khorezm region

Fig. 1.48 Cotton fields have replaced the natural landscape in Jizzakh region

phology, ecology, geographic distribution, economic significance, and species status—two major provinces have been identified:

1. Mountain Middle Asian Province (Western Tien Shan, Fergana, Fergana–Alai, Nurata, Kuhistan, Gissar–Darvaz, Southwestern Gissar, Near-Pjandzh)
2. Turan Province (Upper Syr Darya, Middle Syr Darya, Kyzylkum, Bukhara, South Near-Aral, Aral, Ustyurt)

The major part of the vegetation is constituted by physiognomic types of steppe (Fig. 1.49). In the Aralo-Caspian province, the dominant vegetation consists of hemicryptophytes and chamaephytes, with some rare small trees. These ephemeral plants including some characteristic micro- and nanophanerophytes are prevalent in the arid and semiarid zones of Uzbekistan (Gintzburger et al. 2003; Tojibaev et al. 2017). The native species of the arid and semiarid zones have developed as a result of natural selection under a complex

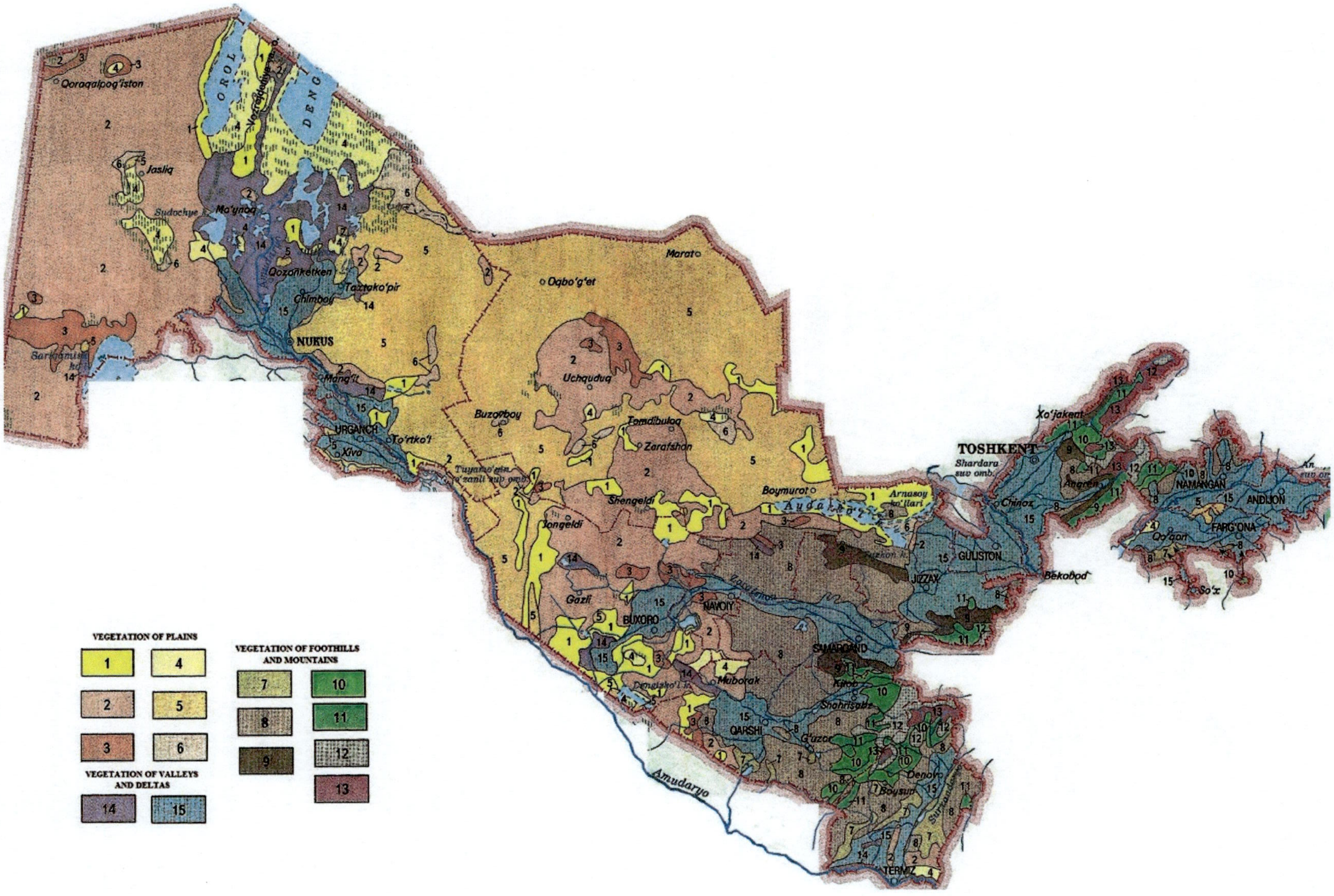

Fig. 1.49 Vegetation in Uzbekistan. *1* indicates sparse desert psamophyte shrub vegetation; *2* indicates desert wormwood and Biyurgun (a saltwater barnacle); *3* indicates saline black saxaul desert; *4* indicates sparse saline desert vegetation; *5* indicates white saxaul, sagebrush, white saxaul, and black saxaul–white saxaul deserts; *6* indicates salt-desert black saxaul; *7* indicates ephemeral semidesert vegetation; *8* indicates semidesert foothill and low-mountain grass and wormwood; *9* indicates mountain dry grass steppes; *10* indicates mountain broadleaf forests, sparse trees, and shrubs; *11* indicates sparse archa trees on the background of dry steppe herbs; *12* indicates high mountain steppes; *13* indicates subalpine meadow steppes and alpine meadows; *14* indicates tugai vegetation; and *15* indicates cultural vegetation. (Courtesy of the State Committee for Land Resources, Geodesy, Cartography and State Cadastre of the Republic of Uzbekistan)

Fig. 1.50 File: Tian Shan Bear (Ursus arctos isabellinus) Rustam Sulaymanov [Creative Commons Attribution-Share Alike 3.0] via Wikimedia Commons

and extreme combination of environmental factors. Researchers (Gintzburger et al. 2003; Tojibaev et al. 2017) note that these species are able to withstand prolonged droughts and nutrient deficiencies, high salinity or sulfate levels in the soil, extremely high summer temperatures and low winter temperatures, and anthropogenic pressure (especially from grazing). Small trees and shrubs represent a small number of species (0.5–7.2% of total).

The vegetation of Central Asia is subdivided (Zakirov 1989) into nine types and 20 subtypes. The plants of Uzbekistan have been studied since ancient times from the perspective of their uses for pharmacy and food. Avicenna (Abu Ali Ibn Sina, 980–1037) and Beruni (Abu Rayhon al Beruniy, 973–1048) contributed to herbal medicine by describing the properties of various species and their uses.

The extremes of the climate, with hot and dry summers and cold winters, have determined food and dietary preferences. Many species have been used as food to supply the needs of people working hard in fields during hot summer days. The need for energy food sources was covered by edible plants. Not all of their properties and ways of using them have been well defined. It was noted by Khojimatov et al. (2015) that 27 species of widely distributed edible plants in Uzbekistan are dicotyledonous and 12 species are monocotyledonous.[15]

[15] Dicotyledoneae (dicots) are plants that have seeds with two cotyledons; monocotyledonae (monocots) are those with one cotyledon.

1.4.8 Fauna

Its geographic position in the center of Eurasia and the relative transparency of its borders for animal migration are major reasons for the richness of the species composition of the fauna of Uzbekistan. The absence of high mountains in the northern part of the country facilitates movement of animal species from the north (Europe) and through Central Asia. Alongside species intrinsic to the desert and steppes, one may find animals such as bears and ermines in the northern regions. Some species originated in Africa or in India. Scientists (Azimov et al. 2009) note that the fauna of Uzbekistan has a unique history and complex genetic relations (Figs. 1.50–1.55).

In terms of distribution, the animal world is divided into groups related to habitats. These are animals of deserts, freshwater, mountains, and oases.

The diversity of landscapes and natural conditions results in diversity of the animal world, too. Another peculiarity of the natural environment is the richness of the fauna, with many endemic species such as snow leopards, mountain sheep, Turkestan geckos, and rock nuthatches.

Deserts and semideserts occupy most of this country (about 80% of its territory); thus, its animals have evolved adaptations to the harsh conditions of the desert environment. The diversity of the fauna depends on the type of desert habitat. The fauna is richer in sandy desert environments than in clay and gypsum desert environments.

Fig. 1.51 Vulpe. Evgeniy Gunenkov [Creative Commons Attribution-Share Alike 2.0 Generic] via Wikimedia Commons

Fig. 1.52 Suslik German Stimban [CC-BY-SA-3.0] via Wikimedia Commons

Fig. 1.53 Trapelus sanguinolentus. BetacommandBot [CC-BY-SA-3.0] via Wikimedia Commons

Fig. 1.54 Eremias regeli. Alexander A. Fomichev [CC-BY-4.0] via Wikimedia Commons

Fig. 1.55 Turtle (Testudo horsfieldi) David Holt (Attribution Share Alike 2.0] via Wikimedia Commons

Special attention should be paid to the fauna of the mountains, which is especially rich and diverse. More than 120 species of mammals, 20 species of reptiles, 20 species of fish, three species of amphibians, and up to 500 species of birds (including migratory species) have been recorded in the Western and the Southern Tien Shan (UNESCO 2015, pp. 41–52) While most species that live in the mountains are similar to those that live on and close to the plains, they are adaptive inhabitants of rocky places, including endemic species.

It has been estimated (UNDP 2015, p. 17) that the fauna of Uzbekistan consists of 14,900 invertebrate species (850 protozoa species, 61 annelid species, 1179 roundworm species, 533 flatworm species, 223 mollusk species, and 12,000 arthropod species) and 714 vertebrate species (84 fish species, three amphibian species, 60 reptile species, 460 avian species, and 107 mammal species).

Scientists (Azimov et al. 2009; UNDP 2015) point out that the Central Asian local origin of the species are Turanian and Turkestan endemics which play significant role in the formation of fauna. According to the *Fifth National Report of the Republic of Uzbekistan on Conservation of Biodiversity* (UNDP 2015, p. 17), 53 species and subspecies of terraneous vertebrate animals in Uzbekistan and Central Asia, including up to 50% of all reptile species in the region, are known to be endemic.

However, the rapidly growing human population, urban expansion, and anthropogenic pressure are decreasing the ranges and populations of many species. Overgrazing is a particular threat to the animal world on the plains and in the mountains. Some species are threatened with extinction or have already become extinct.

A comparison of the data in the fourth edition of the *Red Data Book of the Republic of Uzbekistan* (2009) and the corresponding data in the subsequent (fifth) edition (2019) shows that the total number of species listed in the book increased from 177 to 202, including increases from 73 to 83 invertebrate animal species, from 16 to 20 reptile species, from 48 to 52 bird species, and from 23 to 30 mammal species.

The decline in endemic species (fish, birds, and reptiles) is the result of changes in the water regime in the Aral Sea Basin, deforestation, overgrazing, and chemical pollution of soils and water. Many species lost their habitats as a result of development of virgin lands in desert and steppe areas. A decline in mammal species and the threat of their extinction have been observed in the Western Tien Shan (UNESCO 2015). Human activity and global climate change are resulting in destruction of unique environments and contributing to the imbalance of ecological systems all over the country. Transformation of virgin lands to irrigated lands and mining in desert areas have deprived animals of their means of survival. Decreases in water flow have resulted in shrinkage of the ranges of a number of local endemic pheasant subspecies and other inhabitants of river forests.

Salinity of soils and water has created unbearable ecological conditions for many species. The most disastrous event occurred in the Aral Sea and the wetlands of the Amu Darya delta, where the richest avian diversity has been lost. Not only have fish disappeared but also bird species have declined in vast numbers.

Even in the mountains, human activity has destroyed ecosystems by livestock overgrazing, clearing of trees and bushes, and interference with water catchment in the high mountains by closure of wells and springs. Pollution of springs, their shallowing, fluctuation of water levels in reservoirs, and pollution of soils put rare local endemic invertebrates at risk. All of these actions lead to soil erosion and destruction of the habitats of many species. Nowadays, the rate of wild species extinction continues to grow, following the dramatic decline in species number that started toward the end of 20th century.

Portions of Uzbekistan are included in the Critical Ecosystem Partnership Fund (CEPF) Mountains of Central Asia biodiversity hot spot. Important terrestrial ecological regions include the Gissar–Alai open woodlands, the Alai–Western Tian Shan steppe, the Central Asian southern desert, and the Central Asian northern desert. (Note: For more information on the country's endemic animals and plants, see the Biodiversity Strategy and Action Plan at http://lntreasures.com/uzbekistan.html).

According to the State Committee of the Republic of Uzbekistan on Ecology and Environmental Protection (http://www.uznature.uz/en/activity/view?numer=806), the protective areas include:

- Seven state reserves (Zaamin, Nurata, Gissar, Kitab, Kyzylkum, Surkhan, and Chatkal)
- The Saigachiy complex (landscape) nature reserve
- Three national nature parks (Zaamin, Ugam–Chatkal, and Zerafshan)
- The lower Amu Darya Biosphere Reserve
- Ten Bukhara specialized nursery (Jeyran) natural monuments

The ecological network includes six mountain nature reserves (Gissar, Zaamin, Kitab, Nurata, Surkhan, and Chatkal), three desert–tugai nature reserves (Kyzylkum, Badai Tugai, and Zarafshan), two mountain national parks (Zaamin and Ugam–Chatkal), twelve state natural reserves (Arnasai, Dengizkul, Karakir, Karnabchul, Koshrabad, Kumsultan, Mubarek, Nurota, Oktau, Saigachiy, Sudochie, and Xadicha), and two nature monuments (Vardanzi and Yazyavan). This network contributes to integrated protection and sustainable use of fauna and flora. Threatened animals (Bukhara deer, markhor, Severtzov's argali, Menziber's marmot, the Tien Shan brown bear, the Turkestan lynx, large birds of prey, and many insects) are being conserved within this system of protected areas.

Uzbekistan does not have own national animal or national flower. The fairy-tale bird Humo is depicted on the state emblem. Its open wings symbolize happiness and love of freedom. The bird Semurg (the same as Humo) was mentioned in the ancient book *Zend Avesta*, and Turkic people introduced this bird into their own mythology. A wreath of wheat and cotton branches with open cotton bales accompanies this bird.

1.5 History

It was the Silk Route that united people, and it was the same Silk Route that divided them because of their desire to monopolize and completely control the trade on the Silk Route. The history of Uzbekistan is a sequence of the rises and falls of different civilizations, with revivals after wars and conquests. The ethnic and cultural mosaic of Central Asia, including Uzbekistan, has been determined by historical events for several millennia. One may find traces of the influences of the Persian, Greek, Roman, and Arabic cultures from the west; the Chinese, Indian, and Turkic cultures from the east; and the European and Russian cultures from the north. This mixture of cultures developed the unique culture of the Uzbek nation. The tolerance and friendliness of the people of Uzbekistan is the heritage of being at the crossroads of the Great Silk Route.

The unique geographic conditions of Uzbekistan attracted people to settle there from ancient times. Excavations (Derevyanko et al. 2013; Litvinski 2012; Okladnikiov 1939; Tolstov et al. 1950) demonstrate that people inhabited this region in the Middle Paleolithic (250,000 to 30,000 years ago). A burial of a Neanderthal boy and stone tools found at the Teshik Tash (Teshik Tosh) grotto in the Baysun district of Surxondaryo region (Okladnikiov 1939) and the Obiraḥmat grotto in the Bostanlik district of Tashkent region (Krivoshapkin 2017; Suleimanov 1972) are the most ancient human artifacts in Central Asia. Other discoveries such as the Kulbulak Paleolithic site in the Angren district in Tashkent region (Derevyanko et al. 2013; Kasymov and Grechhkina 1994; Rostovtsev 2006; Frachetti and Spengler 2015), near local flint outcrops, and comparisons of the tools and materials found there with those found at European microindustrial complexes point to great similarities between those early cultures.

According to one of 48 hypotheses based on archeological discoveries, Mousterian technology was probably brought to the region from the west by Neanderthals (Litvinski 2012). This hypothesis assumes that around the middle of the second millennium BCE, movement of primitive agrarian tribes from southern Turkmenistan eastward to southern Uzbekistan represented movement of Indo-Aryans and settling of proto-Iranians in Central Asia. The agrarian tribes lived there alongside local nomadic pastoralists.

According to some sources (Asimov and Bosworth 1998; Lunin 1988; Mayor 2014; Rtveladze 2005; Tolstov et al. 1950), the first human activities in the Central Asian region may have started between 35,000 and 25,000 years ago, in the late Pleistocene epoch. As some ancient and modern historians state, the first identifiable human groups in this region were the Cimmerians and Scythians, who lived in the western part of modern Central Asia in the first millennium BCE (Struve and Snegiryov 1935). In *The Histories*, Herodotus mentioned the Scythians and compared them to Amazons, as Scythian women warriors fought together with men on the battlefield. Herodotus pointed out that Khorezm was the leading political power in Central Asia even before the Persian monarchy was formed. Some arguments have been made in favor of a Khorezmian (Khwarazmian) origin of Zoroastrianism or Mazdeism. The initial development of civilization occurred in regions located along the lower courses of great rivers such as the Amu Darya. Later, cultural supremacy passed to Bactria in the southern part of modern Uzbekistan, where the governors of the Persian Achaemenid resided and from where the cultural influence of India radiated abroad. Central Asia was part of the Persian Empire; Persian culture and language were dominant there. Exchange of goods made in Greek centers started during the Achaemenid period (Litvinski 2012), and the trade between East and West was very intensive.

When Alexander the Great conquered Central Asia in 328 BCE, Khorezm (Khwarazm) was independent of Persia (Lunin 1988; Mayor 2014; Rtveladze 2005). In his book, Baumer (2008) gives a view of the history of the Scythians and Sarmatians, who lived in Central Asia more than 3,000 years ago. His conclusions are based on many years of archeological excavations and research. Baumer (2008) concludes that the arrival of humans in Central Asia occurred during the Paleolithic and that climate change played a great role in the flow of people into and out of Central Asia.

The Greco-Bactrian Kingdom in 256–125 BCE was part of the Hellenistic world. This kingdom was built on the remnants of the empire of Alexander the Great, where two cultures coexisted. The Greek alphabet was used for writing in the Tohar language (Dubrovskaya 2008).

Around 220 BCE, the first known contacts between China and the West were established by the rulers of the Greco-Bactrian Kingdom for technology exchange.

In the second century BCE, this kingdom was destroyed by the Yuezhi (Rtveladze 2005)—nomadic pastoralists from the western part of the modern Chinese province of Gansu, who migrated westward. In the first century BCE, a Greater Yuezhi tribe, the Kushans, played an important role in development of trade on the Silk Route (Dubrovskaya 2008). The Kushan Empire displaced the Greco-Bactrian Kingdom. During their rule, Buddhism entered the region. The Kushan Empire was succeeded by the Hephthalites; the Hephthalite Empire (408–

670 CE) was a confederation of nomadic and settled tribes. Apparently, they were a part of the four major "Hunnic" states.

Historians (Golden 2006; Gumilyov 1967, 2000; Tolstov et al. 1950) have written about the Turkic people who established the Turkic Khaganate in the sixth century. It lasted for two centuries and influenced the subsequent ethnic character of Central Asia and Uzbekistan. This empire was strong enough to defeat all enemies and to take full control of the trade between China and the Byzantine Empire. The Sogdian merchants were traders of silk and jewelry on the Great Silk Route, which passed through the lands of Uzbekistan. Gumilyov (2000) contends that establishment of a tribal confederation in the Western Khaganate weakened it and in 757 it was easily defeated by the Tan Chinese Dynasty.

In the seventh and eighth centuries, Umayyad and Abbasid Arabs conquered Transoxiana (the land between the Amu Darya–Oxus and Syr Darya–Jaxartes rivers) or "the land beyond the river" (*ma wara-al-nahr* in Arabic). The larger part of Transoxiana was conquered in 705–715 by the Umayyad leader Qutayba Ibn Muslim (Golden 2006; Gumilyov 1967, 2000; Tolstov et al. 1950). Islam did not spread widely until Abbasid rule granted non-Arabs equal status. However, the Arab conquest did not mark the end of Buddhism or Chinese influence in Central Asia. Islamization of Central Asia began in the tenth to twelfth centuries and continued until the fifteenth century.

The Uyghurs (also Turkic people) ruled in the eighth century. The Karakhanid Khanate conquered Transoxania in Central Asia and ruled it from 999 to 1211. Historians (Lunin 1988; Rtveladze 2005) suppose that a definitive shift from Iranian to Turkic predominance in Central Asia occurred during that period. The Karakhanids adopted the Persian–Arab Muslim culture and at the same time retained their own Turkish culture. The capitals of this state were in Kashgar, Balasagun, Uzgen, and Samarkand. This khanate was a confederation of people living in the Western Tien Shan and the Western Xinjiang around Kashgar. Few written records of this khanate exist. The people converted to Islam and adopted Muslim names and honorifics in the mid-tenth century (Davidovich 1998; Golden 2006; Gumilyov 1967, 2000; Tolstov et al. 1950). Later, they adopted Arab titles to replace Turkic titles. In the late tenth century, they defeated the Samanids (Persian rulers) and took Isfijab, Fergana, Ilaq, Samarkand, and Bukhara. It was only after the death of the ruler of the Karakhanids that the Samanids came back to Bukhara. However, by 999, the Karakhanids had conquered Tashkent (Chach), Samarkand, and Bukhara once again. The Samanid domains were divided into Khorasan (Persian dominant) and Transoxiana (Turkic dominant) domains. The Syr Darya river was the boundary between these two empires.

According to the rules of the Turkic nomads, the Karakhanid State was divided into parts with four principal urban centers (Qara-khanids 2009). One such center in Fergana was Uzgen, and Samarkand in Transoxiana was another. Historians (Grousset 2004; Soucek 2000) note that after the Turkic Karakhanids converted to Islam, conquest of the Buddhist cities began in the east at Kashgar. The Karakhanid era was marked by the complicated process of Turkization of the local people and Arabization of the Turkic culture. However, Turkic remained the official language (Clark 2010), and Turkic script was used for all documents. At that time, intercultural exchange was very intensive and the influences of the Persian and Islamic cultures on the Turkic culture resulted in the flourishing of Islamic civilization under the Karakhanids. The famous Kalyan minaret in Bukhara is one of the buildings built during their reign (Fig. 1.56).

Additionally, the culture and traditions of the settled population of Transoxiana were adopted by the Karakhanids during their reign (Barthold 1928; Davidovich 1998; Soucek 2000). It was a period in the history of Central Asia when the nomadic tribes and the Turkified settled population formed two major branches of the Turkic language family. According to historians (Barthold 1928; Bosworth and Asimov 2000; Davidovich 1998; Soucek 2000), the combination of the nomadic Turkic culture and settled Islamic institutions spread widely in the region and influenced the cultural development of the nations in Central Asia. Barthold (1928) pointed out that successive phases of Turkish dominance occurred until the arrival of Genghis Khan in the early thirteenth century.

The next wave of Turkic tribes—the Seljuks, followed by the Kara-Khitans—conquered the Karakhanid Khanate. In

Fig. 1.56 The Kalyan minaret in Bukhara

1089, the Seljuks entered and took control of Samarkand, and the Western Karakhanid Khanate became a vassal state of the Seljuks for half a century (Golden 2006; Tolstov et al. 1950).

The Kara-Khitan host invaded Central Asia and defeated the Western Karakhanids in 1137 in Khujand (now a city in modern Tajikistan). The Kara-Khitan became the dominant force in the region, and the Karakhanids remained in power as their vassals. The Kara-Khitan practiced Buddhism, and their rule was marked by religious tolerance in which different faiths coexisted. This Kara-Khitan Khanate was a sinicized Khitan Empire in Central Asia. According to the scant information available from outside sources, it is assumed that this empire took on the trappings of a Chinese state. The rulers adopted many administrative elements from the Liao Dynasty, Confucian administration, the Chinese calendar, and Chinese-style coinage (Rtveladze 2005). At the same time, they adopted local administrative titles. The Khitans were nomads and, after defeating the Western Karakhanids, they took control of the Fergana Valley. In 1141, they gained control over Transoxiana. The Khitans maintained their old customs related to their nomadic lifestyle, including traditional marriages between two clans and practice of a mix of Buddhism and fire worship. The empire was administered in the Persian, Uyghur, Chinese, and Khitan languages (Clark 2010).

This empire was conquered by Genghis Khan in 1218 and became absorbed into the Mongol Empire. In the fourteenth century, they began to lose their ethnic identity. However, the traces of their presence may be found as clan names or toponymy—for example, the Kara-Khitan settlement, the Naiman and Kangli tribe names in the Tashkent region, and the names Kungrad and Kunchji in Surkhandarya (Surxondaryo) region. The Mongols ruled in the thirteenth century.

In about 1400, the era of the Tamerlane (Timur) began (Ibn Arabshox 1992; Mirza Muhammad Haydar 1996). By the end of the fifteenth century, all of the Timurid possessions passed into the hands of Uzbek people, led by Shaybani, who, in 1499, conquered this state and Samarkand (1500–1501) and founded the new centralized Shaybanid State, which was divided in 1599 into the Khanates of Khiva and Kokand (Qoqon) and the Emirate of Bukhara. In the 1860s, after an invasion by the Russian Empire, protectorate status was granted to these states (Rahimov 2001). In 1924, they were united in the form of the Uzbek Soviet Socialist Republic (UzSSR). In 1991, the Republic of Uzbekistan was founded after dissolution of the Soviet Union.

1.6 The Revival of the Great Silk Route

The main part of Uzbekistan is a plain, homogeneous landscape, which made it attractive for establishing trade routes between the Orient and the West—the so-called Great Silk Route—in ancient times (Fig. 1.57).

Today, this idea is also vital, and we see its revival. It is not surprising that because of Uzbekistan's unique geographically central location, it has been central to cultural and economic integration of nations during its long history. All routes between Europe and Asia still go through Central Asian lands, and the capital of Uzbekistan is the pathway to the East, as it used to be many centuries ago.

Uzbekistan is in a less favorable position than other landlocked former Soviet Union (FSU) countries. Its double-landlocked geographic location creates the biggest challenge for bringing its products and services to the world market. In Soviet times, Uzbekistan's foreign economic relations were carried out through three seaports: (1) Ilyichevsk (with access to the Black Sea), (2) St. Petersburg (with access to the Baltic Sea), and (3) Vladivostok (with access to the ports of the Far East). Their respective distances from Uzbekistan—3000, 4300, and 8000 km—significantly increased the costs of transportation.

The double-landlocked location of Uzbekistan is a major barrier to its trade expansion and negatively affects development of its foreign economic relations. Like those of other landlocked developing countries, the economic structures of Uzbekistan are significantly dependent on a narrow range of products for economic activity and export earnings. On the way to the seaports, it is necessary to cross several independent states, and this results in high costs of goods.

International trade and use of cross-border transportation systems depend on transportation through neighboring countries. This is also a political issue related to development of the fundamental transit policy.

This has motivated Uzbekistan's government to strengthen regional cooperation and to initiate programs for infrastructure development. China is also interested in revival of the Great Silk Route, which may expedite the transportation of Chinese goods to the Middle East and Europe (Sookstipaisarnkit and Garimella 2018). An agreement on international road transportation between the Government of the Republic of Uzbekistan and the Government of the People's Republic of China was signed on May 12, 2017, and has been in force since January 1, 2018. This is a new step in reconstruction of the Great Silk Route. In this agreement, mutually acceptable legal conditions have been established. This is not just a positive impetus for further development of bilateral cooperation but also a contribution to formation of a mutually beneficial transportation market. Various countries in the region are interested in this kind of political development. The new Angren–Pop railroad route, constructed through cooperation between Uzbekistan and China, minimizes expenses for transportation of goods and services. A highway and railroad connect the Fergana Valley with other parts of Uzbekistan; this transportation corridor thereby becomes an important part of the China–Central Asia–Europe transit corridor. The length of

Fig. 1.57 In this map there are shown the routes from China and India that connected East and West through lands of Central Asia. The ancient Silk Road existed from the second century BCE until the mid-15th century. It was initiated and spread by China's Han dynasty through exploration and conquests in Central Asia. With the Mediterranean linked to the Fergana Valley, the next step was to open a route across the Tarim Basin and the Hexi Corridor to China Proper. The three main routes spread all over the Xinjiang Uygur Autonomous Region. The Southern Route wandered west along the northern foot of the Kunlun Mountains, passing Ruoqiang (Charkhlik), Qiemo (Cherchen), Hetian, Yecheng (Karghalik), Shache (Yarkand) and reached Kashgar (the last point of the Silk Road in China). [Accessed on April 15, 2021 from https://www.travelchinaguide.com › silk-road › route]

There are four corridors which link India with ancient Silk Roads. (i) The road through high Tibetan plateau and down to the Ganges – to Sravasti. (ii) The Road through valleys and mountains of western Nepal to the fertile valleys of the Ganges. https://en.unesco.org/silkroad/countries-alongside-silk-road-routes/india

Fig. 1.58 The highway through the Qamchyk mountain pass in Tashkent region. [Photo by Magfirat Muminova]

this corridor via the Qamchyk mountain pass in Tashkent province is 19.2 km, and in some parts it reaches an altitude of 2200 m (Fig. 1.58).

The Trans-Asian Railway (TAR) is a project to create an integrated freight railroad network across Europe and Asia. This is a project of the United Nations Economic and Social Commission for Asia and Pacific (ESCAP), which was revived in 1992. According to ESCAP (2010), three components of the project are the Asian Highway, the TAR, and facilitation of land transportation projects through intermodal transportation

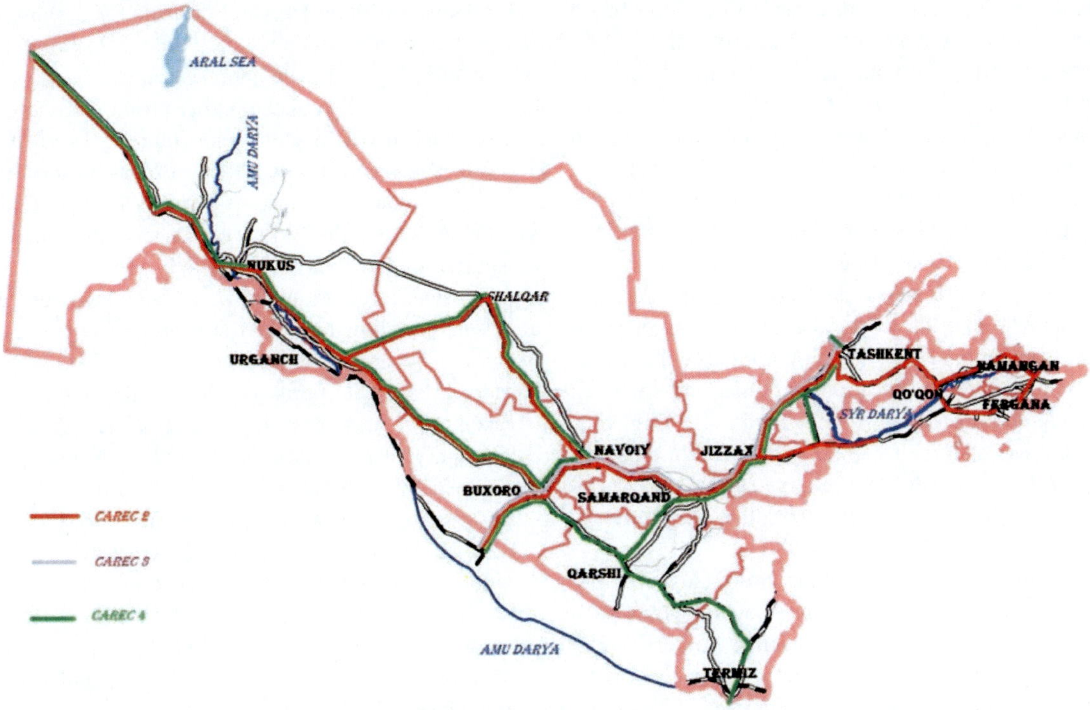

Fig. 1.59 Central Asia Regional Economic Cooperation (CAREC) corridors in Uzbekistan

terminals. This planned network of 141,271 km will cross 32 Asian countries and connect the pan-Asian and pan-European rail networks, as well as highways.

Since three Central Asia Regional Economic Cooperation[16] (CAREC) corridors cross Uzbekistan, this country is a key transit country for trade in Central Asia and beyond (CAREC 2017) (Fig. 1.59).

According to the Asian Development Bank (ADB) (2019), further development of transportation is considered in the ADB country operation business plan (COBP). In accordance with the COBP, electrification of the existing railroad from Samarkand to Karshi (Qarshi), as part of CAREC, will improve the operational performance and institutional effectiveness of Uzbekistan's railroad system. It will also improve regional connectivity, cut transportation costs, lessen greenhouse gas emissions, and boost domestic and regional trade. It is also an integral part of the CAREC Transport and Trade Facilitation Strategy and Action Plan (Asian Development Bank 2011) and is aligned with the Uzbekistan Transport Sector Strategy (Asian Development Bank 2007).

It is clear that all of these projects need enhanced investment.

Uzbekistan is highly motivated to establish a strategy of regional cooperation for development of transportation corridors in Central Asia and an integrated transportation management system. Today, Central Asia connects major international markets. However, without direct access to seaports, this region (especially Uzbekistan) bears significant transit and transportation costs, which reach 70–80% of the cost of export products. Up to 40% of the time required for transportation of goods is lost to suboptimal customs procedures. According to a United Nations (UN) expert analysis cited by the President of the Republic of Uzbekistan in a speech delivered at an international conference (titled the Central Asia: One Past and a Common Future, Cooperation for Sustainable Development and Mutual Prosperity) in Samarkand on November 11, 2017, "the regional GDP [gross domestic product] may grow at least twice in 10 years" if cooperation of Central Asian countries is established. The formation of the Russia–Kazakhstan–Uzbekistan–Turkmenistan–Iran–Oman–India subregional transportation corridor is a result of the Ashgabat Agreement.[17] The Uzbekistan–Kyrgyzstan–China railroad is another construction project, which connects China through Kyrgyzstan and Uzbekistan via the newly constructed Baku–Tbilisi–Kars railroad to the countries of Southern and Eastern Europe, and to Mediterranean ports (Borisova 2012).

[16]Uzbekistan is a member of the CAREC Program.

[17]The Ashgabat Agreement is a multimodal transportation agreement to create an international transportation and transit corridor facilitating transportation of goods between Central Asia and the Persian Gulf.

The Global Development Agenda considers development of international cooperation in the field of transportation as a key element (ESCAP 2017). International partnership for sustainable transportation systems is vital for implementation of the 2030 Agenda (CAREC 2017) and sustainable development goals (SDGs) in Uzbekistan (ESCAP 2017; UNDP 2017).

The involvement of Afghanistan gives an opportunity to form a new transportation corridor with access to the Iranian, Pakistani, and Indian railroad systems and to the Chakhbahar and Gwadar seaports. However, analysis (Safi and Alizada 2018) has shown that these projects require significant improvements in regional transportation infrastructure along with joint actions for simplification of transit and customs procedures. According to the World Bank, the Logistics Performance Index of Uzbekistan ranks 99th out of those of 160 countries (according to data retrieved from the World Bank database (2018) [https://lpi.worldbank.org/international/global/2018] on September 21, 2018).

The best solution is well-functioning trade transportation connecting the whole of Central Asia and beyond with a policy for transit and goods transportation across borders.

The future of this region has been always dependent on its position in world trade. The Russian historian Vasiliy Barthold (1962) noted that the Turkestan's international commerce is entirely dependent on railroads. He compared the cost of trade by sea and land pointing out the importance of integrating Turkestan into world economy through railway construction.

The important development in infrastructure achieved by construction of transportation and energy corridors is evidence of the role played by Central Asian countries, including Uzbekistan, in connecting Asia. One such corridor is the Belt and Road Initiative (also called the Silk Road Economic Belt) proposed by China during a visit to Central Asia in 2013. It stretches from Changán in China through Central Asia to Europe. According to World Bank estimates (World Bank 2018), this project may provide such benefits as a 12% reduction in transportation times along economic corridors, increases in trade volumes of 2.7–9.7%, and an increase in incomes of up to 3.4%, as well as lifting 7.6 million people out of extreme poverty. World Bank experts (Bird et al. 2019, p. 4) have pointed out that "lower trade costs within the region increase the potential for regional specialization of production; reduction in trade costs with the rest of the world increases exposure to external trade, and potential import competition."

Uzbekistan supports all subregional transportation facilitation agreements relating to transit, regional integration, and cooperation. Because of its double-landlocked location, its economic development must prioritize development of infrastructure (railroads, highways, energy, and information and communication technology [ICT]), as well as regional integration initiatives.

Uzbekistan is part of several projects for development of transportation infrastructure, including a 75-km single-rail link connecting Afghanistan, Kazakhstan, Kyrgyzstan, Tajikistan, and Turkmenistan. Other projects are intended to be completed by 2025 within the frameworks of the Asian Highway Network, the CAREC Program, the Eurasian Economic Union, the Transport Corridor Europe–Caucasus–Asia (TRACECA), the TAR network, and other international transportation corridors.[18]

At present, foreign trade in Uzbekistan involves transportation via the following main transportation corridors:

Corridor 1: to the ports of the Baltic States (through Kazakhstan and Russia)—Klaipeda (Lithuania), Riga, Liepaja, Ventspils (Latvia), and Tallinn (Estonia)
Corridor 2: through Belarus and Ukraine (via Kazakhstan and Russia), with border crossings at Chop (Ukraine) and Brest (Belarus), followed by Europe
Corridor 3: to the Ukrainian port of Ilyichevsk (via Kazakhstan and Russia) on the Black Sea
Corridor 4 (also known as TRACECA): the trans-Caucasian corridor (via Turkmenistan, Kazakhstan, and Azerbaijan) to the Black Sea
Corridor 5: to the Iranian port of Bandar Abbas (via Turkmenistan), with access to the Persian Gulf
Corridor 6: eastward via the Kazakh–Chinese border crossing (Dostyk–Alalshankou) to the eastern ports of China and the far eastern ports of Nakhodka, Vladivostok, etc.
Corridor 7: to Chinese ports (via Kyrgyzstan) with access to the Yellow, East China, and South China Seas
Corridor 8: new prospects (following resolution of the conflicts in Afghanistan) for development of alternative southern transportation corridors via Afghanistan to the Iranian and Pakistani ports of Bandar Abbas, Chahbahar, Gwadar, and Karachi

On May 12, 1996, the 320-km Tejan–Sarakhs–Mashhad railroad (with a railroad truck change from a 1520-mm track onto a 1435-mm track at Sarakhs Station) was put into operation. This opened up a new trans-Asian corridor for Central Asian countries to access world markets through Iran and Türkiye. In 1996, Uzbekistan, Azerbaijan, Georgia, and Turkmenistan signed the Agreement on Coordination of Railway Transport and the Agreement on Cooperation in the Area of Management of Transit Transport Between Member States. In 1998, 12 states (including Uzbekistan) signed the Basic Multilateral Agreement on International Transport for Development of the Transport Corridor Europe–Caucasus–Asia.

[18] Source: Ministry of Investments and Foreign Trade of the Republic of Uzbekistan. International Trade. Transportation. Accessed on July 12 2022 (http://www.mift.uz/en/menu/transportirovka).

After construction of the Bafq–Mashhad railroad on the Mashhad–Bandar Abbas route (bypassing Tehran) in Iran in May 2005, the distance to the port of Bandar Abbas was reduced by more than 800 km.

TRACECA[19] is a network of land and maritime routes, which run from Europe to the Black Sea through the Caucasus and the Caspian Sea to the states in Central Asia. This corridor is half length of the transoceanic route from Yokohama to Rotterdam or Hamburg. An international trans-Afghan transportation corridor cuts the distance for transportation of exported goods to the ports of the Republic of Iran by 1500 km.

Transportation of Uzbek goods through Afghanistan will favorably contribute to diversification of routes for exportation of goods via Iranian and Pakistani seaports.

Construction of new railroads such as the 341-km Navoiy–Uchkuduk–Sultanuizdag railroad in 2001, the 220-km Tashguzar–Boysun–Kumkurgan railroad in 2007, and the 75-km Hairatan–Mazar-i-Sharif railroad in 2010 (the first of its kind in Afghanistan) gives more opportunities for transporting goods within the region and beyond. Cooperation between Uzbekistan and Türkiye has made rail transportation of goods in the direction of Türkiye feasible.

The Baku–Tbilisi–Akhalkalaki–Kars railroad became operational in October 2017 and gives Uzbekistan more access to the markets of Southern and Central Europe, as well as states in the Middle East, through the Mediterranean seaport of Mersin. The new Uzbekistan–Turkmenistan–Iran–Oman–Qatar transportation corridor is another reliable communication link for passenger transit and goods transportation between Central Asia and the seaports of the Persian Gulf and the Sea of Oman.

In addition to transportation, Uzbekistan's national priority areas for international cooperation and investment are agriculture, water supply and management, sanitation, energy, health, and education.

1.7 Common Resources, History, and Future

Central Asian countries share not only their history and natural resources but also their future prospects for further development. Historically, ethnically, and culturally tied to each other, these countries are exploring their place in the world and the region. They were politically and economically isolated in Soviet times. Today, they face not only challenges in transitioning from one system to another but also challenges in self-identification and state building. A common idea for unification is in the process of development, and the kind of model that will be chosen depends on many external and internal factors, which influence the policies of every country.

Nature is a common treasure of all nations, and efficient use of all resources should be the priority for regional cooperation. All major rivers in the region start in the glaciers in the mountains of Kyrgyzstan and Tajikistan, and they flow to the lowlands of Kazakhstan, Turkmenistan, and Uzbekistan. Deposits of valuable oil and gas are located in these lowlands. Today, all countries are looking for a model of unification and cooperative solutions to water and energy disputes.

The southwestern part of the region consists mainly of plains and is more or less homogeneous whereas the eastern mountainous part of the region consists of three different sectors: the Western, Central, and Eastern Tien Shan (Fig. 1.60).

The geology of this region is complicated (Fig. 1.61) and represents the results of subduction/collision of various microcontinents, terrains and island arc complexes during Paleozoic and Mesozoic times (Brunet et al. 2017). This history determines the natural resources of the region, which include an extensive orogenic belt, the Tien Shan, and a series of large, economically valuable sedimentary basins, including the South Caspian Basin in the west, the Amu Darya Basin and the Afghan–Tajik Basin in the south, and the Fergana Basin in the east. These places are rich with gas, gas condensate, and oil fields largely located within the Jurassic–Paleogene successions.

Uzbekistan is a member of the world community and a member of many international and regional organizations. Today, Uzbekistan is developing its strategy for widening economic relations with many countries to increase investments in its mining and processing industries.

The aim of this regional cooperation is to improve management of water resources and energy. Development of a renewable energy industry is vital. It was previously estimated that by 2020, Uzbekistan would increase its use of energy from 57.5 billion to 71.8 billion kilowatt hours. The total potential of solar power, wind power, and hydropower combined has been estimated to be more than 530 billion kilowatt hours (see Chap. 4, this volume).

In 2017–2020, the Uzbek government planned to implement a program of measures for further development of renewable energy. According to this program, the share of renewable energy in the energy structure would increase from 12.7% in 2016 to 19.7% by 2025. It was hoped that by implementing this program, the country might achieve several goals, including diversification of its energy structure,

[19]TRACECA is an internationally recognized program aimed at strengthening economic relations, trade, and transportation communication in the regions of the Black Sea Basin, South Caucasus, and Central Asia on the basis of the political will and common aspirations of all member states (http://www.traceca-org.org/en/countries/).

Fig. 1.60 Central Asia [Public domain via Wikimedia Commons User Shauni (note: This image is in the public domain because it contains materials that originally came from *The World Factbook*, a public-domain work published by the US Central Intelligence Agency)]

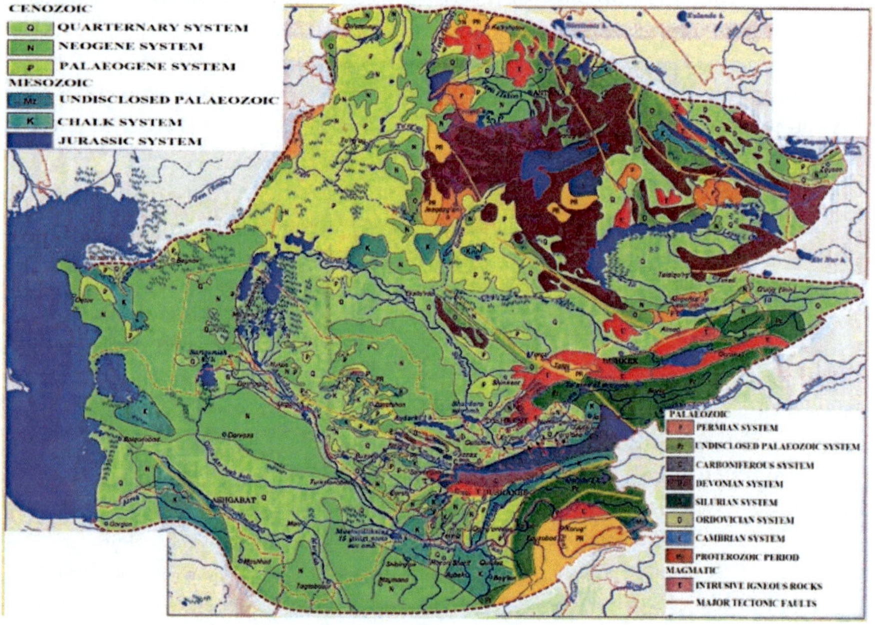

Fig. 1.61 The geology of Central Asia. [From The Physical Geography of Central Asia and Uzbekistan (2017), courtesy of the State Committee for Land Resources, Geodesy, Cartography and State Cadastre of the Republic of Uzbekistan]

reductions in consumption of natural gas and coal, creation of a new industry with new jobs, reductions in atmospheric emissions, and compliance with international commitments to reduce greenhouse gas emissions.

These goals are not only nationally important; they are also of regional and global importance. The natural environment in the Central Asian region is not divided between the states in the region, and the geographic borders of climate, landscape, geology, and topography do not match the state borders. Demarcation is another issue of discussion between states, where it is important not only to define lines on the ground but also to discover solutions for sharing these states' common natural environments and resources.

References

Abdullaev DA, Isaev RI (2005) Resources of solar radiation and wind energies in Uzbekistan and System of Their Combined Utilization. ISESCO Sci Technol Vision 1:76–82 http://www.isesco.org.ma/ISESCO_Technology_Vision/NUM01/doc/D.A.Abdullaev.pdf

ADB (2007) Uzbekistan: Transport Sector Strategy (2006–2020). Accessible from https://www.adb.org/projects/37691-012/main

ADB (2011) Technical Assistance Report: Solar Energy Development. Retrieved February 23, 2019 from https://www.adb.org/sites/default/files/project-document/60309/45120-001-uzb-tar.pdf

Arabshox I (1992) The history of Amir Temur Tashkent Mehnat (In Uzbek ИбнАрабшохАмирТемуртарихиТошкентМеҳнат)

Asimov MS, Bosworth CE (eds) (1998) History of civilisations of Central Asia. Retrieved February 22, 2019 from https://en.unesco.org/silkroad/sites/silkroad/files/knowledge-bank-article/vol_IVa%20silk%20road_the%20karakhanids.pdf

Ayzenshtat BA et al (1983) The Climate of Fergana (In Russian Климат Ферганы. Под редакцией Б.А.Айзенштата, В.Н.Бабиченко, Г.Н.Леухиной Гидрометеоиздат Ленинград 1983) Retrieved February 23, 2019 from http://elib.rshu.ru/files_books/pdf/img-125143636.pdf

Ayzenshtat BA et al (1986) The Climate of Nukus (In Russian Климат Нукуса. Под редакцией Б.А.Айзенштата, В.Н.Бабиченко, Г.Н.Леухиной Гидрометеоиздат Ленинград 1986) Retrieved February 23, 2019 from https://docplayer.ru/37255302-Klimat-nukusa-gosudarstvennyy-komitet-sssr-po-gidrom-eteorologii-i-kontrolyu-prirodnoy-sredy.html

Azimov JA, Umarov NM, Mirabdullayev IM, Khamrayev AS, Mirzayev UT, Chikin YA, Lanovenko EN, Mitropolskaya YO (eds) (2009) The Red Data Book of the Republic of Uzbekistan, vol 2. "Chinor ENK" Tashkent

Barthold W (1928) Turkestan down to the Mongol invasion. Luzac & Co, London

Baumer C (2008) Traces in the desert: Journeys of discovery across Central Asia. I.B. Tauris

Bird JH, Lebrand MSM, Venables AJ (2019) The Belt and Road Initiative: reshaping economic geography in Central Asia? Retrieved from http://documents.worldbank.org/curated/en/836301554729486900/The-Belt-and-Road-Initiative-Reshaping-Economic-Geography-in-Central-Asia

Borisova EA (2012) Aspects of the water crisis in Central Asia. J Hist Modern Times 1(15). Retrieved from http://www.cawater-info.net/bk/7-3_e.htm

Bosworth CE, Asimov MS (eds) (2000) History of civilizations of Central Asia Volume IV. The age of achievement: A.D. 750 to the end of the fifteenth century. UNESCO Publishing, Paris, 700 pp. – (Multiple History Series). – ISBN 92-3-103654-8

Brunet M-F, McCann T, Sobel ER (eds) (2017) Geological evolution of Central Asian Basins and the Western Tien Shan Range. Geological Society, London, Special Publications, 427. https://doi.org/10.1144/SP427.17 # 2017 The Author(s). The Geological Society of London. Publishing disclaimer: www.geolsoc.org.uk/pub_ethics. Retrieved January 20, 2017 from https://hal.archives-ouvertes.fr/hal-01561458/document

CA WATER info (2019) Portal of knowledge for water and environmental issues in Central Asia. Knowledge Base. Degradation of glaciers. Retrieved February 23, 2019 from http://www.cawater-info.net/bk/7-3_e.htm

CAREC 2030: Connecting the Region for Shared and Sustainable Development (2017) CAREC. Retrieved from https://www.carecprogram.org/?publication=carec-2030-connecting-the-region-for-shared-and-sustainable-development

Clark L (2010) The Turkic script and Kutadgu Bilig. Turkology in Mainz, Otto Harrasowitz GmbH & Co, p 96

Climate Zone (2018). Retrieved from https://www.climate-zone.com/climate/uzbekistan/celsius/termez.htm. Portions of this site are based on the CIA World Fact Book, a public-domain work

Davidovich EA (1998) The Karakhanids. In: Asimov MS, Bosworth CE (eds) History of civilisations of Central Asia. Retrieved from https://en.unesco.org/silkroad/sites/silkroad/files/knowledge-bank-article/vol_IVa%20silk%20road_the%20karakhanids.pdf

Derevyanko et al (2013) Results of the Kulbulak (Uzbekistan) Site Research in 2013. In: Problems of archeology, ethnography, anthropology of Siberia and adjacent territories: materials of the annual session of the Institute of Archeology and Ethnography of the SB RAS (In Russian Деревянко А.П., Исламов УИ., Павленок К.К., Шнайдер С.В., Рыбин Е.П., Павленок Г. Д., Раджабов А., Лазарев С.Ю., Мухтаров Г.А.. Результатыис следованиястоянкиКульбулак (Узбекистан) в 2013 году // Проблемыархеологии, этнографии, антропологииСибири и сопредельныхтерриторий :материалыгодовойсессииИнтаархеологии и этнографии СО РАН 2013 г. – Новосибирск, 2013. Retrieved May 20, 2018 from https://cyberleninka.ru/article/n/zubchatoe-mustie-stoyanki-kulbulak-novye-dannye-i-interpretatsii

Dubrovskaya D (2008) The Ghost Empire (In Russian ДубровскаяД. Империя-призрак// Вокругсвета,июнь 2008). Retrieved January 28, 2017 from http://www.vokrugsveta.ru/Империя-призрак_Публикации_Вокруг Света.html

Egamberdieva D et al (2013) Medicinal plants from Chatkal Biosphere Reserve used for folk medicine in Uzbekistan. Med Aromatic Plant Sci Biotechnol 7(1):56–64

Eshkulov FB (2012) Attenuation of solar radiation in desert zones of Uzbekistan. MSc thesis (In Russian: Эшкулов Ф. Б. Ослабление солнечной радиации в пустынных зонах Узбекистана). Retrieved February 23, 2019 from http://library.ziyonet.uz/uz/book/download/5276

Frachetti MR Spengler RN (eds) (2015) Mobility and ancient society in Asia and the Americas. Springer. Retrieved May 10, 2018 from https://books.google.co.uz/books?id=S7E0CgAAQBAJ&pg=PA19&lpg=PA19&dq=site+of+primitive+man+in+Kulbulak&source=bl&ots=OO3u7xWk_K&sig=7WUGt7tZ-zUeeiRfjoQDmI3zcQw&hl=ru&sa=X&ved=0ahUKEwjv062vnKrbAhVLESwKHVe3CZoQ6AEIVTAF#v=onepage&q=site%20of%20primitive%20man%20in%20Kulbulak&f=false

Gintzburger G, Toderich KN, Mardonov BK, Mahmudov MM (2003) Rangelands of the arid and semi-arid zones in Uzbekistan. CIRAD-ICARDA pp. 432. Accessible from May 13, 2016 https://www.researchgate.net/publication/303014592_Rangelands_of_the_Arid_and_Semi-Arid_zones_in_Uzbekistan

Golden PB (2006) Some thoughts on the origins of the Turks and the shaping of the Turkic peoples. In: Mair VH (ed) Contacts and exchange in the ancient world. Hawaii University Press, Honolulu

Grousset R (2004) The empire of the Steppes. Rutgers University Press

Gumilyov LN (2000) The End and the Start Again Moscow Ayris Press (In Russian: Гумилёв Л.Н. Конец и вновьначалоМоскваАйрисПресс)

Gumilyev LN (1967) The Ancient Turks (In Russian ГумилевЛНДревниетюркиМосква: Наука) Retrieved April 20, 2017 from http://gumilevica.kulichki.net/OT/index.html

Ibatullin S (2011) Transboundary water cooperation in Central Asia. Executive Committee of the International Fund for Saving Aral Sea. Retrieved February 23, 2019 from https://www.usea.org/sites/default/files/event-/EC_IFAS_Overview.pdf

Kasymov MR, Grechhkina TYu (1994) The multi-layered Paleolithic site of Karabulak. MaterialyIssledovaniapoArkheologii 185:11–119

Khasanov HH (1981) Travellers-scientists. Uzbekistan Publishing House, Tashkent (in Uzbek) (Сайёҳолимлар, Т., "Ўзбекистон", 1981)

Khasanov HH (1983) Bobur as a Traveller and Naturalist. Uzbekistan Publishing House Tashkent (in Uzbek) (Бобирсайёҳватабиатшунос Т., "Ўзбекистон", 1983)

Khassanov FH (ed) (2016) The Red Data Book of the Republic of Uzbekistan. Volume 1: plants. "Chinor-ENK" Tashkent

Khasanov IA, Gulomov PK (2006) The physical geography of Uzbekistan. Tashkent –'O'qituvchi' (In Uzbek: Ҳасанов ИА, Ғуломов ПК Ўзбекистон табиий географияси 2006 Тошкент Ўқитувчи)

Khojimatov OK et al (2015) Some wild growing plants in traditional foods of Uzbekistan. J Ethnic Food 2(1):25–28. https://doi.org/10.1016/j.jef.2015.02.005. Accessible from http://www.sciencedirect.com/science/article/pii/S2352618115000062

Krivoshapkin AI (2017) Obi-Rahmat variant of the transition from the Middle to the Upper Paleolithic in Central Asia (In Russian Кривошапкин АИ Обирахматскийвариантпереходаотсреднего к верхнемупалеолиту в ЦентральнойАзии. Retrieved May 21, 2018 from https://www.twirpx.com/user/13930987/

Litvinski BAĬ (2012) EXCAVATIONS iii. Central Asia Encycloped Iranica IX(1):96–106. Available online at http://www.iranicaonline.org/articles/excavations-iii. Accessed 30 Dec 2012

Lunin BV (1988) The history of Uzbekistan (Лунин Б.В. ИсторияУзбекистана в источниках. Ташкент, 1988)

Mabberley DJ (2017) Mabberley's plant-book. A portable dictionary of plants, their classification and uses. Cambridge University Press. https://doi.org/10.1017/9781316335581

Makhmadaliev B et al (2008) The Second National Communication of the Republic of Tajikistan under the United Nations Framework Convention on Climate Change. Dushanbe. Retrieved February 23, 2019 from https://unfccc.int/resource/docs/natc/tainc2.pdf

Mayor A (2014) The Amazons: lives and legends of warrior women across the ancient world. Princeton University Press

Mirza Muhammad Haydar (1996) Tarih-I Rashidi Tashkent. The Academy of Sciences of the Republic of Uzbekistan (In Russian МирзаМухаммадХайдарТарих-иРашидиТашкентИздательство АкадемииНаукРеспубликиУзбекистан)

Muhamedjanova MM (2004) Chatkal Nature Reserve as important area for plants's diversity conservation in Western Tien Shan. In: Proceeding book of 4th European conference on the conservation of wild plants, 17–20 September, Valencia, Spain, p 65

National Aeronautics and Space Administration, Science Mission Directorate (2010) The Earth's radiation budget. Retrieved August 10, 2016 from NASA Science website: http://science.nasa.gov/ems/13_radiationbudget

Okladnikiov A (1939) The discovery of the Neanderthal man in Uzbekistan. Bull Ancient Hist (in Russian ОкладниковА. Находканеандертальца в Узбекистане/Вестникдревнейистории. 1939, № 1, стр. 256–257)

Osborn L (2019) Current results. Weather and science facts. Retrieved February 23, 2019 from https://www.currentresults.com/Weather-Extremes/sunniest-places-countries-world.php

Qara-khanids (2009) Encyclopedia of the Peoples of Africa and the Middle East, vol 1 (Stokes J (ed)). Infobase Publishing, p 578

Rahimbekov RU, Dontsova ZN (1982a) The history of study of Central Asian physical geography. Uqituvchi, Tashkent (in Uzbek) (In Uzbek Раҳимбеков Р.У., Донцова З.Н. ЎртаОсиётабиатинигео графикўргаништарихи. Тошкент, Ўқитувчи, 1982)

Rahimbekov RU, Dontsova ZN (1982b) The history of study of the nature of Central Asia. (In Uzbek Раҳимбеков Р.У., Донцова З.Н. ЎртаОсиётабиатинигеографикўргаништарихи.Тошкент, Ўқитувчи, 1982)

Rahimov J (2001) The history of Uzbekistan Tashkent Literature and Arts Publishing House named after GafurGulom (In Uzbek Ўзбе кистонтарихиТошкентҒафурҒуломномидагиАжабиётвасанъат нашриёти)

Rostovtsev OM (2006) Kulbulak is the Stone Age Site of a Primitive Man (In Russian Кульбулак – стоянкаэпохикаменноговека. Retrieved May 20, 2018 from http://arheolog.su/kulbulak/)

Rhodes C (2010) Solar energy: principles and possibilities. Sci Prog 93(1):37–112. https://doi.org/10.3184/003685010X12626410325807

Rtveladze E (2005) Civilizations, states, and culture of Central Asia (in Russian).Цивилизации, государства, культурыЦентральнойАзии. Ташкент, 2005)

Safi M, Alizada B (2018) Integrating Afghanistan into the belt and road initiative review, analysis and prospects FES. Retrieved from http://library.fes.de/pdf-files/bueros/kabul/15587.pdf

Sennikov A et al (2016) The flora of Uzbekistan project. Phytotaxa 282(2):107–118. https://doi.org/10.11646/phytotaxa.282.2.2

Sookstipaisarnkit P, Garimella SR (eds) (2018) China's one belt one road initiative and private international law, Routledge research in international law. Retrieved from https://www.researchgate.net/publication/317317052_China's_'One_Belt_One_Road'_-_Transnational_and_Multilevel_Rule_of_Law_Challenges_from_a_European_Perspective

Soucek S (2000) A history of Inner Asia. Cambridge University Press

Speech of the President of Uzbekistan Shavkat Mirziyoyev at the International Conference «Central Asia: One past and a common future, cooperation for sustainable development and mutual prosperity» in Samarkand – Ministry of Foreign Affars of the Republic of Uzbekistan (2017, November 11). Retrieved from https://mfa.uz/en/press/news/2017/08/13142/

Struve VV, Snegiryov IL (eds) (1935) Turaev BA The history of the ancient East Leningrad OGIZ (In Russian: Тураев В.А.История древнегоВостокаПодредакциейСтруве.В.В, СнегиреваИ.Л. ЛенинградОГИЗ)

Suleimanov RH (1972) Statistical study of the Obi-Rahmat grotto. Academy of Sciences of the Uzbekistan Soviet Socialist Republic, Institute of Archeology Tashkent (in Russian Сулейманов, Р. Х.С татистическоеизучениекультурыгротаОби-Рахмат [Текст] / АН УзССР. Ин-т археологии. – Ташкент : Фан, 1972. – 171 с)

The History of Herodotus. Retrieved June 19, 2016 from https://en.wikisource.org/wiki/The_History_of_Herodotus_(Rawlinson)/Book_4

The History of Uzbek SSR (1953) Tashkent Academy of Sciences of UzSSR (In Russian ИсторияУзбекскойССРТашкент: Издательс твоАкадемииНаукУзССР)

The Physical Geography of Central Asia and Uzbekistan (2017) State Research Institute "Kartografiya" of the State Committee for Land Resources, Geodesy, Cartography and State Cadastre of the Republic of Uzbekistan (in Uzbek)

Tojibaev KSh et al (2017) Botanical geography of Uzbekistan. Korea National Arboretum, Pocheon, Republic of Korea

Tolstov SP, Zahidov VYu, Gulyamov YaG, Nabiyev RN, Trever KV, Yakubivskiy AYu, Voronets ME (1950) The History of People of Uzbekistan Tashkent Academy of Sciences of UzSSR (in Russian Тревер К.В., Якубовский А.Ю., Воронец М.Э. ПодредакциейТолстова С.П., Захидова В.Ю., Гулямова Я.Г., Набиева Р.Н ИсториянародовУзбекистана. Том 1 1950 Ташкент: издательство АН УзССР

UN ESCAP (2010) Transport Division Website. Accessible August 23, 2010, http://www.unescap.org/ttdw/common/TIS/ALTID/Altid.asp

UN ESCAP (2017) Transport and Communications Bulletin for Asia and the Pacific No. 87 Transport and Sustainable Development Goals ISSN: 0252–4392 ST/ESCAP/SER.E/87 Available from https://www.unescap.org/sites/default/files/publications/bulletin87_Fulltext.pdf

United Nations Development Programme (2011) The glaciers of Central Asia: a disappearing resource. Available from http://www.cawater-info.net/pdf/glaciers_of_central_asia.pdf

UNDP (2015) United Nations Development Programme in Uzbekistan Global Environment Facility State Committee for Nature Protection of the Republic of Uzbekistan Fifth National Report of the Republic of Uzbekistan on Conservation of Biodiversity Published in Uzbekistan. Accessed on September 15, 2018 from https://www.cbd.int/doc/world/uz/uz-nr-05-en.pdf

United Nations Development Programme (2017) Uzbekistan. What the global goals mean for Uzbekistan. Available October 25, 2017 from http://www.uz.undp.org/content/uzbekistan/en/home/presscenter/articles/2018/10/26/what-the-global-goals-mean-for-uzbekistan.html

UNESCO (2015) World heritage Centre Western Tien-Shan. Accessed on August 10, 2018 from https://whc.unesco.org/uploads/nominations/1490.pdf

World Bank (2018) Belt and Road Initiative. Accessible March 29, 2018 from https://www.worldbank.org/en/topic/regional-integration/brief/belt-and-road-initiative

World Bank (2019) Climate adaptation and mitigation program for Aral Sea Basin CAMP4ASB. World weather & Climate Information

Worldwide weather forecasts and climate information (n.d.) Retrieved on August 19, 2019 from https://weather-and-climate.com/

Zakirov KZ (1989) Classification of Central Asian vegetation. "Fan", Tashkent (in Russian)

2 Historical Geography

Abstract

This part of the book covers such issues as interrelationships between the human and physical environments and political systems, and the organization of political territories over time. Cultural development is reviewed from the historical and geographic perspectives. The social, ethnographic, and economic influences of the Great Silk Route are explained through the prism of accumulation of traditions associated with clothing, food, and celebrations from many nations.

Keywords

State building · Government · Culture · Traditions · Food · Dress

2.1 State Building

In the modern history of Uzbekistan, several major steps in state building have been clearly defined, and they are reflected in the whole geography of this country. The history of state building is full of achievements and failures of economic and social policies and ideas that have inspired people over time.

2.1.1 Ancient History

The political system in this country is historically connected with the interrelations between humans and the physical environment. According to findings from archeological excavations (Dani and Masson 1992; Koshelenko 1985; Struve 1935; Rtveladze 2005), a transition to irrigated agriculture and a settled lifestyle started in the southern part of Uzbekistan in places with more favorable natural conditions, rich with fertile soil, water, and land resources.

Founded in southern Uzbekistan, the ancient cities and states were conquered by King Cyrus and mentioned in historical chronicles for the first time in the sixth century BCE (Rawlinson 1910) as part of the Achaemenid Dynasty. The state building that occurred under this dynasty resulted in strong governance. The remnants of several provinces of this Empire—such as Bactria, Sogdiana, and the Khorezm (Xorazm) satrapies—were stable over time and existed until the invasion of Alexander the Great in the fourth century BCE (Koshelenko 1985; Rtveladze 2005; Rtveladze and Rtveladze 2017; Struve 1935). The strong state building and governance were praised by Alexander the Great, who used this model in other parts of his empire. After his death in 323 BCE, Sogdiana became part of the Seleucid State until 250 BCE, when the Greco-Bactrian State, with its capital in Bactra, was founded (Rtveladze and Rtveladze 2017; Struve 1935). This age of Hellenism was innovative from the state-building perspective. Later, part of India was joined to the Greco-Bactrian State. Good governance under the umbrella of unification resulted in high levels of prosperity, crafts, agriculture, trade, and town planning, with the state issuing its own coinage. Its division into satrapies is evidence of a widely spread model of the (Persian) Achaemenid Dynasty. The location of Bactria on a trade route from China and India, intensification of infrastructure development, road construction, and establishment of good relations with various countries were the main driving forces for the flourishing of this state.

The Kushan Empire was the next one in this long state-building history (Rtveladze and Rtveladze 2017). It was founded by a nomadic tribe of Kushan (Yuezhi), who united Central Asia, part of present-day Afghanistan, and northern India. At that time, cities played important role as centers of crafts and trade, and contributors to the economy, culture, and art. The central part of Uzbekistan and the Fergana Valley were important parts of the trade network of the Great Silk Route.

L. Gulyamova, *The Geography of Uzbekistan*, World Regional Geography Book Series,
https://doi.org/10.1007/978-3-031-07873-6_2

Part of the Greco-Bactrian State was conquered by the ruler of Parthia, Mitridad, in 170–138 BCE. The area to the north of the Kushan Empire was mentioned in the *Zend Avesta* as Kangha, or Kangju (Litvinsky 1996; Litvinskiy and Antonova 1996; Rtveladze and Rtveladze 2017; Struve 1935). It was praised in Chinese records for its developed economy and system of administration as a confederate state of mainly nomadic tribes. Its strategically important location on the Great Silk Route favored its influence on the whole region.

Between the first and fourth centuries CE, this empire was destroyed by nomadic tribes of Hepthalites, and the Hepthalite State existed until it was invaded and destroyed by Turks in the sixth century. The Turkic Khaganate stretched from China to the Volga River, and Central Asia was part of it (Litvinsky 1996; Struve 1935). Control of the Great Silk Route was the major prize in this state's endless wars. The provinces, including Sogdiana and the Fergana Valley, paid tributes to the Khagan, in the form of products, handicrafts, and other goods.

2.1.2 Medieval Development

In 709–712, the Arab Caliphate was established after invading Arabs conquered the main cities of Sogdiana and Bactria (Litvinsky et al. 1996; Sobti 2005). By the tenth century, the region between the Amu Darya and the Syr Darya rivers (Mawarannahr in Arabic, meaning the lands between the Amu Darya and Syr Darya rivers) adopted Islam after strong resistance. Historians (Barthold 1956, 1966; Gibb 1923) point out that the Arab conquest destroyed the economy and culture dramatically; the governance system also changed, and the governors were appointed by the caliph in Baghdad. The policy of the Arab governors aimed to benefit trade on the Great Silk Route by creating a strong power. To restore life to the destroyed cities, they started work to restore the irrigation system, constructing new canals. For example, by an order of the government, a sum of two million dirhams (Alimova and Filanovich 2017) was released to fulfill the need for irrigation of the new city of Tashkent, which was founded to replace the destroyed city of Shash (the name of a city in the eighth century). Having allocated this sum, Caliph Mutasim (833–842) ordered that a canal be dug to improve the Bozsu–Kalkaus water system in the city (Filanovich 1983). The government of the caliphate also released funds to erect border walls and fortresses for protection against nomadic raids. Under the rule of Caliph Mamun (813–833), the representatives of the old local nobility were appointed as governors of the eastern territories. At the end of the ninth century, an almost independent state emerged here, with its capital in Bukhara. In the tenth century, the rulers of the Samanid Empire created a new economy, based on reasonable tax collection and good governance (Barthold 1966). At that time, Bukhara, Samarkand, and Urgench became major centers of science, culture, trade, and craft on the Great Silk Route.

But the flourishing was stopped by an invasion of tribal remnants of the Turkic Khaganate, who founded the Karakhanid State (Barthold 1956, 1966, 1993; Stokes 2009). This invasion destroyed the culture and economy once again. The Karakhanid Dynasty ruled for 200 years. It was mentioned earlier in this book (Chap. 1, Sect. 1.5) that during their rule, intercultural exchange was very intensive and the influences of the Persian and Islamic cultures on the Turkic culture resulted in the flourishing of Islamic civilization, but this period was not successful for state building when the borders and rules were constantly changing.

The eleventh century was a time of conquest by nomads, and the Seljuk Turk tribes won the fight (Barthold 1966, 1993). Under their rule, there was a flourishing of architecture, arts, and culture. In the twelfth century, however, the Seljuk Turk Empire was defeated by the nomadic Kara-Khitan tribe.

In the thirteenth century (from 1219 to 1225), the Mongol invasion destroyed all major states in Central Asia (Tolstov 1943). New governance was established under Yassa law. In accordance with this law, the legitimate ruler could only be a blood descendant of Genghis Khan. This period in the history of Uzbekistan was marked by very rapid Turkification because of intermixing between the local populations and the Turkic tribes incorporated into the Mongol army. As the result of this ethnic invasion, Iranians became a minority in many states in the whole region. The Mongols damaged wealthy cities such as Bukhara and regions such as Khorezm (Xorazm). The irrigation system destroyed by the Mongols could not be repaired for several generations. Mongol rulers appointed local merchants as governors. Mongol law maintained control of most of the Silk Route. The Mongol Empire was united and strong under orderly succession of direct descendants of Chaghatai (the second son of Genghis Khan).

Competition for influence among the princes of various tribal groups resulted in weakening of the Mongol Empire and establishment of the Timurid State in the 1380s. Although Timur was not a descendant of Genghis Khan, he became the de facto ruler of the lands between western Central Asia, Iran, Asia Minor, and the southern steppe region north of the Aral Sea (Ibn Arabshox 1992; Mirza Muhammad Haydar 1996). The foundation of the Timurid State (1363–1370) became an important step in the history of state building. This state monopolized the Silk Route after destroying such competitors as Urgench and Sarai Berke (the Golden Horde's capital). Timur's governance was based on Yassa law, and he preserved all regulations and institutions of the Mongol Empire, with strong, centralized leadership (Mirza Muhammad Haydar 1996). During his reign and the reigns of his immediate descendants, many construction projects were implemented in the major cities of Uzbekistan. Cities such as Samarkand and Bukhara became major trading centers and also centers of art, crafts, literature, culture, and science. It was during the time of the Timurid Dynasty that the Chaghatai dialect became a literary Turkic language, and it is the basis of the modern Uzbek language. This language was used alongside Persian, which had previously been the primary written and spoken language. However, after the death of Timur, this state was broken up and quickly lost its strength.

This state was conquered in 1499 by the Dasht-i-Kipchak nomadic tribes. In 1501, despite resistance, the capital, Samarkand, fell under their onslaught (Mirza Muhammad Haydar 1996). By 1510, a new, centralized Shaybanid (Uzbek) state was founded (Adle and Habib 2003; Mirza Muhammad Haydar 1996), with a new structure of governance. According to sources (Mirza Muhammad Haydar 1996), the Bukhara Khanate, with its capital in Bukhara, controlled Mawarannahr, the region of Tashkent, the Fergana Valley, the east, and northern Afghanistan. A second Uzbek state was established in the oasis of Khorezm (Xorazm) (Mirza Muhammad Haydar 1996). At the end of the sixteenth century, establishment of ocean trade routes from Europe to China and India resulted in a decline of trade through the Central Asian region. The Silk Route lost its advantage. Wars between the Uzbek states and wars with Iran and nomads from the northern steppe also weakened the Uzbek states of Bukhara and Khorezm (Xorazm). The major cities on the Silk Route began to steadily decline. In 1599, the Uzbek state was divided by the Ashtakhanids into the Khiva and Kokand (Qoqon) Khanates and the Emirate of Bukhara (Adle and Habib 2003). These states lost their role as actors on the Silk Road. Despite the separation, they were economically, socially, and culturally closely related to each other. In the sixteenth to nineteenth centuries, these states, because of isolation as the result of the Silk Road declining and of their landlocked geographical location, faced economic challenges and stagnation.

2.1.3 Colonization

Later, in the nineteenth century, these states became victims of "the Great Game"[1] between the Russian and British Empires. They aimed to control the heartland of Eurasia, which, according to a theory proposed by the British geographer Halford J. Mackinder (Mackinder 1904; Megoran and Sharapova 2014), lay at the center of the "world island" stretching from the Volga River to the Yangtze River and from the Himalayas to the Arctic. According to Mackinder's theory, the power that ruled this heartland would command the world island and hence rule the world (Blouet 2005; Mackinder 1904). The players of this Great Game tried to add Central Asia to their empires.

Despite attempts by the Kongirad Dynasty in Khiva, the Manghits in Bukhara, and the Mins in Kokand (Kastel'skaia 1980; Rahimov 2001) to establish centralized states with standing armies and new irrigation systems, these states were not able to win their fight against the Russians.

In the 1860-is, the Russian Empire invaded Central Asia, and protectorate status was granted to the Emirate of Bukhara and the Khiva and Kokand Khanates (Rahimov 2001; Sergeev 2013; Zagorodnikova 2005). All power was in the hands of the Russian governor general. The political and economic interests of the Russian Empire comprised a desire to control trade in Central Asia, to protect the borders of the empire, and to establish a secure source of cotton other than the southern USA (Kastel'sk ia 1980). The three major cities of the khanates—Tashkent, Bukhara, and Samarkand—were captured in 1865, 1867, and 1868, respectively. In 1868, the Khanate of Bukhara signed a treaty with Russia, making Bukhara a Russian protectorate. Khiva became a Russian protectorate in 1873, and the Kokand Khanate was finally incorporated into the Russian Empire, also as a protectorate, in 1876. Thousands of people were killed in the fighting against the Russians. The treaties establishing the protectorate status of Bukhara and Khiva gave Russia control of the foreign relations of these states and gave Russian merchants important concessions in foreign trade (Kastel'skaia 1980; Rahimov 2001; Zagorodnikova 2005); the khanates retained control of their own internal affairs. Tashkent and Kokand fell directly under a Russian governor general. The colonial policy of the Russian Empire was aimed at increasing cotton production. The main center of cotton cultivation was the Fergana Valley. To export the cotton from this region, the first railroads were constructed between major cities on the Great Silk Route: Tashkent, Samarkand, and Bukhara. In the early 1870s, Andijan (Andijon), Kokand, and Margilan were linked to Samarkand and Tashkent. In 1906, the opening of a new railroad between Tashkent and Orenburg (a city in Russia) created new opportunities for trade.

Two different governance systems existed at the same time in the country. The governance system of the Russian Empire was set up for Russians, and governance under shari'a[2] law applied to the local population. To protect the borders of the empire, the Russians founded several cities; for example, the city of Fergana was founded in 1876, although it was called Skobelev (after the Russian general Michael Skobelev) until 1924 (Fig. 2.1).

In 1893, a Russian fortress and garrison and a military border fortification were built on the right bank of the Amu Darya river, several kilometers east of the city of Termez (Fig. 2.2).

[1]The term "the Great Game" is used by historians to describe the nineteenth-century political and diplomatic competition between the British and Russian Empires for territory and influence in the states of Central Asia.

[2]Shari'a law, or Islamic law, is the religious law forming part of the Islamic tradition. It is derived from the religious precepts of Islam, particularly the Koran and the hadith. Islamic law refers to the idea of divine law as imagined by God (*shari'a*), to Islamic moral epistemology and jurisprudence (*usul al-fiqh*), to doctrines of positive law as formulated by Islamic legal schools (*fiqh*), and in general to the applied legal order of Muslim polities when it makes some claim to religious legitimacy (*sulta*, *siyasa*, *qanun*).

Fig. 2.1 The building of the Officers' House in the city of Fergana, with the date of its construction written on the facade

Fig. 2.2 The fortress of the Russian Empire in the city of Termez

The Amu Darya river port was also built there. This port was strategically important because it was located on the border with the British Empire. For many years, it served as a major port for cargo transportation via the Amu Darya river flotilla and linked the Lower Amu Darya region with other regions.

At the end of the nineteenth century, migration of Russians to Central Asia led to increased Russian vigilance in the region and their intrusion into the internal affairs of the Khanates and the Emirate (International Business Publications 2017). A political resistance movement by local young elites was formed under the pan-Turkish movement, or Jadidism (Bazarbayev et al. 2013; International Business Publications 2017). The modern and secular ideas of Jadidism were met with fear by both the Russians and the Uzbek khans. According to research (Bazarbayev et al. 2013; International Business Publications 2017), the Jadids believed that their society and religion must be reformed for independence to be regained. The leaders of this movement were forced underground or into exile. Some of them became the leaders of Soviet Uzbekistan in the 1920s–1930s.

Before the October Revolution in 1917, Russian rule brought innovations in sectors of the economy directly connected with cotton production and its primary processing (International Business Publications 2017). The transportation network was also improved to transport the cotton out of the region and deliver it to the textile industry in central Russia and other parts of the Russian Empire, such as Finland and Poland. Development of the textile industry in the region was a priority for the Russians (International Business Publications 2017); the czarist government expanded cultivation of cotton and changed the balance between cotton production and food production. To some extent, this policy was implemented in Soviet times when the entire agricultural economy of Uzbekistan was converted to cotton production and industries related to cotton. Today, the consequences of this policy are some of the major issues that Uzbekistan faces with regard to environmental protection, food security, and an unbalanced economy.

2.1.4 The Soviet Period

The Soviet period started in 1917, when the Russian revolutionary events were quickly repeated in Tashkent, where the czarist administration of the governor general was overthrown (International Business Publications 2017). A dual system was established by the Russians without involving the local Muslim population and without considering their opinion. Resistance to that unfair system started in the city of Kokand, where local leaders attempted to set up an autonomous government. The Basmachi (Olcott 1981; International Business Publications 2017) revolt against Soviet rule had the support of almost all sectors of Turkestani society. Olcott (1981) notes that an attempt by the Soviet regime to restructure Muslim society was one of the reasons why the Basmachi were in power for more than a decade. The Soviet leaders then promised political and economic autonomy, with more opportunities for the poor. This policy resulted in withdrawal of Central Asian populations from the Basmachi movement (Rahimov 2001).

In 1924–1925, new states (including Uzbekistan) were founded after failure of the struggle against the Red Army and the Soviet system (Rahimov 2001). The complicated patterns of the spatial distribution of different ethnic groups were not considered during the demarcation in the 1920s (Krasnovskiy 1924). Before 1929, Tajikistan was an autonomous republic in Uzbekistan. The policy of the leaders in the late 1920s and the early 1930s was aimed at reforms in agriculture and establishment of collective farms. In the late 1930s, the political and cultural elites were arrested and executed during Stalinist purges (Rahimov 2001). The Moscow government dictated all aspects of the social and economic lives of all countries in the Soviet Union, and local government was filled with people loyal to Moscow (International Business Publications 2017). Uzbekistan was the primary source of cotton for the rest of the Soviet Union.

During World War II, Uzbekistan contributed to the victory over the fascist regime of Germany, but, in doing so, it lost more than 450,000 of its sons. Sixty-five Uzbeks were awarded the title of Hero of the Soviet Union, and 118,000 were awarded various honorary orders and medals (Gulyamov 1952). Today, on May 9 each year, Uzbekistan marks the Day of Memory and Honor—a holiday of light, sadness, reverent respect, pride, and deep gratitude.

In 1941–1942, many industrial plants in European Russia were relocated in Uzbekistan and other Central Asian countries. However, the local people were primarily agricultural workers and did not have the required skills and knowledge to work in industrial plants. Consequent immigration of Russians workers to staff these plants resulted in further Russification of the major cities in Uzbekistan. Crimean Tatars, Chechens, Volga Germans, and Koreans were exiled to Central Asia, and the ethnic mosaic became more colorful. According to census data from 1989 (the last census in the Soviet Union), more than 100 nationalities lived in Uzbekistan at that time.

Uzbekistan is one of few places on Earth with such a rich history of national state building, experiments and changes in the concept of nationahood 1 and self-identification, destruction of states and institutions, loss of national identity, and revival on a new basis. Researchers (Koshelenko 1985; Koshelenko and Filanovich 1999; Megoran 2006; Reeves et al. 2014; Patnaik 2016) agree that modern political borders dissect historical cultural regions that were formed and existed for a long historical period as united entities.

2.1.5 Independence

After dissolution of the Soviet Union in 1991, the Uzbek Soviet Republic became the Republic of Uzbekistan.

Nation building is one of the major challenges faced by the modern states in Central Asia. After dissolution of the Soviet Union, the search for identity was combined with the need to find ways to join the international community, as well as means of development and approval of ideas for regional unification. Differences between topographies, peoples, customs, and beliefs are difficult to ignore when constitutional outcomes in Central Asian states are compared (Denison 2003). An old pan-Turkism concept that was considered in the twentieth century was the idea of Eurasia, with many ideas for regional unification and cooperation. The leaders of this region's states clearly understand the urgent need for economic cooperation. The major challenge is finding a political umbrella of conditions that fulfill the autocratic interests of the current leadership. The main challenges are rooted in the forced reshaping of state-building institutions several times in the history of Uzbekistan and other countries of the region, destroying ideas for bringing together the nations of the region. This resulted in creation of entirely new categories of nationality, and dissolution of the Soviet Union caused this region to lose its economic and political infrastructures (Reeves 2014). The difficulties of national state building at this stage were closely connected to the configuration of the national idea and the understanding of democracy that were shaped under Soviet ideology and implementation of state building (Denison 2003). They brought centralized and hierarchical rule (Merry 2004), inherited from the Soviet model of state building. It is important also to accept the fact that state building is influenced by multiple different conceptualizations of national identity, which shape people's understandings and the meanings given to nationhood (Isaaks 2015).

2.2 Governance

2.2.1 Government

In 1991, after the collapse of the Soviet Union, Uzbekistan declared its independence. The first year of the newly established state was a year of intensive discussions about the model of governance. According to the country's new Constitution, which was adopted in 1992, a unitary presidential constitutional republic was established. This model of governance provides legislative, executive, and judicial branches of government, dominated by a strong executive. New political parties (except for nationalist or religious political parties) can now be founded. However, although personal liberties are protected by law, the government is given the right to restrict some of them in certain circumstances.

A legislative chamber and a senate form the legislature (the Oliy Majlis, or Supreme Assembly) with authority to amend the Constitution, enact legislation, approve budgets, and confirm presidential appointees. Members of the Oliy Majlis are indirectly elected to five-year terms.

The president is elected for a maximum of two consecutive seven-year terms, with the power to be the head of state and of the government. The president appoints high court justices, the prime minister, and the cabinet, confirm contingent on parliamentary approval. The president also has the authority to issue binding decrees and repeal legislation passed by local administrative bodies. The executive branch holds almost all of the power. The president selects and replaces provincial governors.

The legal system is of a civil law type. The judicial branch is represented by constitutional and supreme courts; the latter comprise administrative, civil, criminal, economic, and military sections.

Administrative Divisions

Uzbekistan is divided into the autonomous Republic of Karakalpakstan and 13 provinces or regions (*viloyats*). The regions are divided into 160 districts (*tumans*) (Fig. 2.3). The Republic of Karakalpakstan has been part of Uzbekistan since 1936 and has its own formal sovereignty. According to the Constitution, relations are "regulated by treaties and agreements," and any disputes are "settled by way of reconciliation." The Republic of Karakalpakstan's right to secede from the Republic of Uzbekistan may be applied "on the basis of a nationwide referendum held by the people of Karakalpakstan" (according to Article 74 in Chapter XVII of the Constitution).

Information on the different regions of Uzbekistan is given in Table 2.1. (Note: The numbers on the map relate to the districts).

2.2.2 The Electoral System

Uzbekistan adopted a new electoral system model after gaining its independence. In this two-round electoral system, a run-off election is held to choose between the two highest-placed candidates. The major political parties are the Liberal Democratic Party of Uzbekistan (LDPU), the Uzbekistan National Revival Democratic Party (UNRDP), the Justice Social Democratic Party (JSDP), and the People's Democratic Party of Uzbekistan (PDPU). The largest party is the LDPU, with a platform primarily focused on such economic issues as doubling the country's gross domestic product (GDP) by 2030, creating new jobs, and eliminating poverty.

In 1991, Islam Karimov was elected as the first president of the independent Republic of Uzbekistan, and he won three consecutive elections.

On December 4, 2016, after the death of Islam Karimov, Shavkat Mirziyoyev was elected to the presidency in a popular vote. On October 24, 2021 he was elected as the President of Uzbekistan for the 2nd term.

2.2.3 After Karimov

The policy of the first president was aimed at adopting the new Constitution and laws, and restructuring the governance system by taking the first cautious steps away from an economy dominated by the state. Starr (2018) notes that "Uzbekistan's early strategy was thoroughly defensive in character, it was focusing single-mindedly on the protection of sovereignty and the avoidance of foreign entanglements."

Fig. 2.3 The administrative divisions of the Republic of Uzbekistan

However, although Uzbekistan gained its independence in 1991, the reforms in governance did not change the institutions of government. In many instances, they remain similar to those that existed in Soviet times. It is not surprising that the government policy emphasizing the needs for stability and a gradual approach to change during the transition period has found credence among the population. It was approved in a time of economic crises in 1998 and 2008, which caused minimal damage to the isolated economy of Uzbekistan. Starr (2018) notes that "a policy to shield away from foreign borrowing helped Uzbekistan to survive the 2008 financial crisis almost unscathed." In the longer term, however, this approach has become an obstacle to further development and regional integration. Curtis (1996; International Business Publications 2017, p. 67) argues that "the first years of independence saw more resistance than acceptance of the institutional changes required for democratic reform to take hold." The result of the Karimov presidency is "inertia of the remaining Soviet-style strong centralized leadership" (International Business Publications 2017, p. 67). Today, after Karimov's era, this style is still in the policy of the government. History teaches us that in countries such as Uzbekistan, after many years of oppression, political conservatism and indifference are stronger than the aspiration for change and reform.

Recent initiatives of the Government of Uzbekistan have been aimed at "construction of new transportation infrastructure, economic deregulation to simplify business entrepreneurship, liberalizing of national currency controls, and other market-oriented reforms in pursuit of Uzbekistan's goal of becoming a regional transportation and investment hub" (Voices on Central Asia 2018). These should help to leverage the country's natural advantage—its pivotal central location—more efficiently. Strong ties with China for infrastructure investment across Eurasia provide opportunities for developing economic connections and advancing social and economic reforms to benefit the citizens and maintain geographic pluralism in the heart of Eurasia (Weitz 2018).

In his domestic policy, President Shavkat Mirziyoyev pledged to implement populist measures such as forcing state bureaucrats and local leaders to be more responsive to the people's concerns. His foreign policy has been marked by openness to regional cooperation and efforts to pull Afghanistan into regional security discussions and reintegrate it into the region through trade, energy, and infrastructure linkages, all of which will contribute to regional stability (Uzbekistan… 2018). At the same time, it is a challenging issue because of the historical tense and competitive relations between the main actors on the Great Silk Route and poor coordination of management of vital common natural resources such as water and energy.

Table 2.1 Information on the different regions of Uzbekistan

Region and area (km^2)	Population	Administrative center	Districts	Location
Karakalpakstan				
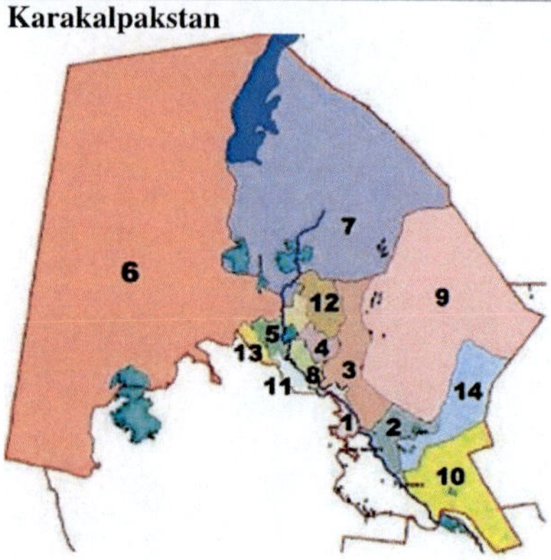				
166,600	1,500,000	Nukus	1. Amudarya, 2. Beruniy, 3. Kegeyli, 4. Muynaq, 5. Nukus, 6. Takhtakupir, 7. Turtkul, 8. Khodjayli, 9. Chimboy, 10. Shumanay, 11. Ellikqala, 12. Qonlikul, 13. Qoraozaq, 14. Kungurat	Karakalpakstan borders Kazakhstan in the north and west, the Navoiy region in the east, the Khorezm and Bukhara regions in the southeast, and Turkmenistan in the south
Andijan (Andijon) region				
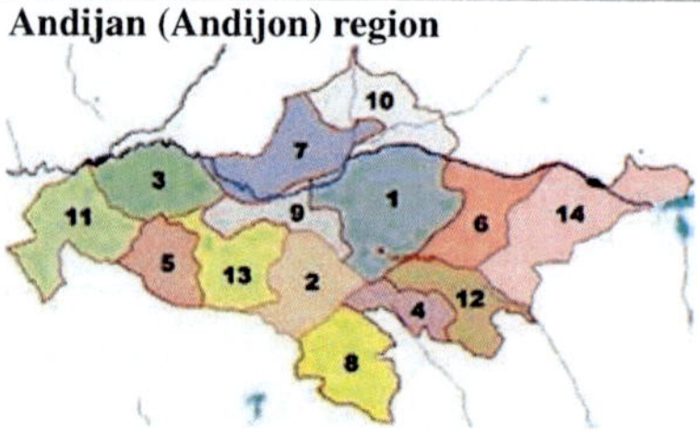				
4300	2,400,000	Andijon	1. Andijon, 2. Asaka, 3. Baliqchi, 4. Buloqboshi, 5. Boz, 6. Jalalkuduk, 7. Izboskan, 8. Markhamat, 9. Oltinkol, 10. Pakhtaabad, 11. Ulygnor, 12. Khodjaobod, 13. Shakhrikhon, 14. Kurgontepa	The Andijon region is located in the easternmost part of Uzbekistan and borders the Namangan region in the north and northwest, Kyrgyzstan in the east, and the Fergana region in the south and southwest
Bukhara (Buxoror) region				
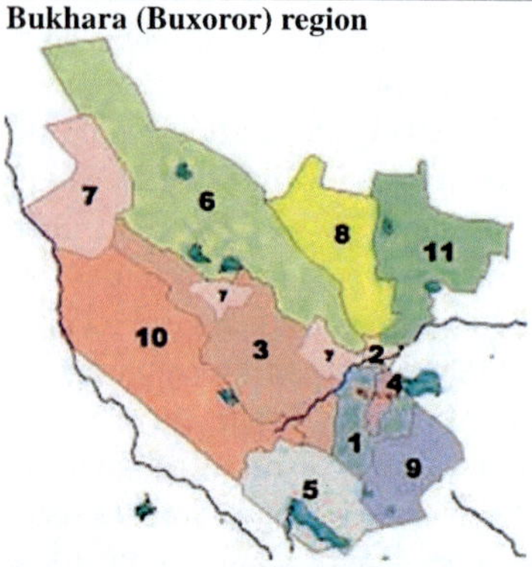				
40,320	1,500,000	Bukhara	1. Bukhara, 2. Vobkent, 3. Djondor, 4. Kogon, 5. Olot, 6. Peshku, 7. Romitan, 8. Shofirkon, 9. Karakul, 10. Karaulbazar, 11. Gijduvan	The Bukhara region is located in the west of the central part of Uzbekistan and borders the Navoiy region in the north and the east, the Kashkadarya region in the southeast, Turkmenistan in the south, and the Khorezm region and Karakalpakstan in the west

(continued)

Table 2.1 (continued)

Region and area (km²)	Population	Administrative center	Districts	Location
Jizzakh(Jizzax) region				
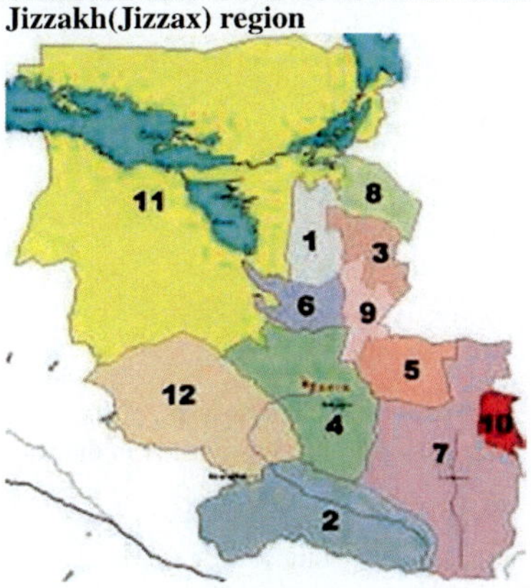				
21,200	1,100,000	Jizzakh	1. Arnasoy, 2. Bakhmal, 3. Dustlik, 4. Jizzakh, 5. Zarbdor, 6. Zafarobod, 7. Zomin, 8. Mirzachul, 9. Pakhtakor, 10. Forish, 11. Yangiobod, 12. Gallaorol	The Jizzakh region is located in the central part of Uzbekistan and borders Kazakhstan in the north, the Syrdarya (Sirdaryo) region in the east, Tajikistan in the south and southeast, the Samarkand region in the southwest, and the Navoiy region in the west
Kashkadarya (Qashqadaryo) region				
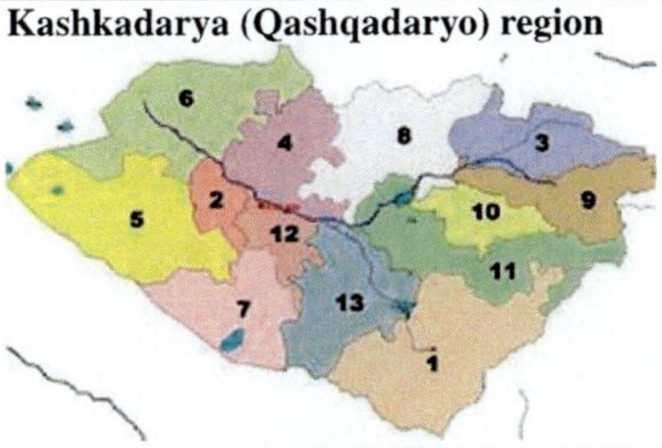				
28,600	2,400,000	Qarshi	1. Dehkanabad, 2. Kasby, 3. Kitob, 4. Koson, 5. Myrishkor, 6. Muborak, 7. Nishon, 8. Chiroqchi, 9. Shakhrisabz, 10. Yakkabog, 11. Kamashi, 12. Qarshi, 13. Guzar	The Kashkadarya region is one of the southern regions of Uzbekistan and borders the Samarkand region in the north, Tajikistan in the northeast, the Surkhandarya (Surxondaryo) region in the east and southeast, Turkmenistan in the south, and the Bukhara region and (partly) the Navoiy region in the northwest
Navoiy region				
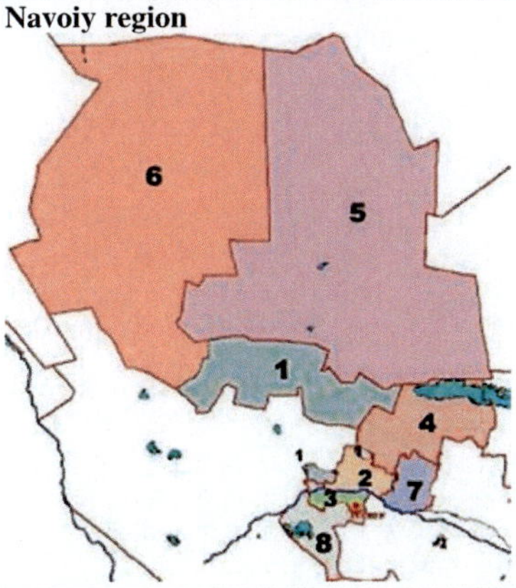				
110,900	850,000	Navoiy	1. Karmana, 2. Konimekh, 3. Navbakhor, 4. Nurota, 5. Tomdi, 6. Uchquduq, 7. Khatirchi, 8. Qiziltepa	The Navoiy region is located in the central part of Uzbekistan, predominantly in the large territory of the Kyzylkum Desert, and borders Kazakhstan in the north, the Jizzakh region in the east, the Samarkand region in the southeast, the Bukhara region and (partly) the Kashkadarya region in the south, and Karakalpakstan in the west

(continued)

Table 2.1 (continued)

Region and area (km^2)	Population	Administrative center	Districts	Location
Namangan region 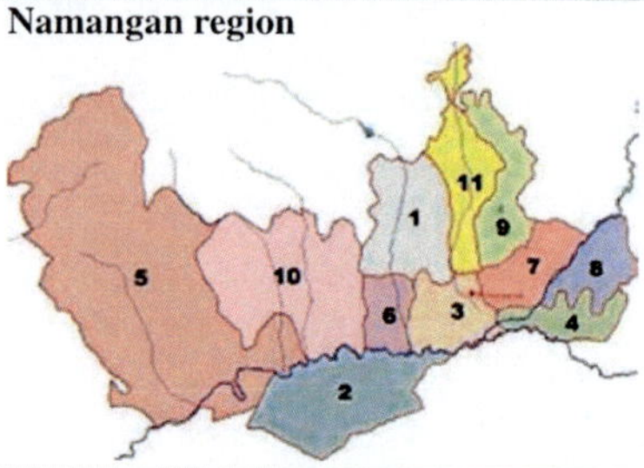				
7400	2,300,000	Namangan	1. Kosonsoy, 2. Mingbulaq, 3. Namangan, 4. Norin, 5. Pop, 6. Turakurgan, 7. Uychi, 8. Uchkurgan, 9. Chartaq, 10. Chust, 11. Yangikurgan	The Namangan region is located north of the Fergana Valley and borders Kyrgyzstan in the north and northeast, the Andijon region in the east and southeast, the Fergana region in the south, Tajikistan in the southwest, and the Tashkent region in the west; the major cities in the Namangan region are Chust, Chortoq, Kosonsoy, Uchkurgon, Khakkulobod, Turakurgan, and Pop
Samarkand region 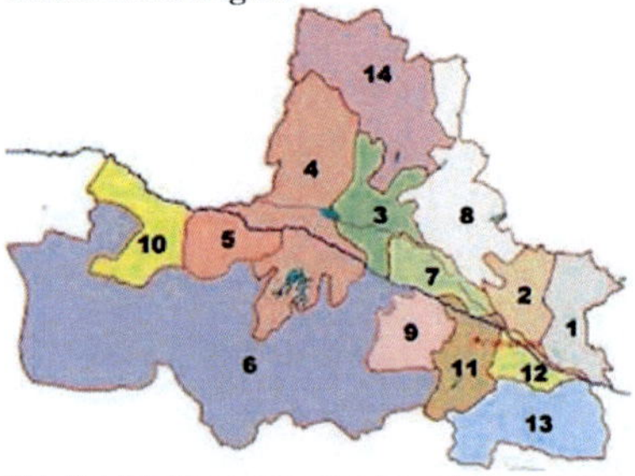				
16,800	2,900,000	Samarkand	1. Bulungur, 2. Jomboy, 3. Ishtikhon, 4. Kattakurgan, 5. Narpay, 6. Nurobod, 7. Oqdarya, 8. Payariq, 9. Pasdargom, 10. Pakhtachi, 11. Samarkand, 12. Toyloq, 13. Urgut, 14. Koshrabot	The Samarkand region is located in the Zarafshan Valley in central Uzbekistan and borders the Jizzakh region in the northeast, Tajikistan in the east, the Kashkadarya region in the south, and the Navoiy region in the west and northwest
Surkhandarya (Surxondaryo) region 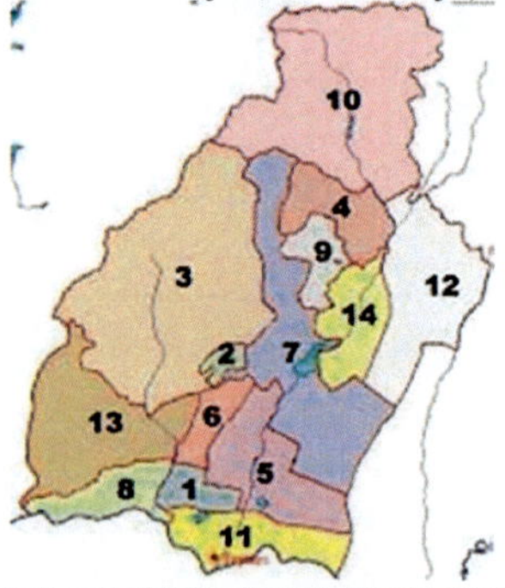				
20,100	1,900,000	Termez	1. Angor, 2. Bandikhon, 3. Boysun, 4. Denou, 5. Djarkurgan, 6. Muzrabod, 7. Oltinsoy, 8. Saryasia, 9. Termez, 10. Uzun, 11. Sherobod, 12. Shurchi, 13. Qiziriq, 14. Qumkurgan	The Surkhandarya (Surxondaryo) region is the most southern region of Uzbekistan and borders Tajikistan in the north and east, Afghanistan in the south, Turkmenistan in the west, and the Kashkadarya region in the northwest

(continued)

Table 2.1 (continued)

Region and area (km^2)	Population	Administrative center	Districts	Location
Syrdarya (Sirdaryo) region 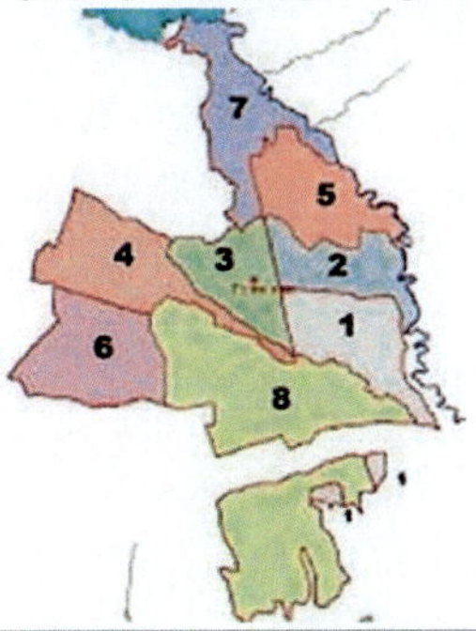				
4280	700,000	Gulistan	1. Boyawut, 2. Gulistan, 3. Mirzaabad, 4. Oqoltin, 5. Saykhunabad, 6. Sardoba, 7. Syr Darya, 8. Khovos	The Syrdarya (Sirdaryo) region is located in the eastern part of central Uzbekistan, to the south of the Syr Darya river (which gave its name to the region), and borders Kazakhstan in the north, the Tashkent region in the south, Tajikistan in the southeast, and the Jizzakh region in the west
Tashkent region 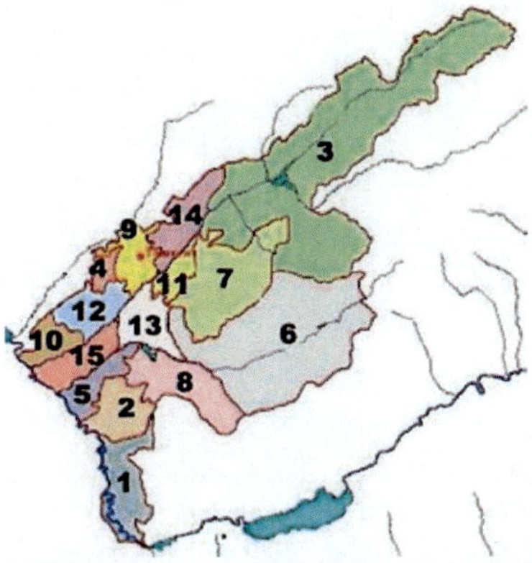				
15,200	2,500,000	Tashkent	1. Bekabad, 2. Buka, 3. Bustonliq, 4. Zangiata, 5. Oqqurgan, 6. Okhargaron, 7. Parkent, 8. Piskent, 9. Tashkent, 10. Chinaz, 11. Yuqori Chirchiq, 12. Yangiyol, 13. Orta Chirchiq, 14. Qibray, 15. Quyi Chirchiq	The Tashkent region is in the eastern part of Uzbekistan, in the valley of the Chirchiq and Akhangaran (Ohangaron) rivers, and borders Kazakhstan in the north, Kyrgyzstan in the northeast, the Namangan region in the east, Tajikistan in the south, and the Syrdarya (Sirdaryo) region in the southwest
Fergana region 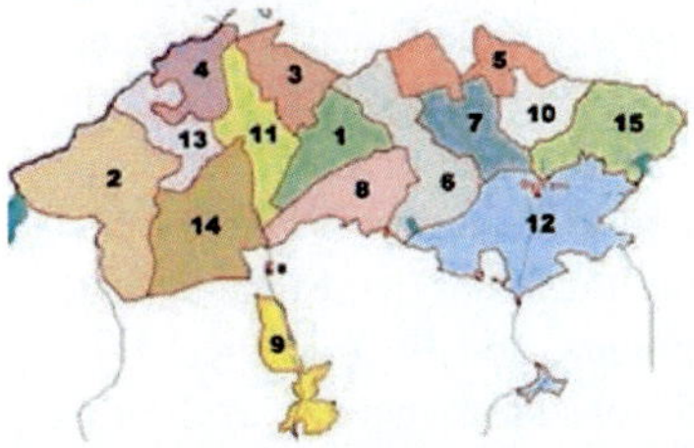				
6800	2,900,000	Fergana	1. Baghdad, 2. Beshariq, 3. Buvayda, 4. Dangara, 5. Yazyavan, 6. Altyariq, 7. Ohkunboboev, 8. Rishton, 9. Sokh, 10. Tashlaq, 11. Uchkuprik, 12. Fergana, 13. Furqat, 14. Uzbekistan, 15. Quva	The Fergana region is located in the southern Fergana Valley and borders the Namangan region in the north, the Andijan (Andijon) region in the southeast, Kyrgyzstan in the south, and Tajikistan in the west

(continued)

Table 2.1 (continued)

Region and area (km^2)	Population	Administrative center	Districts	Location
Khorezm (Xorazm) region 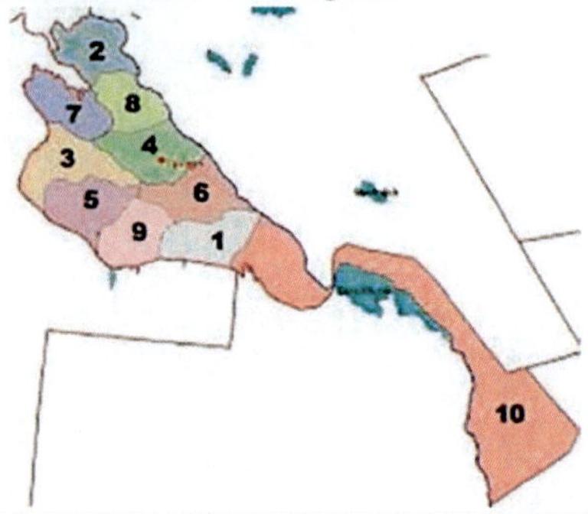				
6100	1,500,000	Urgench	1. Bogot, 2. Gurlan, 3. Urgench, 4. Khazorasp, 5. Khiva, 6. Khonqa, 7. Shovot, 8. Yangyariq, 9. Yangybozor, 10. Kushkupir	The Khorezm (Xorazm) region is one of the northern regions of Uzbekistan and borders Karakalpakstan in the north and northeast, the Bukhara region in the southeast, and Turkmenistan in the south and west
Tashkent (city) 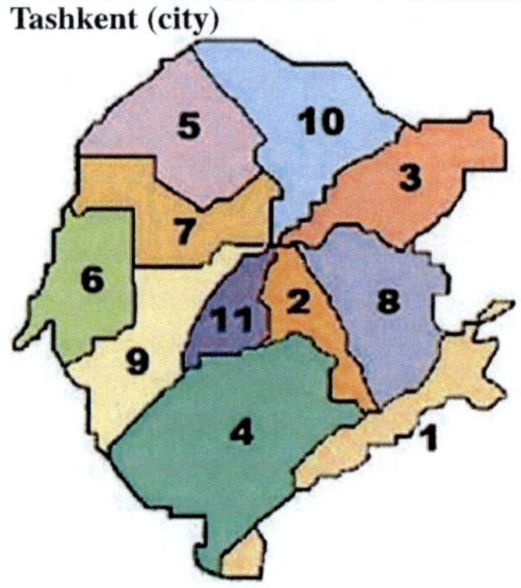				
334.8	2,300,000	Tashkent (capital of Uzbekistan)	1. Bektemir, 2. Mirobod, 3. Mirzo Ulugbek, 4. Sergeli, 5. Sobir Rakhimov, 6. Uchtepa, 7. Shayxontoxur, 8. Khamza, 9. Chilanzar, 10. Yunusabad, 11. Yakasaray	Uzbekistan's capital city, Tashkent, is in the eastern part of Uzbekistan, is the largest city in Central Asia, and is regarded as a main gate to the East

2.3 Cultural Geography

2.3.1 Historical Roots

The culture of Uzbekistan is based on its historical heritage, and the modern concept of its national identity is a reflection of old traditions and new concepts. Today, despite apparent homogeneity, several different cultural regions are defined on the basis of their different natural conditions and prevailing habits and lifestyles. The differences between the beliefs and cultural practices of settled people and nomadic people, inherited from ancient times, are reflected in everyday life; literature, arts, and crafts; and music, songs, and dancing. The culture and traditions have developed over many generations and date back to different religions and ideas of nation building. Invasions and destruction were not able to destroy all cultural foundations, and now the culture of Uzbekistan is a modern extension of the cultural values of the ancient states of Khorezm (Xorazm), Sogdiana, and Bactria. Pre-Islamic and Islamic culture and traditions have added to each other; for that reason, the culture of Uzbekistan is a unique one. For many centuries before the Arabic invasion and adoption of Islam, Zoroastrianism was the main religion of the Persian Empire. Its basic concept of monotheism—which proclaimed the existence of only one God, the right to make choices, and the responsibilities associated with choices—

are also reflected in Islam. Today's traditions and practices include these concepts, and this inclusion is stronger in the regions that were formerly Sogdiana and Bactria. Among the widely spread traditions is a rite of fire purification during wedding ceremonies in many regions of the Zarafshan Valley, particularly in the Bukhara and Samarkand regions. The newlyweds walk three times around a sacred fire, and only after this ritual does the groom take the bride in his arms. This is a good example of remnants of Zoroastrianism beliefs and of the combination of pre-Islamic and Islamic traditions with no sense of contradiction between the two different religious concepts.

Independence and nationalism, as nation-building ideas, were exploited after dissolution of the Soviet Union. In some instances, they were realized by adoption of the new Constitution, declaring Uzbek as an official language, while other languages are used as well. At the same time, the religious issue has still not been finally resolved, and Islam has become a natural rallying point for national identity although other religions are practiced by minorities (Olsson 2011).

The traditions and customs of local people are carefully maintained and preserved, bringing together different generations, and strengthening community and family ties. Many rites and rituals were established in ancient times and are based on hospitality, respect for elders, and cooperation.

The people are socially organized in a form of *mahalla*. The meaning of this word is broader than "community." This institution is based on the corporation principle of people belonging to the same tribe, ethnic group, profession, or blood relationship. The *mahalla* is a repository of such traditions and culture, serving as the principal structure for regulating public life, and has inherited the legacy of medieval cities with common organizations of trade and craft guilds. Being a form of social organization for uniting people of the same ethnicity or profession, the *mahalla* protects and represents their rights and interests in accordance with the principles of mutual aid, consent, respect, and public involvement in decision making (Gulyamova 2012). The safeguarding aspect of the *mahalla* is included in the national list of intangible cultural heritage assets.

Uzbekistan's independence also created challenges for formation of a new culture free of Soviet ideology. During this transformation, strong state control of the shape of the culture created a Soviet legacy of culture production by removing the socialist ideological content of Uzbek identity and leaving a Soviet construction of place-centered heritage with elaboration of traditional Uzbek values (Adams 2010).

2.3.2 Cultural Life

From the historical perspective, the cultural life of Uzbekistan has been transformed several times during the last century. Two different models of urban cultural life existed after the conquest by the czarist Russian Empire. One model was applied in the parts of cities and towns that were built by the Russians for Russian officials, military personnel, their families, and their employees. Another model existed in the neighboring communities of the local population. These cities were religiously, racially, and culturally segregated. The cultural innovations in the Russian parts of the cities were localized, with limited access for the local population. This segregation existed until the beginning of the twentieth century. However, under the influence of European culture, such novelties as theaters, publishing houses, circuses, libraries, societies, and new-method schools changed the cultural life of different population groups (Mustafayeva 2015).

During the Soviet period, cultural life was influenced by socialist and communist ideologies. The national idea as a concise form of people understanding their place in the world and history impacts different forms of culture; it played and still does play a big role in the social and cultural lives of Uzbekistan. After the country gained its independence, the Soviet–Russian social life model was reconstructed into a new Uzbek model. This reconstruction still faces difficulties in avoiding Soviet clichés and stereotypes. Modern culture in Uzbekistan is a combination of socialist and postsocialist ones, or, as Adams (2010) calls it, "postcolonial civic nationalism," which claims to revitalize Uzbek culture, and cultural life under strong state control is "a mirror of official visions and opinions" (Borbieva 2013) and "the relationship between the state, people, and culture" (Walsh 2011).

Public policy in Uzbekistan, with emphasis on development of information and communication technology (ICT) gives the country an opportunity to be a player on the regional and global scales, recognizing the Internet as an important element of the national information network (Kasymova and Usmanova 2015). This policy changes the social and cultural lives of people through modification of their cultural preferences and standards. The absence of an "iron curtain" and new opportunities for communication with other parts of the world make the country's cultural life more international through new forms of expression in music, arts, films, painting, and dancing. Today, the country's cultural life absorbs new trends into the habits, rituals, traditions, and lifestyles that have been established over centuries.

2.3.3 Religion

Most Turkic khanates converted to Islam in the tenth century. This had a profound impact on the culture of the local populations. Islam was introduced to the local nomadic tribes by blending of a new form of Islamic practice into their culture. The Samanid Dynasty played an important role in spreading Islam among the nomads (Mirza Muhammad Haydar 1996).

Fig. 2.4 The mausoleum of Imam Muhammad Al Bukhari in the Samarkand region

Fig. 2.5 The mausoleum of Hakim at-Termezi in the Termez region

Before the thirteenth century, Samarkand, Bukhara, and Urgench were flourishing centers of Islamic learning, culture, and art in Central Asia. The Mongol invasion halted this process for half a century.

Uzbekistan was a major center for dissemination of Islamic thought and culture in Central Asia. Its cities contributed a lot to Islamic religion, culture, and social life. For many centuries, Bukhara, Khiva, and Samarkand were places where many Islamic scholars, poets, writers, and scientists—whose heritage is of global value—were born and spent their lives. Imam Muhammad Al Bukhari (who was born on July 19, 810, and died on September 1, 870) authored the hadith[3] collection known as *Sahih al-Bukhari*, regarded by Sunni[4] Muslims as one of the most authentic (*sahih*) hadith collections (Fig. 2.4). Another prominent author of a number of religious philosophical works is Al-Hakim al-Tirmidhu (Hakim at-Termezi), who was a theorist of Islamic mysticism and a founder of the dervish order[5] (Fig. 2.5).

Among others, it is worth mentioning Bahauddin Naqshband and Khwaja Ahmad Yasaviy, who contributed to development of the Islamic Sufi[6] philosophy. The list of contributors is long and includes leaders in the fields of science, art, culture, architecture, and engineering.

The culture of Uzbekistan is as diversified as the variety of nationalities and cultures of its people. At least 16 different faiths are legally registered in this multinational country. The majority of the population practice Sunni Islam, and their cultural preferences are limited in accordance with its main laws and rules. The culture has historically also been influenced by tolerance, a tradition of pluralism of religion,

[3] In Islam, "hadith" refers to the record of the words, actions, and silent approval of the Islamic prophet Muhammad. The "great bulk" of the rules of shari'a (Islamic law) are derived from a hadith rather than from the Koran (see https://en.wikipedia.org/wiki/Hadith#History).

[4] Sunni Islam is the largest denomination of Islam; Sunni Muslims constitute about 90% of the world's Muslim population.

[5] A dervish is someone guiding a Sufi Muslim ascetic down a path (*tariqah*); dervishes are known for their extreme poverty and austerity.

[6] Sufi philosophy includes schools of thought unique to Sufism, a mystical branch within Islam, also known as *Tasawwu* or *Faqr*, according to its adherents. Sufism and its philosophical traditions may be associated with both Sunni Islam and Shia Islam.

and practice of different religions (Christianity, Judaism, Buddhism, and others) (Fig. 2.6).

During the Soviet period, the state ideology was aimed at complete eradication of religion from the minds of the people. However, because of close ties between religion and traditions, these attempts were unsuccessful, prompting the Soviet State to "make compromises with reality" (Olsson 2011).

Today, Uzbekistan is a secular state, and representatives of different religions and faiths live together, sharing common human values and respecting each other's traditions and beliefs. High tolerance of different religious beliefs is a particular characteristic of social life in Uzbekistan.

Modern Uzbekistan differs from Soviet Uzbekistan by not proclaiming atheism. At the same time, a weak nation-building idea helped Islamic traditions and rituals to fill the ideological vacuum during that era. In the early 1990s, the government was able to strike down Islamic extremists and had to develop a balanced policy on freedom of conscience. Today, Islam has taken a new role in the nation-building process by being embraced as an important part of the national heritage and "moral guidelines" (Olsson 2011), with strong central control by the government.

Islam is the major religion in Central Asia and in Uzbekistan. Today, about 90% of the population are Muslims, mostly Sunnis. In Soviet times, the policy of the government was duplicitous, promoting Islam but attempting to eradicate it at the same time. Many mosques were closed. Since the country gained its independence, post-Soviet religious freedom has filled the ideological vacuum of self-identification. The form of Islam practiced in Uzbekistan is related more to the population's traditions and culture than to politics.

Mosques are located in almost every *mahalla*. Cathedral mosques for Friday prayer are also evenly distributed in urban and rural places (Fig. 2.7).

Today, many new mosques are being constructed and old ones are being reconstructed all over the country. For exam-

Fig. 2.6 Representatives of the three major religious faiths in Uzbekistan discuss issues of common interest. [Image courtesy of Shavkat Boltaev, Bukhara]

Fig. 2.7 The Islom Ota Cathedral Mosque in the Yunus Abad district of the city of Tashkent

Fig. 2.8 The Jome Ali Masjid undergoing reconstruction in the Uchtepa district of the city of Tashkent

Fig. 2.9 An ensemble of Iranian folk musicians in Bukhara. [Image courtesy of Shavkat Boltaev, Bukhara]

ple, the mosque shown in Fig. 2.8 is undergoing reconstruction to create more space for prayer inside it.

Shi'a Islam is also represented in Uzbekistan in the form of a small minority sect composed almost entirely of ethnic Iranians in Samarkand and Bukhara (Fig. 2.9) (see also Chap. 5, Sect. 5.1.3, this volume).

Along with Islam, the population practices Christianity, Judaism, and Buddhism. These religions were also oppressed in Soviet times.

Christianity was known in Central Asia before the Arab invasion. At that time, it was not widespread, since the main religion was Zoroastrianism. In the eighth to tenth centuries, Islam replaced all religions, including Christianity, in Central Asia. To some extent, Christianity returned to the region at the end of the nineteenth century after annexation of Central Asia by the Russian Empire. Today, Christianity is still present in the region, with most adherents following one of three Christian traditions: Orthodox, Protestant, or Roman Catholic.

The majority of Christians in Uzbekistan are Orthodox; 35 Orthodox churches are registered and provide services. The first Russian Orthodox Church in Uzbekistan was founded in Tashkent in 1871; the Cathedral of Holy Dormition is now the main church of the Tashkent eparchy. Other churches include the Church of Alexander Nevsky, the Church of Patriarch Ermogen, and the Church of the Great Prince Vladimir in Tashkent (Fig. 2.10).

As of the first half of 2014, 38 Orthodox churches were registered in Uzbekistan. On February 22, 1996, the Holy Trinity–St. George Monastery was opened by a decision of the Holy Synod of the Russian Orthodox Church, in association with the parish of St. George's Church in the city of Chirchiq in Tashkent province (Fig. 2.11).

Fig. 2.10 The Saint Alexander Nevsky Cathedral was built in 1902–1905 and is the oldest Orthodox Church in Tashkent

Fig. 2.11 The Holy Trinity–St. George Monastery in the city of Chirchiq, Tashkent region

The women's Orthodox community of Tashkent was established in 1893. In 1894, the first church of the Nikolayev Convent was consecrated as a house church for the women's community; the community received the status of a convent in 1901. In 1920, the convent was closed, but it was subsequently re-established in 1990–1992.

There are also representatives of the Catholic, Protestant, and Lutheran churches in Uzbekistan.

The History of the Catholic Church (2018) states that Catholic metropolises and a diocese started emerging along the Great Silk Route during establishment of diplomatic ties between the Holy See and China in the twelfth and thirteenth centuries. The number of Catholics in Uzbekistan exceeded 11,000 at the turn of the twentieth century, when Europeans moved to Central Asia from czarist Russia.

Like many religious institutions, the Catholic Church was oppressed by the Soviet regime. This is why a Catholic church that started being constructed in the capital of Uzbekistan in 1912 was not completed and consecrated until 2000 (Fig. 2.12). It was only after Uzbekistan gained its independence in 1991 that this church was returned to the Catholic parish in Tashkent. In 1976, in accordance with a resolution of the government, the building was restored and handed over to the Ministry of Culture of the Uzbek Soviet Socialist Republic (UzSSR); in 1981, it received the status of a landmark and a historical site in Uzbekistan.

Fig. 2.12 The Catholic Church of Holy Jesus in Tashkent

Fig. 2.13 The German Kirche in Tashkent

In 2005, Pope John Paul II formed an apostolic administration in Central Asia (Cavanaugh 2016). Catholics in Uzbekistan are descendants of Poles, Lithuanians, Germans, French, Latvians, Hungarians, and Czechs who came as military men with the czarist army or were exiled and became refugees in the nineteenth century, or those who were exiled to Central Asia during Stalin's tenure in the 1930s and 1940s.

The Catholic Church's apostolic administration in Uzbekistan represents 3500 Catholics and five parishes, including those in Tashkent (the Sacred Heart of Jesus Cathedral), Samarkand (the Roman Catholic Church of St. John the Baptist), Fergana (the Roman Catholic Church of Holy Mary), Bukhara (the Roman Catholic Church of St. Andrew Apostle), and Urgench (the Roman Catholic Church of Holy Mary, Mother of Mercy) (The History of Catholic Church 2018).

A monastery has now been opened at the Sacred Heart of Jesus Cathedral. Masses are regularly held there in four languages: Russian, English, Korean, and Polish. Catholic clergymen and monastic Franciscan Brotherhood members, as well as missionaries from the Order of Divine Love (the Sisters of Mother Teresa of Calcutta), serve permanently in Uzbekistan.

An Evangelical Lutheran church in Tashkent was built by a project involving the famous architect and artist Alexander Benoit in 1899 (Fig. 2.13).

The story of survival and recovery of this church is familiar. Incorporation of Central Asia into the Russian Empire and establishment of garrisons led to movement to this region of high-ranking officers and soldiers, who, in terms of ethnic origin, were mostly German. Lutherans and Calvinists (members of the Reformed Church) were present in Uzbekistan from the late nineteenth century. Akiner (2018) estimates that in 1897, more than 2800 Protestants (predominantly Lutherans) lived in Central Asia and points out the particular vulnerability of the Protestant community in Central Asia, "not only because they professed a minority faith, but being of German origin, they were inevitably associated with the wartime enemy and thus regarded with suspicion." In the 1880s, a "Bible depot" (a Protestant initiative for distribution of the Holy Scriptures) was established in Tashkent, but it subsequently closed in 1918. It continued to fulfill its mission until 1937. In September 1993, the Bible Society of Uzbekistan was registered with the Ministry of Justice. Today, church services are conducted in Russian and German.

Friendly and mutually respectful relations between different faith communities are a particular characteristic of social life in Uzbekistan. The Armenian Apostolic Church and Jewish synagogues are also represented in Uzbekistan.

According to data from the World Jewish Congress, two distinct Jewish communities exist in Uzbekistan: a more religious and traditional Bukharian Jewish community, and a more progressive Ashkenazi community made up primarily of Jews of European origin. In the 1989 census, 94,900 Jews were recorded as living in Uzbekistan, but fewer than 5000 remained in 2007 (most of them in Tashkent). According to an estimate by the Hebrew University demographer Sergio Della Pergola, the Jewish population in Uzbekistan was 4500 in 2010. Twelve synagogues are registered in Uzbekistan (Fig. 2.14).

The Jewish community in Uzbekistan is affiliated with the World Jewish Congress. The first documented Jewish presence in Uzbekistan dates back to the fourth century. Jews were living in Tashkent, in the cities of Bukhara and Samarkand, in Kattakurgan in Samarkand region, in the cities of Qarshi and Shakhrisabz in Kashkadarya region, and in the cities of Kokand and Margilan in the Fergana Valley. Discriminatory edicts that had existed in the Emirate of Bukhara with respect to Bukharian Jews (referred to as "indigenous Jews") were canceled in the areas annexed by the Russian Empire. After the region came under Russian rule, Ashkenazi Jews appeared there as well. At the same time, the term "Bukharian Jews" emerged—used to define Jews arriving in Russian-ruled areas from the Emirate of Bukhara. At the end of the nineteenth century, there were approximately 16,000 Bukharian Jews. According to the 1926 census, the number of Jews living in Uzbekistan had increased to 38,200 by that time. During World War II, more than a million Jewish refugees from Nazi-occupied Eastern Europe passed through Uzbekistan. The state's first legal Jewish secular organizations emerged in the years 1988–1999. There are now smaller Jewish communities in Samarkand and Bukhara, and quite small ones in Fergana, Andijan, Namangan, Margilan, Kokand, and Navoiy. The communities contain both Bukharian and Ashkenazi Jews. Most of the Jews in Tashkent are Ashkenazi, Bukhara has more Bukharian Jews, and the Jewish community in Samarkand is more or less equally divided between Ashkenazi and Bukharian Jews. Jewish quarters, traditionally called *mahalla*, still exist in Samarkand, Bukhara, and smaller cities in the Fergana Valley. There, Jews continue to follow a traditional way of life .[7]

During Soviet times, as in times of ancient invasions and destruction, traditions and culture were preserved by all nationalities and religious groups. This nation has always been committed to preserving its cultural heritage, including caring for historical figures, places, and shrines. In today's social life in Uzbekistan, "both traditionalism and modernity" are evident (Warikoo 2015).

Fig. 2.14 A synagogue in Bukhara. [Image courtesy of Shavkat Boltayev, Bukhara]

2.3.4 Gender Issues

The roles of women in political, social, and cultural life is one of the indicators of development in any society. Gender[8] equality is not just a human right but also an indicator and precondition of sustainable development. It is accepted that gender equality is a situation in which "all human beings are free to develop their personal abilities and make choices without the limitations set by strict gender roles; that the different aspirations and needs of women and men are considered, valued and favoured equally" (Holzner et al. 2010).

The Constitution of Uzbekistan declares equal rights for all citizens. Women have a right to receive equal pay. According to the United Nations Development Programme's (UNDP's) 2019 Human Development Report, Uzbekistan has a gender development index (GDI)[9] of 0.939, ranking 106th out of 189 countries (UNDP 2020), and a gender

[7]Information from the World Jewish Congress (http://www.worldjewishcongress.org/en/about/communities/UZ), which represents Jewish communities in 100 countries across six continents.

[8]Gender is a basic marker of social position, identity, and order in society (Northrop 2004).

[9]The gender development index is the ratio between the female and male human development indices (UNDP 2020).

inequality index (GII)[10] of 0.288, ranking 62nd, placing it among countries with high human development (UNDP 2014). Uzbekistan's GII is higher than those in neighboring Kyrgyzstan (0.369), Tajikistan (0.314), and the countries in the Caucasus: Armenia (0.245), Azerbaijan (0.323), and Georgia (0.331). In Uzbekistan, 99.9% of women aged 25 years and older have at least secondary education. However, the labor force participation rate for women is just 48.3% versus 76.2% for men (UNDP 2014).

According to the UNDP (UNDP 2014), 16.4% of Uzbekistan's parliamentary seats were held by women in 2019. According to the World Bank (2017), this rate of participation in decision making is lower than the average rate in Europe and Central Asia (27% in 2017).

The stereotype of the role of women as housewives—whose occupation is caring for their families, managing household affairs, and doing housework—still influences women's choices in education and employment. Women are prevalent in public health and education, while men are dominant in industry, construction, transportation, communication, economics, and law. As one study (Kim 2003) has shown, the "glass ceiling" effect results in lower representation of women in management and at high levels in professions, and women have far less relative influence in the field of politics than in families. The traditional notions of family relations and the role of women in society are dominant. In contrast to the larger share of men who are prepared to have a career and professional development, women prefer family stability, devoting more time to childcare instead of reaching the top ranks of management. Women still represent a minority within the government and legal sectors because of limited professional and career growth opportunities, insufficient remuneration, and cultural constraints.

In 2022, the female population of Uzbekistan was estimated to be 17,322,806 (Countrymeters.info 2022), or 50.3% of the total population. However, there is clearly gender inequality in social and economic participation. Only 36.5% of students entering universities are women, and the gap is wider in technical universities, where only 18–19% of students are girls. This inequality creates gender imbalances in public health, education, and the arts, although women represent more than 70% of employees. Women hold only 27% of jobs in public administration and management, 19% in local government, 11% in transportation and communications, and 8% in the construction industry (UNDP 2014).

However, increasing numbers of women are now participating in small businesses, as confirmed by growing numbers of women establishing and heading business entities (35%), services (37%), and trades and public catering services (54%).

The role of women in social and public life has changed over time. In Uzbekistan, such a change occurred dramatically in the late 1920s and the early 1930s, with implementation of a series of policies and actions called the Hujum ("attack" in Arabic), aimed at establishing Soviet power and involving women in public and social life, to give them equality by creating literacy programs and bringing them into the labor force. Basically, the Hujum was part of a larger ideological goal to create a Soviet nation with the same education, culture, and identity.

In a form of an "attack" on female veiling and seclusion, the Communist Party of the Soviet Union pursued Muslim women's liberation. This purpose of this class and cultural revolution was to rapidly change the lives and social status of Uzbek women and their attitudes to the political and social reforms involved in building socialism. This policy was seen by many Muslims as a threat to their religious freedom and Islamic culture, because the veil was considered a marker of cultural identity. The unveiling campaign in Muslim Soviet countries faced difficulties in changing stereotypes and patriarchal attitudes regarding gender equality. Many women were murdered for their attempts to rid themselves of the veil.

In pre-Soviet Uzbekistan, cultural and religious norms promoted female seclusion as a way of asserting male superiority over women. The policy of the Russian Empire was aimed at preserving the status quo in social life without consideration of gender equality and did not attempt to change the social roles of the Uzbek population. The demarcation of gender division was more apparent in settled communities than in nomadic ones. Historically, veiling was related to ethnicity and religious practice. In nomadic communities, women usually covered only their mouths, and this veiling was rooted in tribal custom. Tatar women who emigrated from Russia were unveiled. More strict veiling practices existed among women in settled communities. The form of the veil was dependent on social status and varied geographically. Face and body veiling were common for urban women, and rural women who worked in fields alongside men wore a long robe that could be pulled up to cover the mouth in the presence of men. Limited access to education for girls and women, polygamy, traditions of payment of a bride price, and arranged marriages without the brides' consent to marriage were intrinsic to Uzbek society, marked with gender inequality. The success of the unveiling campaign was mainly ensured by the active participation of Jadids, who, as representatives of the upper ranks of settled Uzbek society, stood for women's enlightenment and modernist cultural reforms. Jadid women contributed to feminism and prepared the ground for equal rights for women in Soviet Uzbekistan.

[10]The gender inequality index is a composite measure reflecting inequality in achievement between women and men in three dimensions: reproductive health, empowerment, and the labor market (UNDP 2014).

Fig. 2.15 A women's club headed by a Jadid woman, Risolat Saidalieva, in the Ohun Guzar *mahalla* in Tashkent in 1925. [Image from the archive of Rabija Axmedova]

The unveiling campaign had been carried out by the Women's Division (Zhenotdel in Russian), while their workers were educated Russians and Jadid educated women, mostly Tatars. The state policy encouraged unveiling through a private initiative, and it was not a state-driven campaign when it started. In 1927, these initiatives became a means for increasing the number of Soviet allies via propaganda promoting gender equality. A mass throwing-off of the veil in public as an act of female liberation was officially launched on March 8, 1927 (International Women's Day) in Tashkent. The activists supposed that this movement would transform the cultural and political outlooks for women. The mass unveiling was supplemented with additional liberation institutions such as a women's club (Fig. 2.15), women-only stores, and schools for women (Fig. 2.16).

For many women, it was a difficult decision, and they were trapped between the coercive methods of the Soviet State and their own society. Fear of public opinion was a major reason for resistance against state pressure. This was a fight in which murders of unveiled women were designed to demonstrate the strength of traditional cultural and religious norms and to remind women of their position in the social hierarchy. Reveiling of unveiled women showed the weakness of the unveiling campaign by the end of the first year. This cultural revolution failed because of the strong position of non-Soviet traditions, with Islamic shari'a being rooted in cultural traditions. In 1928 and 1929, new laws were introduced to treat murder for unveiling as a "counterrevolutionary act" and as a "terrorist act." These laws were designed to protect women and to strengthen the local Soviet authorities.

The Hujum changed the public life of women, whereas their private lives were left without significant changes. By the 1950s, large loose scarves replaced body veils, the literacy rate of women reached 75%, employment rose rapidly, and the health care system was improved significantly.

Fig. 2.16 Young women and girls after throwing off their veils in Tashkent in 1927. [Image from the archive of Rabija Axmedova]

Family matters—including marriage, divorce, and the births of children—came under the aegis of the secular civil administration. Religious marriage became illegal. However, the Islamic marriage ceremony is still considered part of cultural tradition and identity rather than religion.

Since Uzbekistan gained its independence, gender issues have played an important role in the nation-building process. Male dominance of decision-making systems prevent active

Fig. 2.17 The Kukeldash Madrassa in the Shayxontoxur district of the city of Tashkent

participation of women in formal networks of politics and political parties. Women's participation is considerably more active in nongovernmental nonprofit organizations (NGOs), particularly in women's NGOs, which are a new type of organization established to deal with women's issues. The Committee of Women is an NGO integrated into the system of public and public administration at the national and local levels for addressing issues of social, economic, and cultural development.

A survey conducted by the Center for Public Opinion (Ijtimoiy Fikr) (The women of Uzbekistan 2018) revealed that another important niche (for 20.2% of respondents) is entrepreneurial activity, and 24.2% of respondents said they would like to organize their own business. The major obstacle for rural women is lack of knowledge and skills in entrepreneurship. Gender equality is becoming not only a social goal but also a political and economic goal in the development of Uzbekistan today.

2.3.5 Education

From ancient times, and especially in medieval times, Central Asia was a center of science and culture. Education has always played an important role in the scientific and spiritual development of the society.

At the beginning of the twentieth century, with establishment of modern science and enlightenment, the process of formation of the educational system faced contradictions between the ideologies of "Russification" and nationalism, and between Islamic and secular education.

Before Uzbekistan was conquered by the Russian Empire, its system of education was well established and traditionally consisted of two stages: (1) elementary school (*maktab*) and (2) secondary and higher education (*madrassa*). The education system was separated by gender.

Madrassas were scientific and educational centers in many cities of Central Asia. Around the turn of the twentieth century, many madrassas that had operated since the Khan period continued to operate. In the city of Tashkent, for example, according to Tangirov (2015), about 20 madrassas had 1000 students. One such madrassa in Tashkent was the medieval Kukeldash Madrassa[11] (Fig. 2.17), located in the ancient center of the city. It was built around 1570 by one of the rulers of the Shaybanid Dynasty.

Over the centuries, however, some madrassas lost their educational and scientific functions and served the goals of governments that did not give priority to science and education. Like other madrassas, the Kukeldash Madrassa served as a caravanserai because of its close vicinity to the biggest market in the city, as a fortress during the time of incorporation of Tashkent into the Kokand Khanate in the eighteenth century, and as a museum in the Soviet period.

The most famous madrassas (Figs. 2.18 and 2.19) are in "holy" Bukhara. One of them, founded in the sixteenth century, is Mir-i-Arab (Fig. 2.18), which attracted students of Islamic studies from different parts of the world even in Soviet times. Today, it serves 100 students.

In Bukhara, one of the religious centers is the Po-i-Kalyan complex, which contains many mosques and madrassas, and dates back to 713 (Fig. 2.19). In the pre-Islamic era, a central cathedral for fire worshippers was located there. Over many centuries, several religious edifices were built and subsequently razed, restored after fires and wars, and moved from place to place (Fig. 2.20).

[11]The *kukeldash* was an important court position in the time of Genghis Khan.

Fig. 2.18 The Mir-i-Arab Madrassa in Bukhara

Fig. 2.19 The Po-i-Kalyan Islamic religious complex in Bukhara

Fig. 2.20 The Ulugh Beg Madrassa in Bukhara. [Source: Ymblanter – own work, CC-BY-SA 4.0, via Wikimedia]

Fig. 2.21 A madrassa for girls and women in the Kagan district, Bukhara region

Girls were taught separately from boys. Madrassas for girls and women existed in many cities (Fig. 2.21).

Education in maktabs comprised Arabic, calculus, Islamic ethics, and law. Many communities had an elementary school, which was supported by a *waqf*[12] and did not have state funding. Some statistics demonstrate that the population was involved in education in accordance with the tradition of respecting literacy and cultural and scientific values. In the 1860s, 12,000 students were taught in 118 maktabs in Tashkent and 46,000 students were taught in about 5000 maktabs in the Syrdarya (Sirdaryo), Fergana, and Samarkand regions. By the beginning of the twentieth century, 7047 maktabs were functioning in Turkestan, including 1500 in the Khiva Khanate (Mustfayeva 2015). Located in such major cities of Turkestan as Andijan, Bukhara, Kokand, Margilan, Namangan, Samarkand, and Tashkent, the madrassas were of different sizes, depending on their financial support. This education was widely available; in 1890–1893, 68 madrassas were listed as operating in the Fergana Valley and 214 in Fergana, Samarkand, and the Syrdarya regions of Turkestan, including 22 in Tashkent, 19 in Samarkand, and 37–39 in Kokand. Fifteen madrassas in Bukhara, Samarkand, and Tashkent were renowned for their education throughout the Islamic world. This education was religious, with the main priorities being teaching in Arabic and Persian. The libraries of the madrassas were centers of culture and science, and people visited them to attend frequent presentations by poets and writers. Rulers also founded own libraries of collections of manuscripts in Chinese, Uyghur, Persian, and Arabic. However, these libraries did not survive plundering by the czarist Russian Empire (Ziyaeva 2015).

[12]A *waqf* is a charitable endowment under Islamic law.

Reform of the education system in the form of establishment of new-method (Jadid)[13] schools was started at the end of nineteenth century by leaders of the Jadid movement: Munawwar Qari in Tashkent, Mahmud Khoja Behbudiy in Samarkand, Sadriddin Ayni in Bukhara, and Ashur Ali Zahiriy in Kokand. Such schools were established first in Kokand, Samarkand, Tashkent, and Andijan. By 1913, there were 20 in the Fergana Valley alone. The popularity of this reform may have been due to its adoption of a system including the 4 years primary and 4 years of secondary education and a teaching program that included secular subjects as well as programs of traditional education. In 1917, more than 100 new schools with 10,343 students were established in different cities and towns of Uzbekistan. However, this movement was not welcomed by the czarist Russian Empire, which strengthened its control over the schools and closed some of them.

Publishing houses and public libraries played important roles in education. Before establishment of Uzbekistan's first publishing house in 1908 by leaders of the enlightenment movement, no literature was printed within Uzbekistan. The opening of public libraries with collections of modern literature and textbooks made a significant contribution to the spread of knowledge among the local population. The ideas of the Jadids influenced education during the initial stage of the Soviet era and have re-emerged since Uzbekistan became independent. However, the informal Jadid system was destroyed by the communist authorities in the early 1920s.

During Soviet times, education was delivered under the socialist and communist ideologies. However, schools and universities did offer a full range of courses in hard sciences and

[13]Jadids were reformers in the Russian Empire in the late nineteenth and early twentieth centuries, who aimed to introduce a new kind of knowledge and modernist, European-modeled cultural reform.

Fig. 2.22 The National University of Uzbekistan, named after Mirza Ulugh Beg, in Tashkent

humanities. Education at primary schools initially lasted for 7 years and then, from the mid-1960s onward, 8 years. Students could choose to extend their education for another 2–3 years at a public secondary school or at a specialized secondary institution. Undergraduate study at universities took 4.5 or 5 years, and students received a diploma as an engineer, as a teacher, or as another qualified specialist. Postgraduate study at a university to prepare and defend a dissertation took 3 years.

The next stage of reforms in education started in the mid-1980s, when modernization of primary, secondary, and higher education was linked to the idea of self-identification. This resulted in replacement of Russian by Uzbek as the primary language of instruction and communications, making many textbooks in history and other social sciences useless.

Education in Uzbekistan is now run by the state under direct management by the Ministry of General Education and the Ministry of Higher and Specialized Education. At the bachelor's degree level, 69% of students study on a fee-paying basis (with an individual contract), and at the master's degree level, 75% do (Bertelsmann Stiftung 2018).

Today, the population of Uzbekistan is highly literate; the younger generation (including students with disabilities) is totally covered by public education. According to data from the UNDP, 2022 (data.worldbank.org/indicator/ SEADT. LITR.ZS?locations=UZ) Human Development Report, in 2019, the literacy rate of the adult population (aged 15 years and above) is 99.9%.

Officially, in 2015, about 10–12% of Uzbekistan's GDP was spent on the education sector (which is twice the requirement stipulated by the United Nations Educational, Scientific and Cultural Organization [UNESCO]). There are 79 higher-education institutions in Uzbekistan: 20 universities, 36 institutes, 2 academies, 14 university branches in provinces, and 7 branches of international universities. In 2017–2018, a total of 66,586 students started education at the bachelor's degree level at all universities, including international ones. This number was just 9.1% of all applications (Kobilov 2018). According to UNESCO data, the gross outbound enrollment ratio[14] for both sexes in Uzbekistan improved from 0.8 in 2011 to 1.0 in 2017. For comparison, the corresponding 2017 ratios were 6.6 in Kazakhstan and 2.1 in Kyrgyzstan. The UN education index ranks Uzbekistan 78th–80th out of 188 countries, with an index of 0.71.

In 1918, the National University of Uzbekistan was the first higher-education institution to be founded in Central Asia. It was called the Central Asian University until the end of the 1950s and served as a major center of science and education in this region. In 1960, its name was changed to the V.I. Lenin Tashkent State University, and after Uzbekistan gained its independence, it became the National University of Uzbekistan, named after Mirza Ulugh Beg (Mirzo Ulugbek nomidagi O'zbekiston Milliy Universiteti). Today, it is the largest university in Uzbekistan (Fig. 2.22).

Many other universities (for example, the Tashkent State Technical University) were previously part of the Central Asian University (Fig. 2.23).

[14]UNESCO describes the gross enrolment ratio as the total enrolment within a country "in a specific level of education, regardless of age, expressed as a percentage of the population in the official age group corresponding to this level of education" (see https://en.wikipedia.org/wiki/Gross_enrolment_ratio).

Fig. 2.23 The Tashkent State Technical University's main building

Various international universities (the Westminster International University in Tashkent, the Inha University in Tashkent, the Management Development Institute of Singapore in Tashkent, the Turin Polytechnic University in Tashkent, the Moscow State University, the Russian State University of Oil and Gas in Tashkent, and the Russian University of Economics in Tashkent) provide international diplomas and are relatively better equipped and of higher quality than the other universities. Education at these universities is conducted in English or, in the Russian universities, in Russian (Figs. 2.24 and 2.25).

The Academy of Sciences was established in 1942. However, after Uzbekistan became independent, the academy lost a lot of subsidy funding. By 1992, many scientific entities had closed. Different structural reforms negatively impacted the academy. Today, this policy is undergoing revision.

According to the statute of the Academy of Sciences, its "highest governing body is the General Assembly, which includes in its membership the full members (Academicians) and Directors of scientific-research institutions affiliated with the Uzbek Academy of Sciences. The General Assembly is convened at least once per year. In between the activities of the General Assemblies, the Presidium (Hayat) consisting of the President, Vice Presidents as per science directions, General Scientific Secretary, and the chairmen of the Regional Departments of the Academy of Sciences of Uzbekistan carries out administration and management of the Academy". Under the Statute, "the candidacy of President of the Academy of Sciences of the Republic of Uzbekistan is submitted to the General Assembly of the Academy of Sciences on behalf of the President of the Republic of Uzbekistan. The President of the Academy of Sciences of the Republic of Uzbekistan shall be elected from among the members of the Academy of Sciences on the majority-vote-basis of the General Assembly by open vote" (Academy of Sciences of the Republic of Uzbekistan 2018).

The Academy of Sciences consists of 28 research institutions, two regional departments (Karakalpak Department and Khorezm (Xorazm) Mamun Academy), and four state museums, which are organized into three organizational complexes based on the following scientific areas:

- Physical and mathematical sciences and engineering
- Chemical, biological, and Earth sciences
- Social and human sciences

2.3.6 Music

The music of Uzbekistan originated in the traditions of many nationalities and the heritage of previous civilizations, and it includes classical singing forms (*shashmaqom*, *maqom*, *doston*, and *katta ashula*) and popular singing forms (*lapar*, *yalla*, and *qo'shiq*). The singing tradition of *shashmaqom* was more popular in Bukhara and Samarkand, the tradition of *katta ashula* in the Tashkent and Fergana regions, *maqom* in Khorezm (Xorazm), and *suvora* in Khiva and Urgench. The tradition of singing a cappella spread in the Fergana Valley, while traditions of instrumental musical accompaniment developed in Bukhara and Samarkand. In many cases, the performers presented their own songs and music (Ziyaeva 2015), and this tradition continues today, when modern singers are the composers and poets creating their own songs. The musical instruments of Uzbekistan are the *ghijjak* (a local musical instrument resembling a violin), the *dutor* (a two-stringed instrument), the *surnay* (a kind of flute), the *nay* (another kind of flute), the *qoshnay* (a double flute), and the *doira* (drum), and they are used to entertain people at festivals and wedding parties (Fig. 2.26).

Fig. 2.24 The International Westminster University in Tashkent

Fig. 2.25 The Inha University in Tashkent

Fig. 2.26 Wedding musicians. [Image courtesy of Shavkat Boltaev, Bukhara]

2.3.7 Theater

Before the twentieth century, theaters did not exist in the region. In the nineteenth century, instead of theaters, popular live shows were an important element of urban cultural life. The most popular type of show was puppetry, due to the ease of construction in markets and the simplicity of creating an impression and a performance for personification of heroes, animals, and depiction of scenes in everyday life. Closely related to the tradition of oral poetry in the Turk culture, the art of the comic show and jokes were also popular and widespread. *Askiya* (joking with puns) is still one of the traditional types of comic show in the Fergana Valley. Rope walkers, stone-lifting wrestlers, musicians, acrobats, and tricksters performed their art in cities during holidays and market days. The first Coliseum Cinema was built by the Russian czarist administration in Tashkent in 1910. The national Uzbek Theater "Turon" was established by Abdulla Avloniy in Tashkent in 1914 and later also in the cities in the Fergana and Zarafshan Valleys. At the national theaters the plays were translations of musical dramas and comedies, and they were a new element in cultural life. In Soviet time Uzbek theater was formed into a school of original musical and stage art. In the difficult economic situation of the 1990s, however, loss of government subsidies led to a drastic decline in theatrical activity, and cinema and television further emptied the seats in live performance theaters. Today, however, theaters play an important role in the cultural life of major cities in Uzbekistan. Concerts by popular singers, dance performances, and other forms of entertainment attract people who come to the big cities.

2.3.8 Literature

Many great poets and writers who lived in the territory that is now Uzbekistan have contributed to the culture of humanity. The scholars of the medieval era were influenced by an encyclopedic spectrum model related to the knowledge-ordering texts of the ancient Greek and Roman cultures and traditions (König and Woolf 2013). They usually possessed various skills and knowledge, and their lyrics and writings on ethical and moral issues are still impressive. Al-Biruni lived in the tenth century, and his geographic works in the natural sciences were combined with writings in humanities. Alisher Navoiy advanced Turkic-language literature and, like other scholars in the fifteenth century, he created interesting musical compositions. His poetic works are part of the priceless heritage of humankind.

Many forms of literature and poetry were developed in this country and spread beyond the region. They may be recognized in the modern cultures of other neighboring countries. The authors of the Alpamysh are unknown. As the most important example of Turkic oral literature, this heroic epic, or *dastan*, dates back to the sixth to seventh centuries and was spread by Oghuz Turks (Paksoy 1989).

At the beginning of the twentieth century, such writers and poets as Abdalrauf Fitrat, Sadriddin Ayni, Abdullah Qodiriy, Cholpon (Abdulhamid Sulayman Yunus), Elbek (Mashriq Yunus O'g'li), and Batu created fresh and modern stories, novels, rhymes, and poems on issues associated with common human values and everyday life. Their works became very popular because of their new form and language, and because they were different from the verse composed in the traditions long employed by the poets of the region. The new traditions of writing about the lives of ordinary people and the excellent psychological portraits in historical novels such as *Days Gone By* and *Scorpion from the Altar* (written by Abdullah Qodiriy) made them very popular among the young generation of the 1920s. However, these writers (Fitrat, Abdullah Qodiriy, Cholpon, and others) were executed, exiled, or declared enemies of the people in 1937–1938 under Stalin's purges. As a result, Uzbekistan's intellectual and cultural life suffered trauma for decades to come (Allworth 2017). This created a vacuum in the country's literature and poetry. Even an increase in the number of writers in the second half of the twentieth century did not fill the gap. Soviet Uzbek writers and poets were very cautious and produced their works in line with official Communist Party themes. A new generation of writers is now looking for advanced forms of expression and new themes in a changing Uzbekistan. Several of these new writers have contributed striking dramas and comedies to the theater of Uzbekistan (Allworth 2017).

2.3.9 Newspapers and Magazines

Freedom of speech and absence of censorship, which were granted by a manifesto from the Russian emperor in October 1905, provided an opportunity for establishment of a national press industry. The first newspaper, *Taraqqi*, was published in Tashkent in 1906 but was immediately banned, as were others. Despite such strict conditions, other newspapers replaced banned ones and increased public information and awareness. During Soviet times, newspapers were financially supported by the government. Today, many newspapers and magazines share their content via the Internet and other ICT channels.

2.4 Architecture

The cultural history and heritage of Uzbekistan are reflected in its magnificent monuments. The national architectural traditions are diverse and depend on the natural conditions of the region, the lifestyles of its inhabitants, and access to natural resources. In Uzbekistan, one may find interesting examples of architectural monuments constructed in different periods. The pre-Islamic architectural history is not well known, and many monuments were destroyed by invaders over the centuries.

2.4.1 Ancient Monuments

In contrast to many other places, the city of Bukhara has always stood in the same place, and new structures have been built on top of earlier ones there. Some evidence about early Bukhara architecture and arts has been obtained from excavations on the site of Varakhsha, 30 km from Bukhara. Building of the city of Bukhara was begun in the sixth century and by the eighth century, it covered an area of 87 km^2, making it a fairly large city for its time. In the tenth century, it became the capital of a large state under the Samanids. That period of economic and cultural renaissance turned Bukhara, which was no longer Arab dominated, into a center of science and arts. Architects in Bukhara learned to manipulate burnt brick techniques. The famous mausoleum of the Samanid ruler Ismail in Bukhara (built in the ninth and tenth centuries) is constructed of bricks made of loess (Fig. 2.27). It is the only surviving building from that period.

The composition of the mausoleum is symmetrical, with equal-sized facades that widen toward the bottom. The cubic volume, with four open arches on all sides and massive columns at the corners, is crowned by a light arcature gallery. The overall composition was made logical and complete by installing four miniature domes at the corners, which conveyed finish to the semispherical main dome. One of the facades has a door, while the others are blocked by brickwork grating. Monumental and airy, this building, well lit and with interplay in the folds of its relief masonry, is a classic work of its time. Its features revived the remote past by retaining closeness to the spirit of Sogdianian art. All of these elements are vividly alive in the crowning arcature, the small columns of the trompe belt or vaulted structure over an external angle of a building, and the decorative elements, which replicate wooden pieces. The ideas of dignified tranquility, security, and firmness are manifested in the down-to-earth manner of the execution of this mausoleum (Voronina 1969).

The monuments of the eleventh and twelfth centuries are examples of incredible skill in using burnt bricks as an important decor element. The brick layers have been ground and fitted without any mortar to secure them to the masonry, captivating spectators with a gamut of yellow and pink tones and combinations of geometric shapes. The Kalyan Minaret of the cathedral mosque is a masterpiece of medieval architecture. Its tremendous height necessitated ample experience and uncommon skill on the part of the builders to consider the forces of the wind, the risk of seismic quakes, and the properties of the soils in towns with a thick cultural layer. The Kalyan Minaret is 45.6 m high and was previously taller, since its foundations are now buried in more recent strata. Ten decorative belts of relief brickwork and intervening borders are highlighted by an on-rib manner of bricklaying (Fig. 2.28).

The Chashma Ayub (Job's Well) has an outline with a well-developed plan inscribed with a rectangular contour and a conic dome. This outline is uncommon for the architecture of the fourteenth century, with an absence of glazed facing (Fig. 2.29). It is supposed that this building was constructed by architects from Khorezm (Xorazm) (Vorornina 1969).

Fig. 2.27 The mausoleum of Ismail Samani, built in the tenth century, in Bukhara

Fig. 2.28 The Kalyan Minaret and Madrassa in Bukhara

Fig. 2.29 The Chashma Ayub in Bukhara

A remarkable example of town planning with consideration of local climatic conditions is the Lyabi-Khauz architectural complex, with three large monumental buildings: the Kukeldash Madrassa (Fig. 2.30) in the north, a *khanqah* (Sufi lodge) in the west, and the Nodir Divan-begi in the east, surrounding a water reservoir.

The name Lyabi-Khauz means "at the reservoir." Such reservoirs are a common element of the town landscape and serve as water sources in the dry summertime, as well as natural humidifiers and air coolers. The air temperature is usually several degrees cooler at sites around reservoirs. The rectangular pool (*khauz*) measures 36 by 46 m and is shaded by ancient trees, with steps down to the water, made from massive blocks of yellow limestone. In the old times, there was a tea bazaar, and sweets, delicacies, bread, and other food was sold there. Today, Lyabi-Khauz is a center of the cultural life of Bukhara (Fig. 2.31).

The city of Samarkand is the most ancient urban community in Uzbekistan. Its pre-Islamic history is buried under the lifeless hills of the Afrasiab stronghold. After the Mongol invasion in 1220, Afrasiab was deserted. Under Tamerlane rule, this city became the capital of the empire at a time when it flourished and major construction projects were undertaken. Architects from Khorezm (Xorazm), India, Iran, and Syria contributed to the tremendous construction efforts. In the construction, Timur's concepts foreshadowed his daring political exploits, with his attempts to construct gigantic buildings in many parts of his empire, particularly in Shahr-i-Sabzi and Samarkand.

Fig. 2.30 The Kukeldash Madrassa in Bukhara [This file is licensed under the Creative Commons Attribution-Share Alike 4.0 International license via Wikimedia] public domain [Ymblanter https://commons.wikimedia.org/wiki/File:Kukeldash_Madrasa_Bukhara.jpg]

The Samarkand architectural style came into its own under Ulugh Beg (1409–1449). His reign was beneficial for science and arts. He himself was a talented scientist and considered a great astronomer. However, the observatory constructed by Ulugh Beg was subsequently destroyed; only the foundation and part of the sextant survive (Fig. 2.32).

In the fourteenth and fifteenth centuries, many great and fine tombs, mosques, palaces, madrassas, and mausoleums were constructed in Samarkand. The Ulugh Beg Madrassa was completed in 1420 and was the latest word in contemporary architecture. The portal displayed a carved glazed mosaic of elements of vegetables and geometrical designs. In contrast to the architecture of Bukhara, these mosaics were faced with light blue, dark blue, green, white, black, and yellow glazes. The finesse of their execution and their delicate designs are impressive.

The Shir-Dor Madrassa and the Tillya-Kari Madrassa were constructed by the Shaybanid rulers. The Shir-Dor was started in 1619 and completed in 1635–1636. It is across the square from the Ulugh Beg Madrassa and replicated the latter from top to bottom by repeating its composition and size. The glazed facade of the Shir-Dor is of comparatively high quality. A distinguishing feature is its images of a lion (or tiger) attacking a doe and a sun with human features. The Tillya-Kari Madrassa was built in 1646–1647 in the northern section of the square. The decoration work continued until 1660, but some parts of the building were never finished. In terms of its composition, the building of the Tillya-Kari Madrassa is different from the Shir-Dor and the Ulugh Beg buildings, and it recalls the Bukhara madrassa style. Registan

Fig. 2.31 Lyabi-Khauz in Bukhara. [Image courtesy of Magfirat Muminova]

Fig. 2.32 The Ulugh Beg Observatory in Samarkand

Fig. 2.33 Registan Square in Samarkand. [Modifications made by MrPanyGoff Creative Commons Attribution 2.0 Generic license via Wikimedia]

Square is a gem of the medieval architecture of Central Asia (Figs. 2.33, 2.34, and 2.35).

Muslim rebuilders have now accurately reconstructed mosques and mausoleums that were damaged during the Soviet era, as part of a comprehensive drive to recreate the Islamic life suppressed by the communists between 1920 and 1990 (Allworth 2017).

2.4.2 Modern Architecture

The architecture of Uzbekistan is the result of development of urban planning and architecture of Uzbek cities from fundamental Islamic orientation through Russian colonization and Soviet modernization up to the search for a new identity in a world characterized by globalization (Gangler et al.

Fig. 2.34 The Shir-Dor Madrassa in Samarkand

Fig. 2.35 Interior of the Shir-Dor Madrassa in Samarkand

2002). The modern architecture and town planning share three periods historically related to state building and revision of the national idea.

The invasion by the Russian Empire at the end of the nineteenth century influenced the national architecture through insertion of new elements into the country's social and economic life, such as the founding of new types of institutions and construction of buildings such as banks, train stations, gymnasiums, hospitals, and schools. The architecture was planned in several styles, combining eclecticism, gothic style, and modernism. Burnt bricks were used as a primary building material.

The residence of the Grand Duke Nicholas Constantinovich Romanov (the first cousin of the Russian Emperor Nicholas), who was exiled to Central Asia in 1881, is an example of the architecture of the end of the nineteenth century. This building was built in 1891 (Fig. 2.36) and has survived over time. Its functions changed several times: it was the Lenin's Young Pioneers' Palace in Soviet times, and in 1980 it became a museum housing a fabulous jewelry collection. It is now the property of the Ministry of Foreign Affairs of the Republic of Uzbekistan (Fig. 2.36).

Traditional architecture and town planning coexisted with Russian and European architecture (Fig. 2.37). In contrast to the "additive" growth of oriental cities, new streets were laid out to converge in a radial star-shaped symmetry, while standardized building blocks formed the new cities. The old and new parts of cities contrasted markedly with each other in terms of their development and town-planning concepts. In the old parts of the cities, irregular one-storied buildings

Fig. 2.36 The residence of the Grand Duke Nikolas Romanov in Tashkent

Fig. 2.37 The former Women's Gymnasium building was constructed in the traditional style of the 1890–1900s

were situated in labyrinths of streets. In contrast, the model of European cities was used by Russian architects and engineers in new cities, with the idea of regular plans of straight, planted, and paved streets oriented in accordance with the four winds. The building of the Coliseum Cinema in Tashkent is another example of the architecture of the 1900s. This building has survived many reconstructions and is now used for offices (Fig. 2.38).

Between the 1930s and 1991, the architecture was developed under the idea of socialist culture in three stylistic directions. One of them was constructivism, another was imitation of classical building styles, and the third used motifs of Uzbek national architecture (Figs. 2.39 and 2.40).

In the post–World War II years, new planning principles and industrial methods of mass building were implemented without sufficient consideration of local natural and social

Fig. 2.38 The former Coliseum Cinema, constructed in Tashkent in the 1900s

Fig. 2.39 This building in Tashkent is an example of the style and design of the 1950s

Fig. 2.40 An administrative building of the 1950s

Fig. 2.41 The Municipal Building in Tashkent

Fig. 2.42 The Circus in Tashkent

conditions. In the 1960s, mass construction of residential buildings, using industrial methods, resulted in inexpressive design and architecture (Fig. 2.41). However, the designs used for some public buildings (such as cinemas and theaters) were combinations of modernism and local traditional design (Figs. 2.42, 2.43, 2.44, and 2.45).

Since 1991, new forms and ideas have been realized in Uzbekistan's architecture. Traditional motifs such as arches and domes are widely used in new construction, as well as decoration reminiscent of mosques and madrassas. Independence created new conditions for residential buildings; in particular, privatization of residential building activities resulted in widespread building demolition, building conversion, and new building projects both in the oriental–Islamic old cities and in the large settlements and apartment blocks constructed using prefabricated slabs during the socialist era. Newer public buildings erected after 1990 differ considerably from those built in the 1970s–1980s. Gangler et al. (2002) note that they reflect the idea of a new understanding of architecture as a carrier of state representation.

Uzbekistan's eclectic architectural heritage ranges from glittering turquoise-domed mausoleums to the showy white marble structures favored in post-independence Uzbekistan, the rusty-hued brick structures of the czarist period, and the grandiose civic buildings of the Soviet era.

In residential areas, the design and planning of houses has also changed from the traditional style to an eclectic style combining various ideas (Fig. 2.46).

Fig. 2.43 The Hotel Uzbekistan in Tashkent

Fig. 2.44 The State Museum of Arts of Uzbekistan in Tashkent. Grand Duke Nicholas Romanov's large and very valuable collection of works of art is now the centerpiece of this museum

Fig. 2.45 The Turkiston Theater in Tashkent

Fig. 2.46 New houses in a residential area in the Shayxontoxur district of the city of Tashkent

Fig. 2.47 Exclusive residential areas in the Uchtepa district of the city of Tashkent

In various parts of the country, gentrification[15] has resulted in increasingly exclusive residential areas. The chosen architectural style is usually a combination of a courtyard and a villa, with a mixture of traditional and modern stylistic elements (Figs. 2.47 and 2.48). As in other postsocialist cities, the modern architecture style and forms clearly perform the function of social–spatial segregation (Gangler et al. 2002).

Askarov (2015) points out that in the twenty-first century, Soviet-style design concepts are being replaced by designs that aim to replicate traditional culture and crafts—alongside classicism and corporate design of new hotels, banks, business and commercial centers—to reveal Uzbekistan's culture to the world (Fig. 2.49). This idea is implemented in combinations of traditional culture and new design and engineering (Fig. 2.50). Since 1991, the idea of self-identification has been reflected in the new style and forms (Fig. 2.51). The reconstruction of the central part of the city of Tashkent, as well as other major cities, is a manifestation of strong gentrification in which residential areas are replaced by commercial and business facilities.

In 1993, as a sign of honor and a new symbol of the Uzbek nation, a sculpture of Amir Temur was unveiled (Fig. 2.52) in Amir Temur Square in Tashkent. This place had originally been developed in 1870 as the central park of the new Tashkent by Russian Empire architects. At that time, it was called the Square of Constantine. Since then, the main monu-

[15]Gentrification is a process in which a place, especially part of a city, changes from being a poor area to being a richer one.

Fig. 2.48 As land becomes expensive, multistory buildings replace traditional houses in the Mirzo Ulugbek district of the city of Tashkent

ment in the center of the square has been changed several times. The first was a monument to General Kaufman, which was later replaced by the Free Workers' Monument (1917), then a monument to the tenth anniversary of the October Revolution (1927), a monument to Stalin (1947), and a monument to Karl Marx (1968). Similar sculptures were also erected in Samarkand and Shakhrisabz. In Tashkent, Amir Timur's famous motto "Power is in Justice" is engraved in four languages on the monument's plinth.

The styles and designs of sculptures (Fig. 2.53) and mosques have also changed (Figs. 2.54 and 2.55).

Fig. 2.49 The Xalq Bank building in the center of Tashkent

Fig. 2.50 The UzAuto company building in the center of Tashkent

2.5 Food

Uzbek food unites various cuisines of the nations of the Silk Route. Different food preferences in the different regions of this country are based on the various traditions and histories of these places, the natural conditions, and the available resources. Persian and Mediterranean influences on the cultural life in the Zarafshan Valley and Chinese and Turkic influences in the Fergana Valley are evident in their different food preferences. Many Uzbek dishes incorporate the food of the people's nomadic and settled ancestors, making them very delicious. In contrast to the traditional dishes of the Turkic ancestors, Uzbek cuisine is a reflection of the long journeys taken to transport rice, noodles, spices, and tea from the East to the West and from the West to the East.

2.5.1 Bread Is Sacred

An Uzbek proverb says "Respect for bread, respect for nation." The most serious oaths are made with the breaking

Fig. 2.51 The Business Center building in Tashkent

Fig. 2.52 The sculpture of Amir Temur in Tashkent replaced a sculpture of Karl Marx. [LBM1948 / CC-BY-SA (https://creativecommons.org/licenses/by-sa/4.0) [public domain] via https://commons.wikimedia.org/wiki/File:Tamerl%C3%A1n,_Tashkent_1.jpg]

of bread. To break such an oath is the worst sin. As a sign of the exceptional respect reserved for bread, it is broken into pieces by hand and never cut with a knife (Figs. 2.56 and 2.57).

All important family celebrities are marked by parties with traditional dishes. At the beginning of any party, bread and tea are shared, and in this way, the host welcomes the guest. In Uzbek culture, bread is a metaphor for basic necessities and living conditions in general; it has social significance. Different types of Uzbek bread (known as *non* or *nan*) are made using different types of fancy, flaky dough, with different appearances and tastes. The methods used for leavening and the techniques used for baking are as varied as the water, land, and people of this country.

Fig. 2.53 The sculpture of the poets Zulfiya and Hamid Alimdjan in Jizzakh

In Uzbekistan, bread is made from a wheat flour dough or cornflour. It may be cultured with yeast, allowed to rise, and baked in a tandoor oven (*tandir*), or it may be baked without being cultured with yeast, as in the making of matzo bread (Fig. 2.58).

Many rites and beliefs refer to bread. When departing on a long journey, it is common to carry half a loaf of bread and keep it for the return home. Bread serves as a mascot. During an engagement ceremony, bread is broken by the parents of the bride and groom, and this ritual confirms the agreement between both parties. The bread is a sign of consolidation and hope for a happy future for the young couple.

Uzbek bread is usually round and flat (as a reminder of the sun) and is baked in loaves of different sizes and designs. According to Uzbek tradition, it is strictly forbidden to place the bread or its broken pieces face down (with the flat side up). If one drops bread on the ground, it is traditional to lift it up, kiss it, put it to one's eyes, and apologize. Throwing of breadcrumbs on the ground or in the garbage is considered an insult to the bread. All of these traditions were inherited from the pre-Islamic period and strengthened by Islamic laws and regulations.

Breads of different sizes, forms, and recipes, baked in different ways, tell the stories of their origins. Two different types of bread are festive bread (*patyr*) and everyday bread (*obi-non*) (Figs. 2.59 and 2.60). The bread varies depending

Fig. 2.54 The reconstructed Komolon Mosque in the Shayxontoxur district of Tashkent

Fig. 2.55 The Kokcha Mosque in the Shayxontoxur district of Tashkent

Fig. 2.56 (Xorazmian) bread of Khorezm is broken into pieces by hand

Fig. 2.57 In Jizzakh province, bread is never cut with a knife

on its taste, appearance, ingredients, design, and cooking technique. The famous Samarkand bread is made in 15 different ways, and the *shirmoy-non* bread of the Fergana Valley is made with chickpeas and aniseed. All types of bread are decorated.

2.5.2 The Tea Ceremony

On hot and dry summer days or on windy and cold winter days, a cup of tea is the greatest pleasure. It may be cold green tea (*yahnakuk choy*), hot black tea, or tea boiled with milk and butter (*shir choy*). The host checks the tea by filling

Fig. 2.58 Matzo bread—a flat, usually brittle piece of unleavened bread, which is eaten by Bukharian Jews especially during Passover—is baked in a *tandir* in Bukhara. [Image courtesy of Shavkat Boltaev, Bukhara]

Fig. 2.59 *Festive bread (Patyr)*

Fig. 2.60 Bread with onion (piyozlinon)

their own cup, carefully pouring it three times into the kettle and back again (Fig. 2.61), and then, only after this procedure, offering the tea to the guest. The cup (*piala*) is only half filled in order to allow the guest to experience the beauty of the drink. When passing a cup of tea to the guest, it is acceptable to put one's left hand on one's heart, emphasizing respect for the guest, and the guest taking this cup of tea should also put their left hand on their heart, emphasizing respect for the host. It is not as complicated as the Chinese tea ceremony. However, it recalls the origin of tea, and this procedure emphasizes respect for both sides.

2.5.3 Uzbek Cuisine

Many Uzbek dishes have a long history. The recipes and rituals of cooking are passed on from generation to generation. Traditional Uzbek food is diverse and bears the imprint of

Fig. 2.61 A tea ceremony starts in Urgench, Khorezm (Xorazm) region

Fig. 2.62 Kebab

Fig. 2.63 (**a**) *Lagman* (noodles with or without soup). [Source: Kim Sergey at Russian Wikipedia, CC-BY-SA 3.0, via Wikimedia]. (**b**) Manti (steamed dumplings). [Source: Ramón from Llanera, España; brightened by Off-shell, CC-BY-SA 2.0, via Wikimedia]

the conditions and lifestyle of people living in a particular locality. Historically, a settled or nomadic lifestyle influenced the nature and variety of dishes, the components, the recipes, and the cooking techniques. A particular feature of Uzbek cuisine is a combination of recipes inherited from nomadic ancestors for cooking meat dishes and other recipes inherited from settled ancestors. More vegetables and spices are used in cooking meat dishes, which makes them different from the cuisine of other Turkic nations. A common feature in Turkic cuisine is roast (kebab) (Fig. 2.62). Flour-based dishes such as dumplings or ravioli, steamed dumplings (*manti*), steamed meat roulette (*hanum*), noodle soup made with beef chuck, garlic, tomato paste, vegetables, and kumin (*lagman*) (Figs. 2.63a, b and 2.64), and a pasta dish made with fresh hand-rolled noodles and meat with herbs (*norin*) (Fig. 2.63) have much in common with Chinese

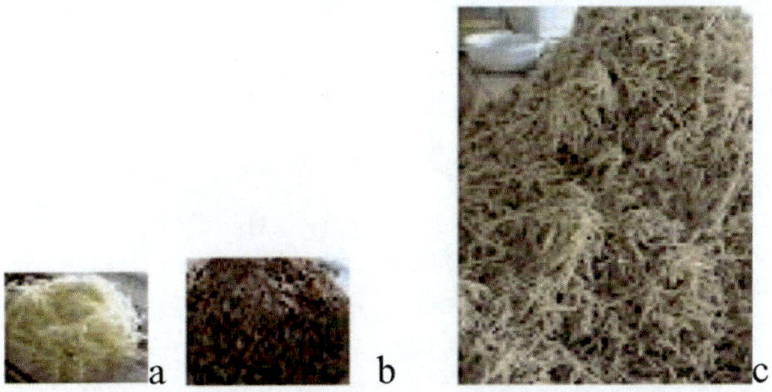

Fig. 2.64 Noodles (**a**) and meat (**b**) for *norin* (**c**)

Fig. 2.65 *Samsa* (traditional pastry patties with different fillings)

Fig. 2.66 *Tukhum barak* (dumplings with egg or pumpkin)

and Uyghur cuisines. Among the most popular Uzbek flour-based dishes, one may find the dough that may be a simple bread dough or a layered pastry dough (*samsa*) (Fig. 2.65), or traditional patties prepared using flaky pastry with different fillings (meat, pumpkin, potato, or greens). The ravioli with an egg or pumpkin filling (*Tukhum barak*) that widespread in Khorezm and Bukhara regions (Fig. 2.66).

The most tangible influence of Persian cuisine is evident in sweets cooked with fruit, vegetables, and even rose petals (Figs. 2.67 and 2.68). The favorite sweets are Halva (*khalva*),

Fig. 2.67 Sweets: cotton candy (*pashmak*) and caramel candy (*parvarda*)

Fig. 2.68 *Crystallized sugar (Novot)*

Fig. 2.69a Sprouted wheat seeds are the main ingredient used for cooking *sumalyak*

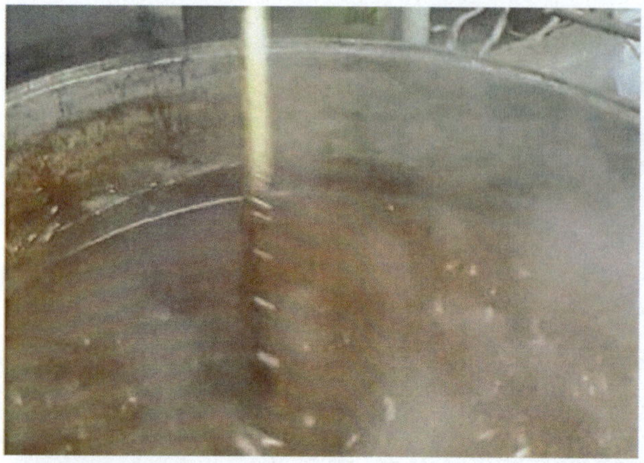

Fig. 2.69b It takes about a day to cook *sumalyak*, which is made using sprouted wheat seeds

Fig. 2.69c *Sumalyak* (made using sprouted wheat seeds) on a Navruz (New Year) festive table, accompanied by dishes of fresh greens

Crystallized sugar (*novot*), cotton candy (*pashmak*), made using beaten egg whites with sugar (*nisholda*), and *made using sprouted wheat seeds (sumalyak). Sumalyak* is a special food used to celebrate the New Year (Navruz) on March 21. This holiday is rooted in Uzbekistan's pre-Islamic past (Figs. 2.69a, 2.69b, 2.69c, 2.70, and 2.71).

The traditional Uzbek pilaf (*plov*) is cooked in different ways in different parts of the country. In Samarkand or Bukhara, *plov* is cooked in accordance with an Iranian recipe, which entails separate cooking of rice and meat with carrot, steaming them. This *plov* is usually white, in contrast to the brown *plov* cooked in the Fergana Valley, where they put rice in boiled soup containing onion, carrot, and meat, with spices (cumin, curcumin, raisin, pepper, barberry, and others). In Tashkent, all ingredients are roasted at the beginning (Figs. 2.70 and 2.71).

Fig. 2.70 Pilaf (*plov*). [GeoO (CC-BY-SA 4.0) (https://creativecommons.org/licenses/by-sa/4.0) via Wikimedia]

Sour milk (Figs. 2.72 and 2.73) is a heritage of nomadic food traditions and serves as a necessary ingredient in soups. Milk dishes are widely consumed, especially during fasting. Sour milk dishes are divided into vegetarian and meat dishes, and into cold (*ayran*) and hot ones. They may be spiced with herbs and are very popular all over the country.

The soups are quite thick and rich with vegetables and spices; one may find many commonalities with Chinese soups. The recipes and cooking techniques vary from the broth-based type with different vegetables, green gram, turnip, spices, rice, and noodles, to the type in which all ingredients are fried and then cooked in boiling broth. The recipes for cooking meat dishes include a variety of techniques and methods, including boiling in water, broth, or milk; stewing; smoking; frying; and frying followed by boiling. These techniques have absorbed the richness of the culinary arts of Western and Eastern cuisines, and some may have come from even more distant parts of the world. Who knows?

2.5.4 Fruit and Vegetables

In Uzbekistan, fruit and vegetables accompany any feast (Figs. 2.74, 2.75, and 2.76). Some fruit and vegetables can be harvested two times a year. The climate allows vegetables to be grown in greenhouses all year round. Uzbek dishes are supplemented by succulent tomatoes, emerald-green cucumbers, yellow carrots, radishes, eggplants, peppers, cabbages, and pumpkins. Grape leaves are used for wrapping fillings

Fig. 2.71 *Uzbek Pilaf.* [Beko CC-BY-SA 4.0 (https://creativecommons.org/licenses/by-sa/4.0) via Wikimedia]

Fig. 2.72 Sour milk (*chakka*) with spices

Fig. 2.73 Sour milk is an ingredient in soups

Fig. 2.74 Apples in the village of Chimgan, Tashkent region

Fig. 2.75 Grapes in the city of Tashkent

Fig. 2.76 Pomegranates in the Sherabad district, Surkhandarya region. Each piece of fruit weighs about 1 kg

Fig. 2.77 Black basil

Fig. 2.78 Green basil

(such as meat and rice) and when cooking dolma stuffed with such vegetables as tomato, pepper, onion, and garlic. These dishes have been known since antiquity and are also common in Mediterranean cuisine.

Spices and herbs are widely used in cooking for flavor strength, as are some vegetables and fruit (for example, quinces, pomegranate arils, fresh and dried plums, raisins, and wild cherries), both as ingredients and as spices. Such spices as black cumin, black pepper, cayenne, barberries, coriander, bay leaves, cinnamon, and sesame seeds were originally brought to Uzbekistan by travelers and merchants. Commonly used herbs are dill, parsley, celery, cilantro, green onion, garlic, basil (*rayhon*), and thyme (*jambil*) (Figs. 2.77 and 2.78).

2.6 Clothing

The colors and designs of clothing are based on their cultural traditions, beliefs, and historical background. The preferred clothing colors are usually bright, like the sun, nature, and the environment. The most popular color is blue, which, in the pre-Islamic religion of the Turks, represented the east and the god of the sky. In combination with green (the traditional color of Islam), it represents fertility, luck, and wealth. In the Surkhandarya region, red is a symbol of well-being. Apparently, this color was inherited from the Buddhists beliefs, in which it symbolizes good luck, joy, prosperity, celebration, happiness, and a long life (Figs. 2.79, 2.80, 2.81, and 2.82).

The national headwear is the skullcap (*tubeteyka*), worn by men, women, and children. A tetrahedral and slightly conical form of *tubeteyka* is commonly worn in many regions of this country. The heights, patterns, and colors are very diverse and reflect traditions and preferences rooted in the past (Figs. 2.81, 2.82, 2.83, and 2.84).

2.7 Traditions and Rituals

In Uzbekistan, traditions and rituals are reflections of beliefs and behavior with symbolic meaning inherited from the ancestors of the different ethnicities, cultures, and religions. Today, they combine various religious and cultural concepts, including a number of ideas from the pre-Islamic past and from Islam. The cultural policy of the Russian Empire and the policy during the Soviet period influenced many rituals by introducing secular forms. For example, some traditional Western customs, such as white wedding dresses and veils, have been adopted by both urban and rural populations in Uzbekistan.

In an agricultural society, the idea of a big family is reflected in the importance of rituals such as weddings and the births of children. They are still celebrated, strictly following the rules of practice transmitted through time by teaching from one generation to the next. The Uzbek saying that "Uzbeks need peace and freedom" is a reflection of the attitude formed during the long history of nation building.

Fig. 2.79 Traditional women's and men's suits in Khorezm

Fig. 2.80 A festive men's suit

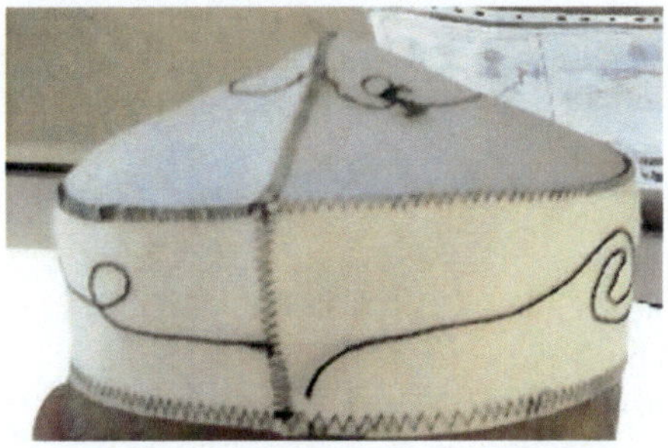

Fig. 2.82 The traditional headwear of Kazakh men is different in design

Fig. 2.81 The traditional skullcap (*tubeteyka*) for men is black and embroidered with an embroidered white pattern in the form of four paprika fruit

Fig. 2.83 The traditional headwear of women in the Fergana region is embroidered with a floral pattern

Fig. 2.84 A geometric pattern in the form of horns, in the Baysun district, Surkhandarya region

Fig. 2.85 The houses of the groom and the bride are usually decorated

Fig. 2.86 Musicians announce the event

Fig. 2.87 The traditional *kelin salom* is a ritual to welcome new relatives and guests and to exchange gifts with them

Uzbek people have several dreams: to build a house, to marry, to have children, and to live a peaceful and healthy life. They work hard and save money to build a house and to celebrate marriage and the births of children.

2.7.1 Weddings

A wedding is one of the most important events for a family. People usually try to make it beautiful and bright, with a lot of guests, music, and food (Figs. 2.85 and 2.86).

Today, the wedding party is a combination of traditional and modern rituals. The wedding ceremony varies in different regions and now incorporates new rituals in addition to ancient ones (Figs. 2.87, 2.88, 2.89, 2.90, and 2.91).

2.7.2 The Birth of a Child

The celebration of the birth of a baby is the next important stage in a person's personal life. It is usually marked by costly festivities in the case of the birth of the first child. It is called the *beshik toi*[16] ("cradle celebration") and is celebrated by the family when the baby is 40 days old.

2.7.3 Consecration of the Mother

The *sallabandon*[17] is a ritual to honor motherhood when a young woman gives a birth to a baby (Fig. 2.92). It is assumed that this tradition, in which the young mother wears a turban

[16] A *beshik* is a type of cradle made of juniper wood.

[17] In Persian, *sallabandon* means "to tie a turban."

Fig. 2.88 A ceremony of purification with water from the well at the Mahmud Pahlavan Mausoleum in Khiva, Khorezm region

Fig. 2.89 A wedding ceremony in Khiva, Khorezm region, in which the young couple releases pigeons into the air

(*salla*), started in the seventh century. This ritual is widespread in the Bukhara and Samarkand regions.

Boys undergo circumcision, and this is usually marked by a festive party (*sunnat*, *ogiltoi*) at which the relatives and guests celebrate this important moment (Fig. 2.93).

2.8 Holidays

2.8.1 New Year

The New Year is a national holiday, and people celebrate New Year's Eve two times. The first celebration is held on December 31, with good wishes for the future. This holiday was established after incorporation of Central Asia into the Russian Empire and was retained during the Soviet period, with all relevant attributes such as Christmas trees, Santa Claus, and gifts. Today, Santa Claus's gifts are welcomed by children in every family (Fig. 2.94).

2.8.2 Navruz

The second New Year celebration is a national holiday (known as Navruz) on March 21. It was established after Uzbekistan gained its independence. It is another manifestation of self-identification. However, this holiday has been celebrated in Central Asia for at least 3000 years and is associated with Zoroastrianism. For an agricultural society, this day is a day of hope for a good year ahead, with a heavy crop, a rich harvest, and peace. Navruz is Persian for "New Day", a day when the seasons begin to change, nature is awakening after winter, and a new agricultural cycle of cultivation and planting is beginning. For ancient people, this day was a symbol of the power of light, which overcomes darkness. By tradition, people clean their own houses and their own souls, since it is a time for reconciliation, forgiveness, and cheerfulness. Turks and other nomadic tribes adopted the celebration of Navruz with festivities, singing, dancing, special markets,

Fig. 2.90 A wedding ceremony of Bukharian Jews. [Image courtesy of Shavkat Boltaev, Bukhara]

Fig. 2.91 A wedding ceremony of young Lyuli (Central Asian gypsies) incorporates Uzbek rituals. [Image courtesy of Shavkat Boltaev, Bukhara]

Fig. 2.92 A *sallabandon* ritual in Bukhara. [Image courtesy of Magfirat Muminova, Tashkent]

Fig. 2.93 A boy wears colorful clothing for his *sunnat* party. [Image courtesy of Magfirat Muminova, Tashkent]

Fig. 2.94 Santa Claus and children on December 31

Fig. 2.95 The traditional dishes of Navruz

horse racing, different forms of entertainment, and parties with the traditional dishes of Navroz (Fig. 2.95). This holiday survived Islam and Soviet restrictions.

2.8.3 Religious Holidays

In spite of restrictions, Islamic holidays were celebrated in Soviet times. Today, they are set officially by the government as holidays of mercy, kindness, and spiritual generosity. Eid-al-Fitr (or Ramazan Hayit) and Eid-al-Adha (or Kurban Hayit) were introduced by Arabs when they invaded Central Asia in the middle of the seventh century. These holidays have social meaning in addition to religious meaning. In Uzbekistan, during these holidays, it is customary to commemorate the deceased, and to visit and help the sick, poor, and elderly. Eid al-Fitr is one of the most festive periods in the Islamic calendar and is often commemorated with large feasts and family time, as well as charitable acts and donations.

2.9 Arts and Crafts

As important trade and craft centers on the Great Silk Route, many cities in Uzbekistan have contributed to arts and crafts through development of a variety of artistic traditions brought by the peoples who formed the Uzbek nation. Different regional centers and schools of folk arts and crafts have been formed under the influence of ancient traditions. These traditions include the making of gold embroidery, *suzani* (embroidered textiles), ceramics, Uzbek pictures on leather, skullcaps of various kinds and purposes, Uzbek national knives (*pichaks*), silk and wool carpets, papier-mâché souvenirs, wooden items, Uzbek jewelry, and wonderful ironwork, which for many centuries have been handmade local craftsmen and masters.

The ethnocultural traditions of the people of Uzbekistan have a common history and ethnographic origins, creating the basis of the national culture and art, which are rooted in ancient Xorazm, Bactria, and Sogdiana, and have also accumulated elements of other cultures on the Great Silk Route.

2.9.1 Cold Steel from Fergana

The availability of natural resources and the lifestyles of the local people have led to regional specialization in particular arts and crafts. One example is metallurgy at sites rich in carboniferous iron ore, such as the Namangan region in the Fergana Valley. The technology for smelting steel and masterly hammering to make it very strong was mentioned in Arabic chronicles in the tenth century. Fire-resistant crucibles made of kaolin could resist a temperature of 1300 °C. For smelting furnaces, local metallurgists used natural fuel—the wood of the juniper tree, which, when burned, can produce high temperatures. The famous knives from Chust in Namangan are descendants of that same "cold steel from Fergana" (Fig. 2.96).

Fig. 2.96 Knives from Chust in the Namangan region of the Fergana Valley

2.9.2 Gold Embroidery

Gold embroidery was mentioned in texts as early as the fifteenth century. Special gold embroidery techniques and national patterns were developed universally in Bukhara, Nurota, Shakhrisabz, Samarkand, Jizzakh, Tashkent, Fergana, and Pskent (in Tashkent province). This kind of needlework is labor intensive and time consuming. Hand-stitched embroidery is a work of art that was highly valued by all nations on the Great Silk Route (Fig. 2.97). Because of its high cost, handmade gold embroidery was available only to high-ranking elites in Central Asia. Outerwear (*zarchapan*) richly decorated with patterns was presented as a symbol of the ruler's special favor at the emir's court. Even today, it is a generous gift bestowed on particularly important occasions, including weddings. Today, machine-made gold embroidery is widely available and is used for many purposes—for example, wallets, outerwear, headwear, decoration, curtains for theatrical stages, and large, thematic picture panels for large halls. Gold-embroidered products are produced in large volumes by a gold embroidery factory in Bukhara.

2.9.3 Carpet Weaving

Carpet weaving is an old tradition of both settled and nomadic peoples. The designs and colors of carpets vary from blue-green in Xorazm (Fig. 2.98) to red-brown in the

Fig. 2.97 Nineteenth-century handmade gold embroidery from the Tashkent region

Fig. 2.98 Handmade carpet weaving in Khiva, Khorezm region

Kashkadarya and Surxondaryo provinces. A dominant geometric shape of an ornament recalling the horns of sheep or cows is a major element of the designs in Karakalpakstan and Jizzakh, where nomadic traditions are dominant. Floral ornaments are more popular in the Zarafshan Valley.

2.9.4 Embroidery

Embroidery is the most widespread type of fancywork and has traditionally been done by settled, seminomadic, and nomadic tribes for decoration of clothing, covers, religious textiles, festive and everyday accessories, and traditional interior design. Several centers of artistic embroidery—including Nurota (in the Samarkand region), Bukhara, Samarkand, Tashkent, and Fergana—are well known for their high-quality artworks. These centers have created their own styles, with various ideas implemented in their designs. In Islam, it is forbidden to portray people, animals, and birds, and that is why stylized images were used to accompany floral designs. At the same time, Zoroastrian influence can be seen in embroidery patterns, where flowers, fruit, and trees with magical meanings were traditionally depicted (Figs. 2.99, 2.100, and 2.101). The embroidery patterns created by settled communities in Fergana, Tashkent, Bukhara, and Samarkand are quite different from those with a stronger influence of nomadic history and preferences in Kashkadarya or Surkhandarya regions.

Thin graphic designs of twigs with large areas of uncovered space on a colored background are a distinguishing feature of embroidery from the Fergana Valley (Fig. 2.102). A common element in the embroidery is beautiful patterns of floral ornaments, with images of pomegranates[18] and colorful elements of various kinds.

[18] Pomegranate arils (seeds) symbolize fertility, prosperity, and wealth.

Fig. 2.99 Nurota in the Samarkand region is famous for embroidery of flowers and floral patterns on white fabric. This is an example of an embroidery pattern created in settled communities, with strong influence from the Persian tradition

Fig. 2.100 Tashkent embroidery is usually composed in the form of a central star or circle with blossomed branches, leaving the greater part of the fabric visible. This example of embroidery with a pattern depicting the sun and snakes symbolizes prosperity and wealth

2.9.5 Ceramics

Ceramic art is the most developed craft in Uzbekistan. Several schools of ceramic art are well recognized for their distinguished work. The Samarkand–Bukhara school of ceramic art has centers in Tashkent, Samarkand, Urgut (in Samarkand region), Bukhara, Gujduvon (in Bukhara region), Shakhrisabz, Kitab (in Kashkadarya region), and Denau (in

Fig. 2.101 An example of embroidery with a geometric pattern of horns, from the Baysun district, Surkhandarya region

Fig. 2.102 A nineteenth-century prayer mat, Fergana valley

Fig. 2.103 Ceramics from the Rishtan School in the Fergana Valley

Surkhandarya region). The Fergana school has centers in Rishtan (in Fergana region) and Gurumsaray (in Namangan region). The Khorezm school has centers in Khiva and Chimbay (in Khorezm region). The ceramics from these centers differ from each other in terms of decoration, shape, and type. The technique used for making pottery varies depending on the quality of the available clay, which differs from region to region, and so too do the local ingredients used for dyes to paint plates and jars. Schools of traditional ceramic arts have developed their own techniques for ceramic firing, with unique designs for painting and glazing. The schools of ceramic arts in Rishtan in the Fergana Valley, Gijduvan in Bukhara region, Urgut in Samarkand region, Shakhrisabz in Kashkadarya region, Khorezm, and Tashkent have existed for many centuries. In Rishtan, a blue glaze is created using a natural dye produced from plant ash.

All centers of pottery production were located in proximity to the Great Silk Route, and the growing demand for glazed dishes and tiles stimulated their development. Even endless raids by nomadic tribes were not able to stop their growth. The ceramic art of Uzbekistan integrates achievements by different cultures and nations on the Great Silk Route. It combines the variety of floral decoration developed by settled people with geometric shapes developed by nomadic tribes. In accordance with Turkic tradition, a blue color is preferred, including all shades of azure with impregnations of noble turquoise under a barely perceptible, thin layer of glaze. The technology of alkali (*ishkor*) glazes which, after glazing (firing) make the finished wares look blue, originated in Iran and Samarkand.

The best examples of this type of ceramic art are ceramics from the Rishtan School in the Fergana Valley, developed under the influence of Chinese masters during the Ming Dynasty (Fig. 2.103). This is considered the starting point for ceramic art in Uzbekistan.

Wares from the Bukhara–Samarkand school are of ochre-yellow and greenish colors, achieved by application of lead glaze (Fig. 2.104).

Ceramics from the Gijduvan school are characterized by floral motifs and natural colors, with an emphasis on greens

Fig. 2.104 Ceramics from the Bukhara–Samarkand school

Fig. 2.105 Ceramics from the Gijduvan school

and browns, and are unusually lustrous because of the clay-heavy paint they are brushed with (Fig. 2.105).

2.9.6 Silk Fabrics

Centuries-old handmade silk fabric, or ikat, is a very ancient way of creating designs on fabric by resist dyeing the threads before the fabric is woven (Fig. 2.106). In Uzbekistan, this technique is called *abr-band*, which means "tying a cloud" in Persian. The fabric products created this way are called *abra* ("cloud"). Legends about the origin of this technique mention depictions of the lunar or solar path on the running water surface of a stream and an image of clouds playing with col-

Fig. 2.106 The ikat technique from the Fergana region

Fig. 2.107 *Atlas* fabric made of pure silk in Tashkent

Fig. 2.108 *Adras* fabrics made of cotton weft and silk wrap in Tashkent

ors. Fabrics made of pure silk are called *atlas*, *khan-atlas*, and *shoi* (Fig. 2.107). Fabrics made of cotton weft and silk warp are called *adras* (Fig. 2.108), *bekasab*, *pasma*, and *banoras*. Ikat patterns are diverse and are named after the shapes they resemble. Common designs imitate almonds, pomegranates, the moon, horns, butterflies, and other Turkic symbols of prosperity and wealth.

Fig. 2.109 Simple and restrained shapes, with typical herbal ornamentation, in embossing from the Bukhara school

2.9.7 Copper Embossing

In places with rich metallic ore deposits, such as Uzbekistan, copper embossing is one of the most ancient applied traditional arts. The main workshops making embossed copper articles were located in Bukhara, Khiva, Kokand, Samarkand, Qarshi, Shakhrisabz, and Tashkent. The embossed and chased metal articles were valued not only for their decorative qualities but also for their roles in everyday life as indicators of social status and wealth.

The different schools developed their own styles of ornamentation, form, and design, as well as different embossing techniques. In the nineteenth century, embossed copper artworks from Bukhara and Khiva became very popular for the great variety of their shapes, the classical balance of their proportions, their ornamental motifs, and their original engraving and embossing techniques (Figs. 2.109, 2.110, and 2.111). The variety of methods of background and surface pattern decoration produced by the Kokand school and the conservative style of large, simple, vegetable-inspired

Fig. 2.110 Embossing from the Samarkand school

Fig. 2.111 Embossing from the Khiva school

patterns produced by the Samarkand school are representations of the environments and traditional beliefs of settled and nomadic peoples.

2.9.8 Wood Carving

Wood carving was widely used in ancient architecture, and later it was performed on many objects used in everyday life.

Fig. 2.112 Wooden pillars in the Itchan Kala in Khiva, Khorezm region

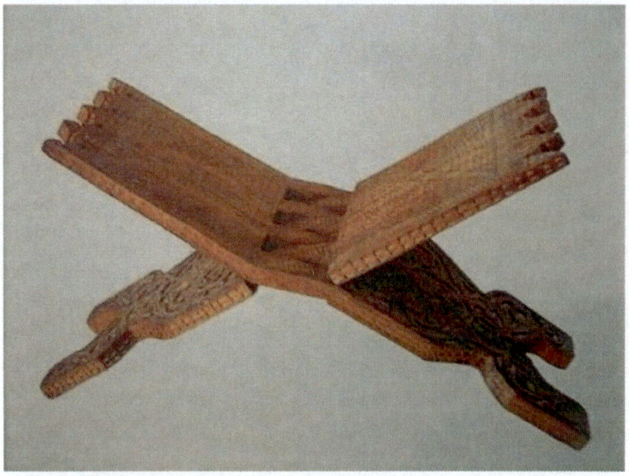

Fig. 2.113 A book holder from Khiva, Khorezm region

Floral and geometric ornamentation is used in carving of jewelry boxes, book holders, tables, and chairs (Figs. 2.112, 2.113, 2.114, 2.115, and 2.116). Woods such as elm, walnut, sycamore, juniper, and mulberry are used by craftsmen. The centers of wood carving are Khiva, Bukhara, Samarkand, Tashkent, and Fergana.

Fig. 2.114 A door in the palace of Said Alimxan, the emir of Bukhara

Fig. 2.115 Elm (*gujum* in Uzbek) is used for wood carving in Khiva, Khorezm region

Fig. 2.116 Wooden pillars in the Jome Masjid in Bukhara

References

Academy of sciences of the Republic of Uzbekistan (2018) About Academy of Sciences. http://www.academy.uz/en/news/23

Adams L (2010) The spectacular state: culture and national identity in Uzbekistan, Duke University Press. Retrieved from https://books.google.nl/books?hl=ru&lr=&id=TtcwIXC0pl8C&oi=fnd&pg=PP6&ots=3_qSjOrxXq&sig=ukAfqQgUZioUSPewG9Zz5WJ1atc&redir_esc=y#v=onepage&q&f=false

Adle C, Habib I (eds) (2003) History of civilization of central Asia, volume V, Development in contrast: From the sixteenth to the mid-nineteenth century. UNESCO Publishing, Paris, 934 p

Alimova D, Filanovich M (2017) The history of Tashkent. From the ancient times to today (In Russian: ИсторияТашкента. С древнейшихвремендонашихдней). Retrieved from https://mytashkent.uz/istoriya-tashkenta-s-drevneyshih-vremen-do-nashih-dney/, on May 27 2017

Allworth E (2017) Uzbekistan. In Encyclopedia Britannica. Last updated 4 May, 2017. Retrieved from https://www.britannica.com/place/Uzbekistan

Askarov S (2015) Design, engineering, and traditional culture in the modern day architecture of Uzbekistan. J Sanat 11(10):4. Retrieved from http://sanat.orexca.com/2014/2014-4/design-engineering-and-traditional-culture-in-the-modern-day-architecture-of-uzbekistan/

Barthold V (1956) A short history of Turkestan in four studies on the history of central Asia Vol. I. E. J. Brill, Leiden

Barthold V (1966) Studies on the history of Islam and Arabic Caliphate Vol VI (In Russian Сочинения том 6 Работы по истории ислама и арабского халифата) Издательство «Наука», Москва. http://publ.lib.ru/ARCHIVES/B/BARTOL'D_Vasiliy_Vladimirovich/_Bartol'd_V.V..html

Barthold V (1993) Turks; twelve lectures on history of Turkish people of central Asia (In Russian Тюрки: Двенадцать лекций по истории турецких народов Средней Азии Алматы: Жалын,) Accessiblae from http://publ.lib.ru/ARCHIVES/B/BARTOL'D_Vasiliy_Vladimirovich/_Bartol'd_V.V..html

Bertelsmann Stiftung (2018) BTI 2018 country report: Uzbekistan. https://bti-project.org/fileadmin/api/content/en/downloads/reports/country_report_2018_UZB.pdf

Blouet B (2005) Global Geostrategy: Mackinder and the defence of the West Frank Cass, London/New York. Retrieved from https://books.google.co.uz/books?id=JYeTJy7Oyj8C&pg=PA165&lpg=PA165&dq=great+game+McKinder&source=bl&ots=tze2Ir7JwE&sig=0Q9_4_fWTDazw-VbWR1BQqI1ocE&hl=ru&sa=X&ved=0ahUKEwj6mIb6urzbAhVFJ5oKHcR2A3kQ6AEIejAQ#v=onepage&q=great%20game%20McKinder&f=false

Borbieva NO (2013) Book review: Laura L Adams, the spectacular state; culture and national identity in Uzbekistan. J Int Sociol 28(2):170. Available on 1 Mar, 2013 from http://journals.sagepub.com.vu-nl.idm.oclc.org/doi/abs/10.1177/0268580913477954a?journalCode=issa&volume=28&year=2013&issue=2

Curtis GE (ed) (1996) Uzbekistan: a country study. GPO for the Library of Congress, Washington DC. Retrieved from http://countrystudies.us/uzbekistan/

Countrymeters.info (2022) Uzbekistan population. https://countrymeters.info/en/Uzbekistan. Accessed 3 Jun 2022

Dani AN and Masson VM (1992) (ed) History of civilizations of Central Asia v I: The dawn of civilization: earliest times to 700 BC. UNESCO, Paris, 535 p

Denison M (2003) Identity politics in Central Asia. J Asian Affair XXXIV(1):58–64. Available from 1 Sept 2011. http://www.tandfonline.com/doi/abs/10.1080/03068370320000054243

Filanovich MI (1983) Tashkent: The Birth and Development of the Urban Culture Tashkent FAN (In Russian Филанович, МаргаритаИвановна. Ташкент :Зарождение и развитиегорода и гор. культуры / М. И. Филанович. – Ташкент: Фан, 1983. – 199, [32] с. : ил.; 21 см.; ISBN В пер. (В пер.) : 3 р.Внадзаг.: АН УзССР, Ин-т археологии. Археологические памятники – ТашкентБ 83-29/211Б 83-29/212)

Gangler A, Nagler H, Schwartze F, Ribbeck E (2002) Tashkent in change – transformation of urban structure Städtebau-Institut Universität Stuttgart Lehrstuhl Städtebauund Entwerfen). Retrieved from: https://www.uni-stuttgart.de/si/stb/stb_forschung/taschkent/Taschkent%2008%2012%2002%20EN_s.pdf

Gibb HAE (1923) The Arab conquests in central Asia Vol II. Royal Asiatic Society, London

Gulyamov X (1952) The communists of Uzbekistan in the great war of the Soviet Union Tashkent Gosizdat of UzSSR (In Russian: Гулямов ХКоммунистыУзбекистанавпериодвеликойотечественнойвойныСоветскогоСоюзаТашкентГосиздатУзССР 1952)

Gulyamova L (2012) Renewable social energy: Mahalla as a traditional institution for sustainability in disaster by design: the Aral Sea and its lessons for sustainability, vol 20. Research in Social Problems and Public Policy Emerald Group Publishing Limited, pp 335–348. https://doi.org/10.1108/S0196-1152(2012)0000020034

Holzner B et al (eds) (2010) Gender equality and empowerment of women: policy document. Federal Ministry for European and International Affairs, Vienna

International Business Publication (Ipb) (2013) Uzbekistan Country Study Guide. Vol 1 Strategic Informatio and Developments World Business Information Catalog IBP Inc

International Business Publications (2017) Uzbekistan: country study guide: volume 1 strategic information and developments. Washington, DC: International Business Publications

Isaacs R (2015) Between "imagined" and "real" nation-building: identities and nationhood in post-Soviet Central Asia. J Natl 43(3):371–382. https://doi.org/10.1080/00905992.2015.1029044. "Imagined" vs "real" nation-building language and identity policies between theory and practice in Central Asia. Available from 10 March 2015. Published online: 13 May 2015. Retrieved from http://www.tandfonline.com/doi/full/10.1080/00905992.2015.1029044?src=recsys

Kastel'skaia ZD (1980). On history of Turkestan region (In Russian Из истории Туркестанского края Москва «Наука»)

Kasymova N, Usmanova K (2015) Information policy in central Asian countries. J Himal Cent Asian Stud New Delhi 19(3/3):194–111. Retrieved from http://search.proquest.com.vu-nl.idm.oclc.org/docview/1722189516/fulltextPDF/432E56C5BAC24748PQ/16?accountid=10978

Kim L (2003) Gender Stereotypes in employment in Uzbekistan. Central Asian Network for Gender Studies. Retrieved from http://www.eldis.org/document/A20698

Kobilov B (2018) Deficient higher education (In Russian: Кобилов БДефицитноевысшееобразование). Retrieved from https://www.gazeta.uz/ru/2018/05/28/education/

König J, Woolf G (eds) (2013) Encyclopaedism from Antiquity to the Renaissance. Cambridge University Press, Cambridge/New York. Retrieved from http://www.cambridge.org/catalogue/catalogue.asp?isbn=1107454387

Koshelenko GA (ed) (1985) The Ancient States of the Caucus and the Central Asia Moscow "Science". (In Russian Кошеленко ГА (отв.редактортома) ДревнейшиегосударстваКавказа и СреднейАзии. М, Наука, 1985)

Koshelenko GA, Filanovich MI (1999) The Central Asia and the far East in the mediaeval Moscow: science. (In Russian Кошеленко ГАФиланович МИ, СредняяАзия и ДальнийВосток в эпохусредневековья М. Наука, 1999)

Krasnovskiy DI (1924) The materials on regionalization of Turkestan Tashkent TES. (In Russian: МатериалыпорайонированиюТуркестана (под редакцией Д.И.Красновского) Ташкент: Изд-во ТЭСа- 1924)

Litvinsky BA (ed) (1996) History of civilizations of central Asia, 3, UNESCO, Paris

Litvinskiy BA, Antonova EV (ed) (1996) Asia-dialogue of civilizations (In Russian: Азия - диалог цивилизаций'; ред. Литвинский, Борис Анатольевич; Антонова, Елена Вадимовна; Изд-во: СПб: Гиперион, 1996)

Litvinsky BA & et al (1996) The Arab Conquest In Litvinsky BA History of Civilizations of Central Asia Vol III: The Crosroads of Civilizations: A.D. 250 t0 750 Paris UNESCO Publishing

Mackinder HI (1904) The geographical pivot of history. Geogr J 23(4):421–437. Available from https://www.jstor.org/stable/1775498?seq=1#page_scan_tab_contents

Megoran N (2006) For ethnography in political geography: Experiencing and re-imagining Ferghana Valley boundary closures. J Politic Geogr Elsevier 25:622–640. https://doi.org/10.1016/j.polgeo.2006.05.005. Retrieved from https://www.staff.ncl.ac.uk/nick.megoran/pdf/megoran_ethnography_pg.pdf

Megoran N, Sharapova S (eds) (2014) Central Asia in international relations. The legacies of Halford Mackinder. Hurst Publishers, Oxford University Press, London. Retrieved from http://www.hurstpublishers.com/book/halford-mackinder-and-the-international-relations-of-central-asia/. https://doi.org/10.1093/acprof:oso/9780199327973.001.0001

Merry EW (2004) Governance in Central Asia: national in form, Soviet in content. J Cambridge Rev Int Affairs 17(2) Available from 21 Oct 2010. https://doi.org/10.1080/0955757042000245906

Mirza Muhammad Haydar (1996) Tarih-I Rashidi Tashkent. The academy of sciences of the republic of Uzbekistan (In Russian МирзаМухаммадХайдарТарих-иРашидиТашкентИздательство АкадемииНаукРеспубликиУзбекистан)

Mustafayeva N (2015) Formation of secular higher education system in Uzbekistan. J Himal Cent Asian Stud. New Delhi 19.3/4 (Jul–Dec 2015): 46-III. Retrieved from http://search.proquest.com.vu-nl.idm.oclc.org/docview/1722189335/abstract/432E56C5BAC24748PQ/5?accountid=10978

Northrop D (2004) Veiled Empire: Gender & Power in Stalinist central Asia. Cornell University Press

Ohlsson H (2011) Islam and secular state in Uzbekistan: state control of religion and its implications for the understanding of secularity. J Cahiers d'Asie Centrale 19-20:485–493

Olcott M (1981) The Basmachi or Freemen's revolt in Turkestan 1918–24. Eur Asia Stud 33(3):352–369

Paksoy HB (1989) Alpamysh: Central Asian identity under Russian rule paperback Assn for the advancement

Patnaik A (2016) Central Asia. Geopolitics, Security and Stability Routledge India. Retrieved from https://books.google.nl/books?id=-wneCwAAQBAJ&pg=PT49&lpg=PT49&dq=The+Heartland+Fallacy:+Central+Asia,+Geography+and+Globalisation+%E2%80%94+Levent+Hekimoglu&source=bl&ots=Xknv8TZMJx&sig=35npkqx_nHU9Oj-AobV7IbghRS8&hl=ru&sa=X&ved=0ahUKEwis7_b5jIbUAhWPZlAKHcrYDBAQ6AEIRjAE#v=onepage&q=The%20Heartland%20Fallacy%3A%20Central%20Asia%2C%20Geography%20and%20Globalisation%20%E2%80%94%20Levent%20Hekimoglu&f=false

Rahimov J (2001) The history of Uzbekistan (2nd half of XIX - beginning of XX centuris) Textbook for 9th grade Tashkent (In Russian: Рахимов Ж. История Узбекистана (вторая половина XIX в. - начало XX в.). Учебник для 9-го класса. Ташкент, "Узбекистон", 2001)

Reeves M, Rasanayagam J, Beyer J (eds) (2014) Ethnographies of the state in Central Asia. Performing Politics. Indiana University Press. Retrieved from http://www.iupress.indiana.edu/product_info.php?products_id=807110

Rtveladze (2005) Civilizations, states, culture of central Asia (In Russian: Цивилизации, государства, культуры Центральной Азии. Ташкент)

Rtveladze E, Rtveladze L (2017) Tashkent: along the roads of history. In J Sanat 2. (In Russian). Available from 1 April 2017. http://sanat.orexca.com/2007/2007-2/history_roads/

Sergeev E (2013) The great game, 1856–1907: Russo-British Relations in Central and East Asia. John Hopkins University Press, Baltimore MD

Sobti M (2005) Urban metamorphosis and change in central Asian cities after the Arab invasions. A Thesis Atlanta University

Starr FS (2018) Change and continuity in Uzbekistan, 1991–2016. In: Uzbekistans New Face. Central Asia-Caucasus Istitute Silk Road Studies Program A Joint Transtlantic Research and Policy Center. Accessible from 27 Sept, 2018. http://silkroadstudies.org/resources/pdf/Monographs/1809-Starr-UZ.pdf

Stokes J (ed) (2009) Qara-khanids, encyclopedia of the peoples of Africa and the Middle East, Vol.1. Infobase Publishing, 578 p

Struve VV, Snegiryov IL (eds) (1935) Turaev BA the history of the ancient East Leningrad OGIZ. (In Russian: ТураевВ.А.ИсторидревнегоВостокаПодредакциейСтруве.В.В, СнегиреваИ.Л. ЛенинградОГИЗ)

Tangirov OE (2015) About activities of the madrassas in the city of Tashkent at the end of XIX and the beginning of XX centuries (in Russian: Тангиров О. Э. О деятельности медресе города Ташкента в конце XIX – начале XX вв). https://web.archive.org/web/20150928172311/; http://rex-history.ru/history-a-a/425-tangirov.html

The History of Herodotus translated by George Rawlinson. Edited by Blakeney EH (1910) London JM Dent. Retrieved from https://en.wikisource.org/wiki/The_History_of_Herodotus_(Rawlinson)

The Women of Uzbekistan: Social Optimism, Strong Confidence in Future (2018) (In Russian: ЖенщиныУзбекистана: социальныйоптимизм, твердаяуверенность в будущем). Retrieved from http://uza.uz/ru/society/zhenshchiny-uzbekistana-sotsialnyy-optimizm-tverdaya-uverenn-24-01-2018

Tolstov SP (1943) The ancient culture of Uzbekistan (In Russian:Толстов, Сергей Павлович.Древняя культура Узбекистана- Ташкент : Изд-во УзФАН, 1943. - 36 с)

United Nations Development Programme (2014) Figures show: teacher, doctor or artist – a typical "woman's job" in Uzbekistan. Retrieved from http://www.uz.undp.org/content/uzbekistan/en/home/ourperspective/ourperspectivearticles/2014/11/250/gender-equality-is/

United Nations Development Programme (2020) Human Development Reports. Table 4: Gender Development Index. Retrieved from http://hdr.undp.org/en/composite/GDI

Uzbekistan's Newfound Foreign Policy and Its Implications for U.S.-Uzbekistan Relations (2018). Retrieved from http://voicesoncentralasia.org/uzbekistans-newfound-foreign-policy-and-its-implications-for-u-s-uzbekistan-relations/ on May 24, 2018

Voronina V (1969) Architectural monuments of middle Asia. Aurora, Leningrad

Walsh P (2011) Book review Laura L Adams, The spectacular state; culture and National Identity in Uzbekistan. J Am J Sociol 116(5):1684–1686. The University of Chicago Press Journals. Available from March 2011 http://www.jstor.org.vu-nl.idm.oclc.org/stable/10.1086/659023?seq=1#page_scan_tab_contents

Warikoo K (2015) Historical cultural heritage and tradition in Uzbekistan. Himal Cent Asian Stud 19(3/4) (Jul–Dec 2015) 16-II. Retrieved from https://vu.on.worldcat.org/search?queryString=ti:culture%20in%20uzbekistan&databaseList=4054,3882,2276,2274,2273,1697,2269,2662,4048,3995,638,3999,1978,3879,3891,3374,3250,2437,2038,2796,1861,1982,2433,2795,3782,2572,2175,3384,3261,1875,3018,3258,1476,3378,3652,3654,3552,3431,2584,2221,3551,2462,1920,3547,2215,2972,2178,3421,3784,3200,2474,3441,2198,2229,1931,3039,3556,3313,3433,2586,3036,2585,3950,2507,4023,3573,3572,3450,283,1941,2513,1666,2237,3205,3567,2236,2233,3962,3967,2375,3582,3218,1953,2369,3336,2005,2401,1672,3575,4026,4028,2409,3976,1834,4044,3195,3903,3229,1842,3227,2259,2897,3589,3586,3909,1847,3988&sortKey=LIBRARY_PLUS_RELEVANCE&scope=&page=3

Weitz R (2018) Uzbekistan's new foreign policy: change and continuity under new leadership. In Silk road paper, The Central Asia-Caucasus Institute and Silk Road Studies Program, Joint Center. Retrieved from https://silkroadstudies.org/resources/pdf/SilkRoadPapers/1801Weitz.pdf on May 28, 2018

World Bank (2017) Women in parliament in Europe and Central Asia. http://www.worldbank.org/en/data/interactive/2018/03/08/eca-data-women-parliaments

Zagorodnikova TN (ed) (2005) The great game in central Asia (I Russian "Большая игра в Центральной Азии Москва изд РАН)

Ziyaeva D (2015) Education and culture in Turkestan (19th–20th centuries). J Himal Cent Asian Stud. New Delhi 19.3/4 (Jul–Dec 2015): 24-III

3 Geography of Resources

Abstract

This chapter describes the influences of such global geological processes as plate motion, continental drift, and development of large-scale features of the Earth on formation of the nature of Uzbekistan. The country's natural resources—including its mineral reserves, groundwater, soil and land, water, and energy resources—are described in the context of their origins and distribution.

Keywords

Uzbekistan · Geology · Earthquakes · Mineral reserves · Gold mining · Loess · Groundwater · Land · Water · Energy resources

3.1 Global Geological Processes and Geology of Uzbekistan

The study of geology helps us understand the past, present, and future of the Earth and of nature. Global geological processes do not stop, and our knowledge of the geological evolution of Eurasia helps us understand the cyclic development of its nature. It is a significant way to explain and determine the geography of such complicated places as Uzbekistan, which is an example of the multiplier effects of global and regional geological processes and the history of the Earth itself as well.

An assumption accepted by scientists (Windley 2015) is that the age of the Earth is roughly 4.6 billion years. The geological evolution of Central Asian started almost 4 billion years ago and is still in process.

3.1.1 Geological Evolution of Asia

Asia, including the Central Asian region, is considered in geological works (Hendrix and Davis 2000; Mossakovsky et al. 1993) to be one of the youngest, most dynamic, and structurally most complicated continents to be studied in depth. More than half of the continent, including Uzbekistan, remains seismically active. Recent earthquakes in Asia are the result of emergence of new land and episodic collisions of island arcs with the mainland. According to some opinions (Heubeck 2001), the polyorogenic character of the Central Asian mountain belts stipulates the complexity of the region and makes it impossible to unambiguously identify, locate, and date the former plate boundary zones. The complexity of the geological evolution of the region creates challenges for precise identification of the extent of geological objects and the role of each of the driving forces. For example, the evolution of the Tien Shan Mountains is characterized by complexity of interrelated geological processes.

As a result of unique geological processes on the Asian continent (Torsvik and Cocks 2016), Central Asia contains some of the morphological extremes of the Earth's land surface. The Himalayas and the Hindu Kush Mountains are located quite close to the region, and the Tien Shan Mountains are part of these highest mountain systems on Earth. All of these mountain ranges are results of plate tectonics; since the process is still ongoing, it plays significant roles, at the global and regional levels, in geological evolution and in forming the natural environment of Central Asia and Uzbekistan.

According to the opinion of many researchers (Hendrix and Davis 2000; Zonenshain et al. 1990; Berzin et al. 1994; Buslov et al. 2001), Asia is the largest mosaic of assembled tectonic elements on Earth, and Central Asia is a block of various fragments derived from sliding of the Indo-Australian Plate beneath the Eurasian one and collision of various microcontinents. Since Asia is a relatively young continent, secondary structures are dominant and define its neotectonic units.

Continental collision is the result of different geological processes such as accumulation of tectonic collages, faulting, and postcollisional disruption. Researchers (Ryabchikov

L. Gulyamova, *The Geography of Uzbekistan*, World Regional Geography Book Series,
https://doi.org/10.1007/978-3-031-07873-6_3

et al. 2018) have indicated that the theory of plate tectonics explains how forces within the Earth propel sections of its crust in different directions and form continents. In their opinion, large structural deformations of the Earth's lithosphere are caused by interactions between tectonic plates. This is the primary mechanism of mountain building, which is a complicated geological process, or orogenesis; the process of orogeny takes tens of millions of years to build mountains from plains and the ocean floor.

From the global perspective, the journey of India's landmass is quite new; according to the US Geological Survey (2009) and to Kious and Tilling (2015), it started about 40–50 million years ago (Ma) when it moved 6000 km to the north before colliding with the Eurasian Plate. Kious and Tilling (1996) argues that the collision of Asia and India took place about 2000 km south of the present location of the line of collision along the Indus–Brahmaputra suture of the main range of the Himalayas. This tectonic event led to uplifting of the mountain ranges surrounding the Fergana Valley (as a result of fracturing at great depths) and to folding over a large radius, examples of which are in the Tien Shan and Gissar Mountains.

The motion of the plates played, and still does play, an important role in geological processes in the region. Ulomov (2004) notes that the region is characterized by high intracontinental seismicity due to intensive geodynamic interaction of several large lithospheric plates: the European, Asian, Iranian, Indian, and Chinese Plates.

The model developed by Natural Resources Canada (NRCan) in Ottawa, Canada, on behalf of the International GPS Service (IGS), shows the complexity of the geodynamics of the Asian continent and the Central Asian region (Ferland 2002). The map shown in Fig. 3.1 was compiled by the author, using data from the Geodesy and Geodynamics Group at the Department of Earth, Atmospheric and Planetary Science at the Massachusetts Institute of Technology (MIT; Cambridge, MA, USA) (MIT 2018). Today, little information is available about the geodynamics of Central Asia. This map represents only a general view of the motion, which is continuing in a northeasterly direction.

Today, the Indian Plate is still moving into the Asian continent. Different methods used to estimate the motion of the Indian Plate show that the movement and collision of these two plates (Eurasian and Indian) in some selected places varies from 37 mm annually (Grechko 2015) to 67 mm annually (Kious and Tilling 1996). Researchers (Chaterjee and Bishop 2015; Gansser et al. 1987) consider that this movement results in formation of a convergent boundary and leads to

Fig. 3.1 The motion of tectonic plates and fault lines (shown as *blue lines*) at the edges of the plates. The *black arrows* indicate the direction of the motion. (Source: Central Asia (public domain) via Wikimedia Commons User Shauni. (Note: This image is in the public domain because it contains materials that originally came from *The World Factbook*, a public-domain work published by the US Central Intelligence Agency)

geological and seismic activity in the Himalayas and in neighboring mountains.

3.1.2 Geological Evolution of Uzbekistan

The geology of Uzbekistan is an example of the evolutionary processes undergone by the components of the terrestrial environment during the times at which each layer was formed. Many researchers (Hendrix and Davis 2000; Mukhin et al. 1989; Nurtaev et al. 2013) note that the basement of the Turan platform was exposed as a result of new tectonic activity. The variety of rock layers in different parts of this country is evidence of this geological evolution (Fig. 3.2). These rocks are also a result of constant erosion, and, adjacent to eroding mountain ranges, the Fergana, Zerafshan, and Surxondaryo Valleys are filled with sediment, which has become buried and turned into sedimentary rocks that are widely spread as sand and clays to the west of the region.

In Uzbekistan, rock formations that undergo orogeny are severely deformed and undergo metamorphism. In mountains, as a result of orogenic processes, deeply buried rocks are pushed to the surface, and this phenomenon is widespread in this area. A combination of various phenomena such as magmatism, metamorphism, crustal melting, and crustal thickening as a result of tectonic forces plays a significant role in forming the natural environment of this country.

The geology of Uzbekistan is rich with a broad range of geological activities and can be considered a "living document" of the history of the land and nature. It includes evolution of ancient oceans, the closure of which formed mountain ranges. Many controversies exist about the tectonic history of the Tien Shan and about the ancient Turkestan Ocean. The geological history of the Turkestan Ocean has not yet been fully established. It is not clear how big this ocean was. The extent of the Paleo-Tethys Ocean as a precursor of the Turkestan Ocean (Zonenshain et al. 1990) is also not well understood.

Today, many researchers (Safonova et al. 2016; Heubeck 2001; Nurtaev et al. 2013) believe that the Turkestan Ocean (also known as the South Tien Shan Paleo-ocean or Paleo-sea) was located between the Kazakhstan and Tarim paleo-continents, which are parts of present-day Eurasia. This suggestion is based on evidence such as certain types of sediments and other rock formations. Recently discovered by a group of researchers (Safonova et al. 2016), ancient oceanic sediments (dated as being from 415–360 Ma) in the Southern Tien Shan Mountains in Kyrgyzstan are additional evidence of the existence of the Turkestan Ocean there. Some assumptions are made by researchers (Safonova et al. 2016; Zonenshain et al. 1990; Zonn and Kostianoy 2013) that the Aral and Caspian Seas are relics of this ocean, which was reduced and closed as the continents approached each other. The relics of the Turkestan Ocean are not just water objects but also mountains.

Researchers (Biske and Seltman 2010) suppose that by late Devonian times, this ocean reached 1500 km in width. A correlation exists between the size and the age of ancient oceans, and the thickness of the crust is evidence of their lifetimes: the bigger the ocean is, the older it is. Geologists (Safonova et al. 2016) believe that the Turkestan Ocean may have been as big as the Indian Ocean, surrounded by two or more continents.

In accordance with this hypothesis (Safonova et al. 2016), as these continents approached each other, the ocean was reducing and closing. Most of the oceanic plate was subducted to the deep mantle, but some of its parts were broken off by the margins of the continental plates and pushed up to the surface. While the continents were moving and thrusting, the mountains were rising and more fragments were pushed upward. The volcanic activity of the ancient plate of the Turkestan Ocean played a role in formation of the mountains. The mantle plume provided energy that caused melting of the Earth's crust and eruption of lava that became the chain of mountains.

The reason for learning the history of the Turkestan Ocean is to better understand this complicated geological process (which has not stopped) and to identify the combination of various geological processes leading to deformation of the Earth's surface and formation of the natural environment. A reconstruction by Nurtaev et al. (2013) of the evolution of a Paleozoic suture zone in the North Nurota fault zone (Fig. 3.3) shows the complexity of events from the Early Silurian Period to the Late Carboniferous Period.

The major tectonic orogenic belts in Asia were formed as a result of the motion of plates and their collision, and their shape tells us about the directions of these driving forces (Fig. 3.4).

Attempts by Heubeck (2001) to illustrate the assembly of Central Asia during the Middle and Late Paleozoic Era show the complexity of the processes and the uncertainty involved in trying to define the width of the Turkestan Ocean (Fig. 3.5) in the Late Carboniferous Period.

The model of the approximate extent of the Turkestan Ocean in the Early Permian Period is one of the assumptions (Fig. 3.6) with many unanswered questions.

Study of the geological evolution of Uzbekistan (Dolgopolova et al. 2017) confirms the age of the Turkestan Ocean at some selected sites: in the Lower Amu Darya river region in the Sultan Uvais Mountains, 505 Ma; the age of rocks in the accretionary[1] complex of the Chatkal–Kurama

[1] In geology, accretion is a process by which material is added to a tectonic plate or a landmass. This material may be sediment, volcanic arcs, seamounts, or other igneous features.

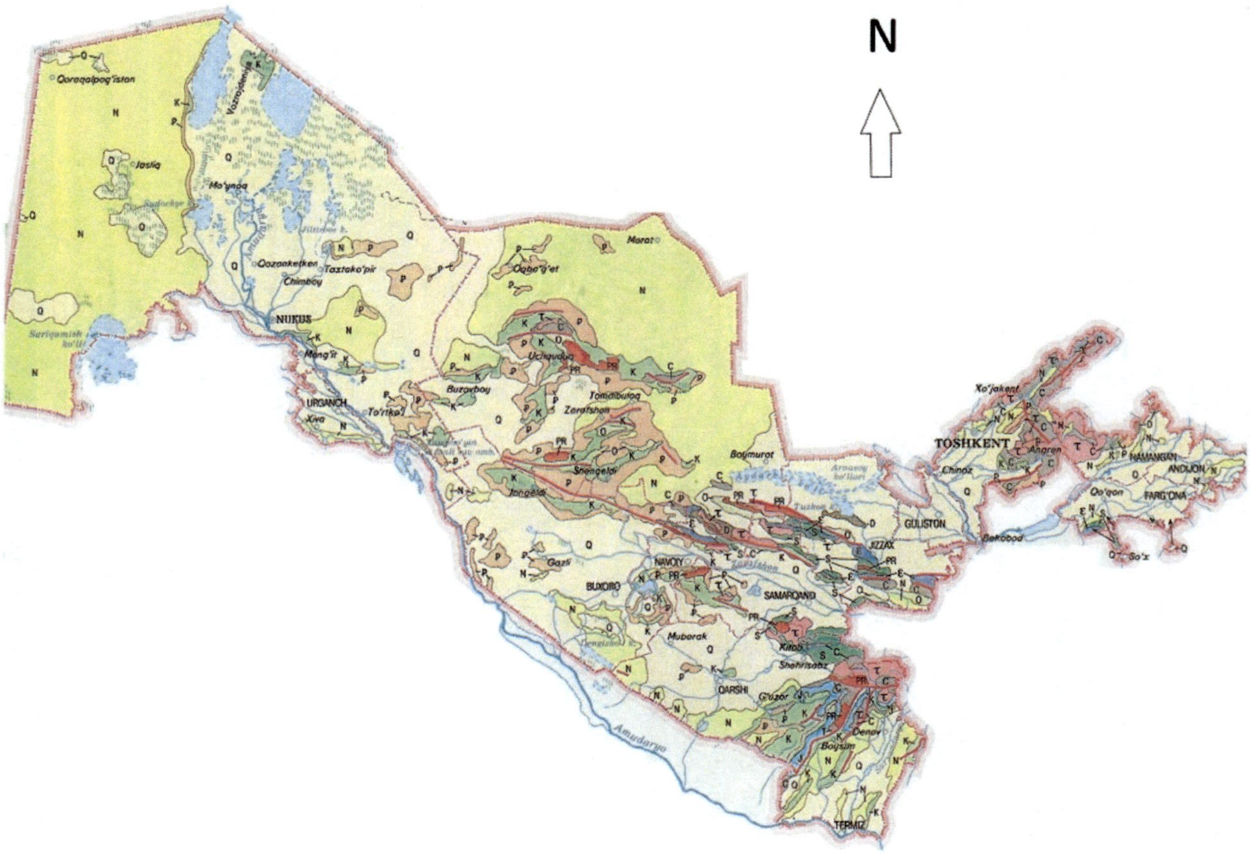

Fig. 3.2 Geological map of Uzbekistan. (Source: Atlas 2017, courtesy of the State Committee for Land Resources, Geodesy, Cartography and State Cadastre of the Republic of Uzbekistan)

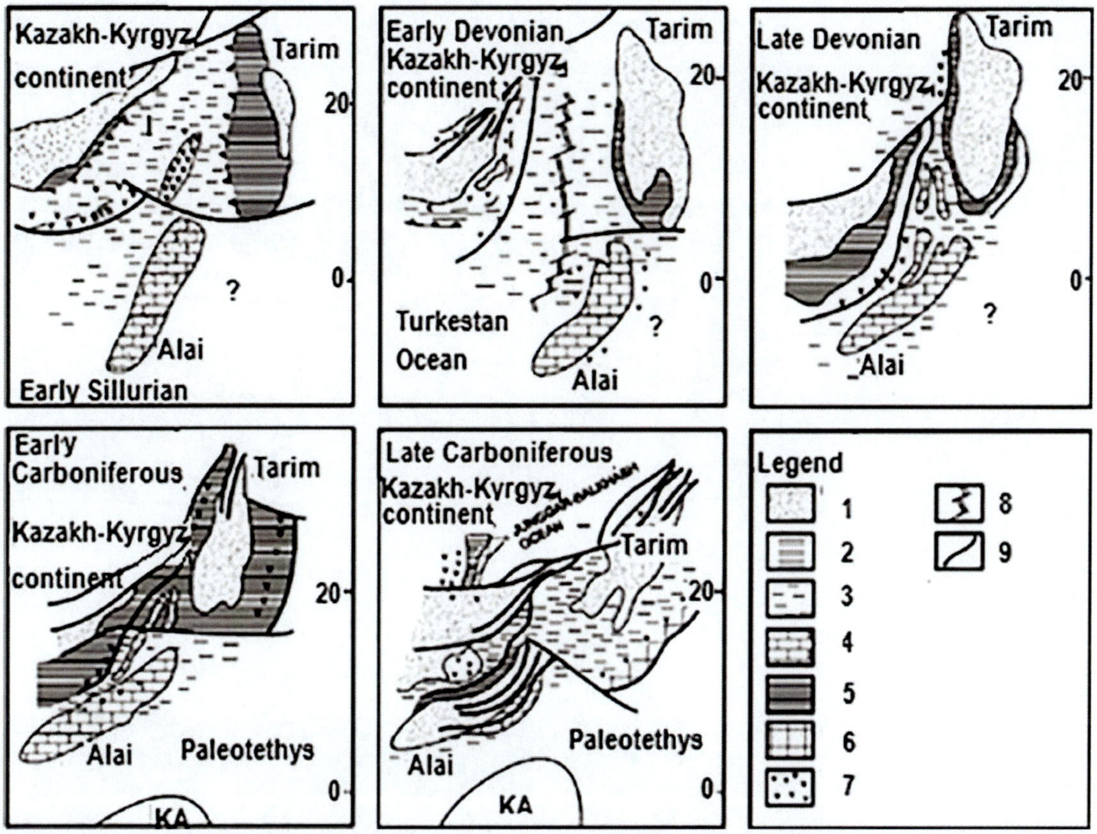

Fig. 3.3 The history of the Turkestan Ocean from Silurian to Permian times. These schematic paleotectonic/paleogeographic maps of the region illustrate the history of the Turkestan Ocean. On the maps, *I* indicates the Turkestan Ocean, *1* indicates the continental area, *2* indicates a shallow marine zone, *3* indicates a deeper marine shelf, *4* indicates limestone (a platform area), *5* indicates Turbidites *6* indicates an accretionary prism, *7* indicates volcanoes, *8* indicates a spreading ridge, and *9* indicates faults. The abbreviation KA means Kyzylkum-Alai "block". (Reproduced from Nurtaev et al. (2013) with permission from Elsevier through the Copyright Clearance Center's RightsLink® service order 4224270481315, November 8, 2017)

Fig. 3.4 The major tectonic orogenic belts in Asia. On the map, *1* indicates the approximate position of modern Uzbekistan. The Altaids are also known as the Central Asian orogenic belt (CAOB). (Reproduced from Heubeck (2001) with permission from the Geological Society of America)

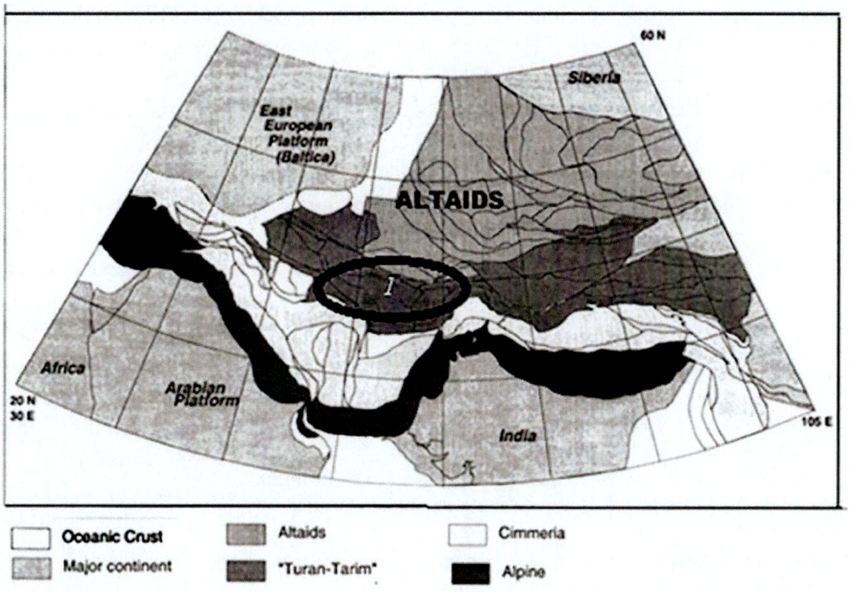

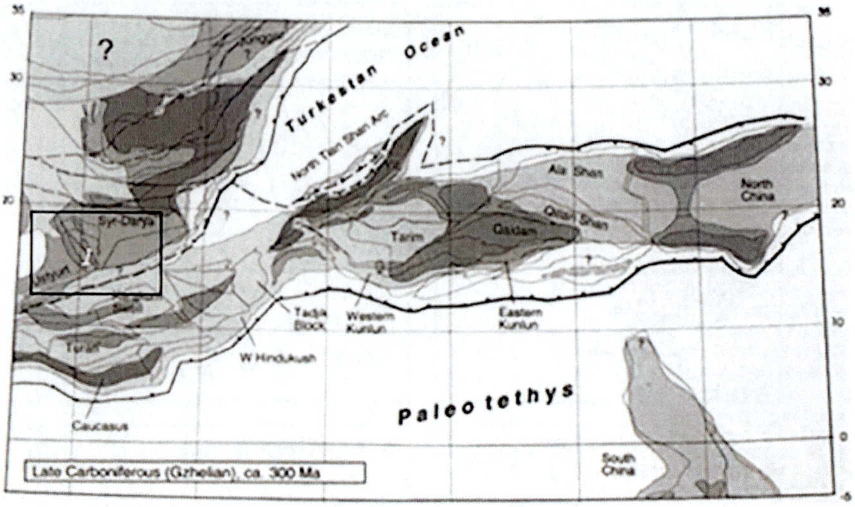

Fig. 3.5 Late Carboniferous paleogeography of Central Asia. On the map, *1* indicates the Northern Tien Shan arc. (Reproduced from Heubeck (2001) and displayed with permission from the Geological Society of America)

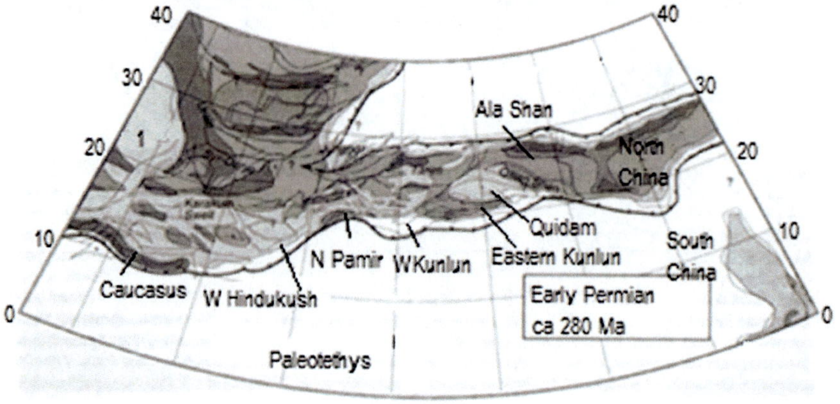

Fig. 3.6 Early Permian paleogeography of Central Asia. (Reproduced from Heubeck (2001) with permission from the Geological Society of America)

terrane[2] in the Western Tien Shan, 429–416 Ma; in the upper allochthon[3] of the Southern Tien Shan in the Tamdytau Mountains in the Kyzylkum Desert, 438 Ma; and the results of the early collisional stages in the Chatkal–Kurama and Gissar segments, around 320–310 Ma. Scientists (Dolgopolova et al. 2017, p. 76) suggest that the westernmost Tien Shan is characterized by "multiple subduction[4] processes responsible for 300 million years of geodynamic evolution history including accretionary collage, crustal growth, with the pre-Mesozoic crust formation concluded by Permian post-collisional extensional magmatism."

The Turan Plateau and Kyzylkum Desert represent another example of the complicated geological history of Uzbekistan. They are places where relics of ancient Turkestan or the Paleo-Asian Ocean, such as the lowest point in Uzbekistan (the Mingbulak depression) demonstrate several transgressions. The Sarygamysh depression is partly located in Uzbekistan and is also an example of a relic of the ancient Turkestan Ocean (Zonn and Kostianoy 2013). Further evidence of the complicated geodynamic evolution comes from the recorded events related to the closure of this ocean and the consequent ones in the central part of Uzbekistan in the North Nurota region (Golonka 2009). Researchers (Mukhin 1998; Safonova et al. 2016; Zonenshain et al. 1990; Zonn and Kostianoy 2013) suppose that the closure of the ocean basin probably did not happen until at least the early Late Carboniferous Period, i.e., about 359–299 Ma. The differing opinions (Heubeck 2001; Natal'in and Celâl Şengör 2005)

[2] In geology, a terrane is a fragment of crustal material formed on, or broken off from, one tectonic plate and accreted or "sutured" to crust lying on another plate. The suture zone between a terrane and the crust it attaches to is usually identifiable as a fault.

[3] An allochthon is a geological formation that was not formed in the region where it is found, and it was moved to its present location by tectonic forces.

[4] Subduction is sideways and downward movement of the edge of a plate of the earth's crust into the mantle beneath another plate.

on the timing of the closure of the Turkestan Ocean from the Ordovician Period through the Early Triassic Period are evidence of the complicated tectonic history of different parts of this region.

The base of Uzbekistan is part of the Turan platform (Fig. 3.7), which belongs to the Ural–Mongolian Late Paleozoic orogenic belt. Separated by the Turkestanian paleo-oceanic structure, the Kazakh and Karakum–Tajik microcontinents constitute the major ancient continental blocks. Mukhin (1998) suggests that when the two microcontinents collided, a large nappe pile intruded by granite was formed in southern Uzbekistan, with a volcanic–plutonic belt in the northern part of the country. The present appearance was formed by Late Neogene–Quaternary secondary orogeny.

The multiple folded basement of the Turan platform is constituted by southwest-facing arc fragments from the Paleozoic Era and the Jurassic Period. These fragments are sandwiched between two boundaries with an east–northeasterly trend (Natal'in and Celâl Şengör 2005), and this is the reason for the complicated stratigraphy and lithology of the region.

The Central Asian geologically young mountain ranges on the Earth's land surface are usually associated with the alpine belt, which includes southeastern Asia through the Himalayas and ends in the Alps. This belt is the result of the closure of the Tethys Ocean (Zonenshain et al. 1990). The Tien Shan is a large system of mountain ranges, which is considered by researchers (Bazhenov et al. 1993) as a fold-and-thrust belt. It crosses Uzbekistan and is associated with the highest mountain range on Earth—the Himalayan orogenic belt.

Being part of the Central Asian orogenic belt (CAOB; also known as the Altaids), the Tien Shan orogen covers the main part of Uzbekistan (Fig. 3.8). The CAOB is frequently considered by geologists (Huang et al. 2013; Widley et al. 2007) to be the one of the Earth's largest accretionary orogens with a complicated structure (Fig. 3.9). An orogen and a plain platform are formed during different periods of mountain formation and volcanic activity, resulting in creation of reserves of many ores and rare and precious metals.

One of the places of geological interest is the Tien Shan gold belt (also known as the Gold Dragon belt), which extends for over 2500 km from Kyzylkum in western Uzbekistan through Tajikistan, Kyrgyzstan, and southern Kazakhstan to western China. Researchers (Şengör et al. 1993; Şengör and Natalin 1996; Yakubchuk 2004) consider that it represents the central part of the Altaid orogenic collage of Central Eurasia and is a part of the Himalayan orogenic belt formed by the collision of the Indian and Eurasian

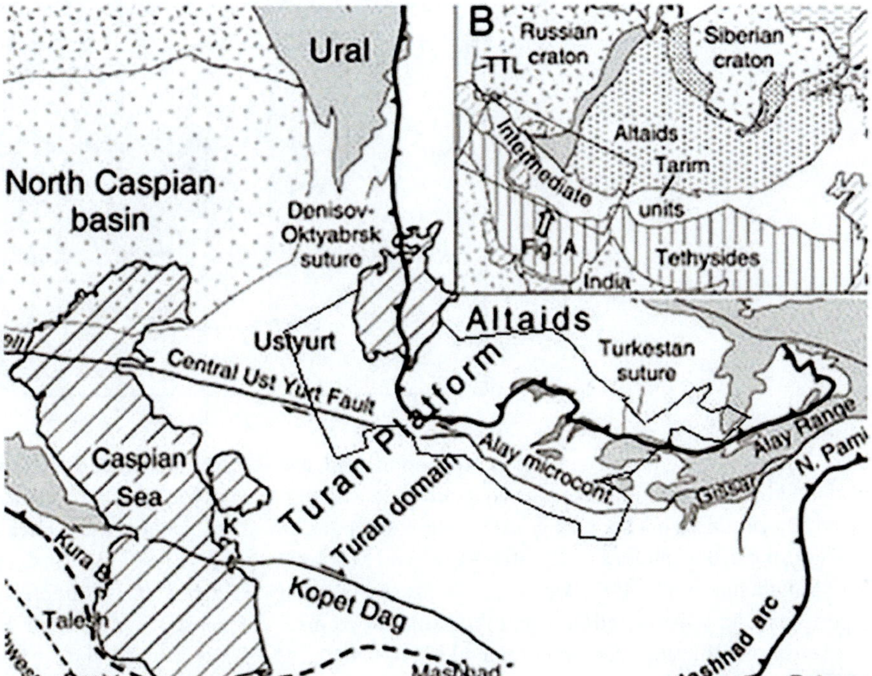

Fig. 3.7 The Turan platform. *B* indicates the location of site of interest on the map of more general view. (Reproduced from Natal'in and Celâl Şengör (2005) with permission from Elsevier through the Copyright Clearance Center's RightsLink® service order 4223730099284, November 7, 2017)

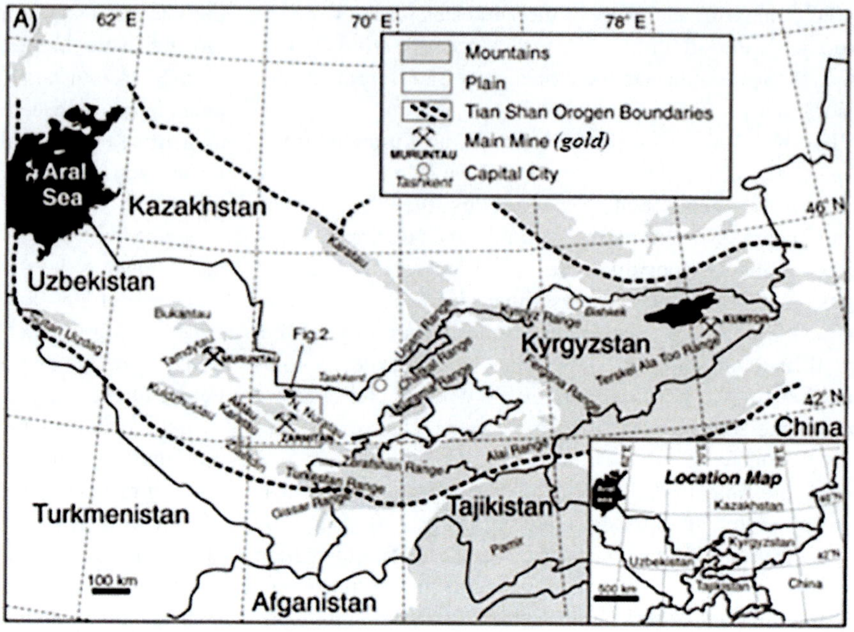

Fig. 3.8 The Tien Shan orogen boundaries. (Reproduced from Abzalov (2007) with permission from the US Society of Economic Geologists)

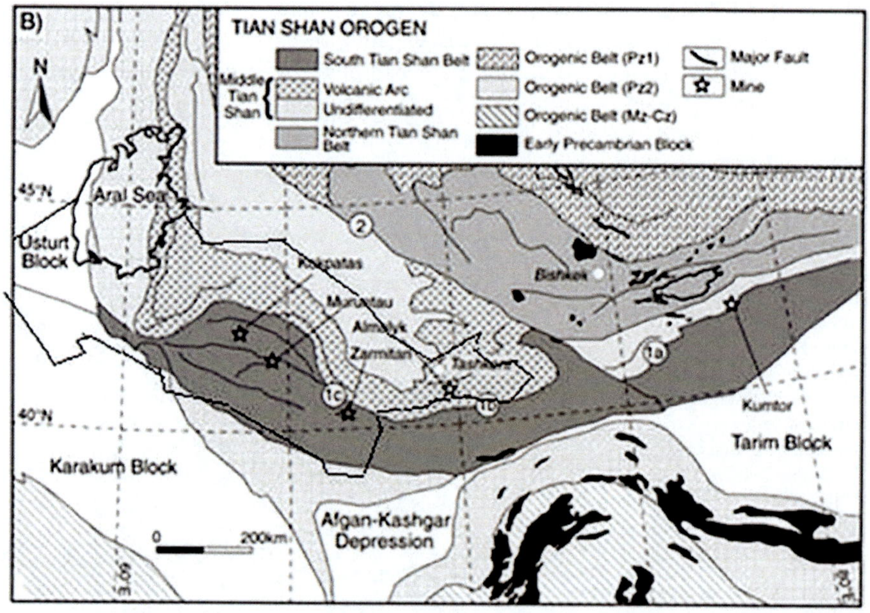

Fig. 3.9 The Tien Shan orogen structure. (Reproduced from Abzalov (2007) with permission from the US Society of Economic Geologists)

Plates during the Cenozoic era. As a Hercynian fold and thrust, it hosts numerous mesothermal gold deposits. It is now fast becoming a hot spot for gold-mining companies.

According to the opinions of scientists (Cole and Seltmann 2000; Yakubchuk et al. 2002; Mao et al. 2004) the peculiar property of gold mineralization is that it occurs in two principal settings within the Tien Shan mineral belt—namely as "porphyry and epithermal systems developed within magmatic arcs, and orogenic-type gold deposits that are structurally controlled, and temporally and spatially associated with late Paleozoic, syntectonic to early postcollisional, highly evolved, I-type granodioritic to monzonitic intrusives in fore- and back-arc terranes."

Because of its complicated geological evolution, Uzbekistan is attractive for geologists and paleontologists. One such site is the Bissekty formation (Bissekt) in the Kyzylkum Desert. It is supposed that it was formed during regression of the saltwater sea. The Bissekt is dated to about 90–85 Ma. Analysis of species from Bissekt has helped define the periods of its formation and ancient sea regressions. According to research (Redman and Leighton 2009), the fossils are a mix of marine, brackish water, freshwater, and terrestrial ones.

The interest in such sites is based on the suggestion that many familiar groups of Cretaceous dinosaurs had their roots in Central Asia. It is the Bissekt where the stratigraphically oldest remains of a ornithomimid theropod dinosaur, *Timurlengia euotica*, were found by a group of scientists (Sues and Averianov 2016). This new fossil helps answer the question of how tyrannosaurs evolved so rapidly at the end of the nearly 70-million-year period during which they lived on Earth. This discovery fills in a 20-million-year gap in the fossil record of tyrannosaurs during the early Cretaceous Period.

3.1.3 Earthquakes

More than half of the Asian continent, including Uzbekistan, remains seismically active, and recent earthquakes are a result of emergence of new land and episodic collisions of island arcs with the mainland.

The complexity of the tectonics may be explained by overlapping of geological processes, and this creates difficulties for the understanding, forecasting, and study of earthquakes.

The complicated northward motion of the Indian Plate under the Eurasian Plate is a major cause of frequent earthquakes in the Central Asian region. The erection of the Himalayan mountain chain is a result of collision of these plates about 50 Ma (Kious and Tilling 1996; Chaterjee and Bishop 2015).

This motion is still going on, and it results in erection of mountains and adds to the instability of the whole region. The principal border of the Indian and Eurasian Plates is quite close to Uzbekistan (Fig. 3.1) in northern India and Pakistan (Kious and Tilling 1996). The Main Central Thrust is also close to the region, and its fault continues along the Himalayan Mountains.

The alpine belt is one of the active areas on Earth, and it is estimated that 15–17% of the total world energy releases in earthquakes occur there. By passing Central Asia, it adds instability and raises the threat of earthquakes (Bolt 2016). Uzbekistan is close to northeastern Afghanistan, which a researcher from the US Geological Survey (Grechko 2015) has called one of the most seismically hazardous regions on Earth. The difficulties of understanding the forces behind earthquakes that have happened 210 km underground in the Hindu Kush range create additional uncertainties for prediction of earthquakes.

All of these factors are important in consideration of the frequency of earthquakes. According to research (Ulomov 2004), many eastern parts of Uzbekistan are defined as areas with a high probability of earthquakes measured or estimated as having a magnitude of 5.0 or more on the Richter scale. In the area of Uzbekistan, the World Bank Group GFDRR (2018) classifies the earthquake hazard as high, with more than a 20% chance of a potentially damaging earthquake (Figs. 3.10 and 3.11).

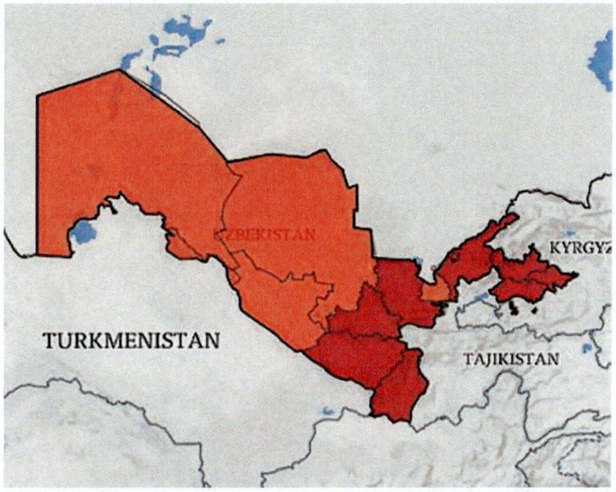

Fig. 3.10 Earthquake hazards. On the map, *dark red* indicates a high earthquake hazard and *red* indicates a medium earthquake hazard. (Source: CC-BY-SA, via ThinkHazard)

The Chatkal–Fergana, Southern Fergana, Southern Tien Shan, Central Kyzylkum, Piskom–Karjantag, Gissar–Alay, and Southern Uzbekistan overlapping faults are considered the most seismically active zones (Khasanov and Gulomov 2006). The epicenters of earthquakes are mainly 5–40 km deep and overlap the faults.

Earthquakes measured or estimated to have had magnitudes of 7–9 points have occurred in all of these zones. According to the data (Earthquake Hazards Program 2016), earthquakes with a magnitude of about 5 points occur at least once a week in neighboring Kyrgyzstan and Tajikistan, and the seismic waves reach many eastern regions of Uzbekistan.

In eastern Uzbekistan, the Chatkal–Fergana seismic zone includes the Piskom–Karjantag subzone, which overlaps the Ugam–Karjantag fault (Fig. 3.12).

In Eastern Uzbekistan, Artikov et al. (2020) identify three mega zones: (1) Tashkent; (2) Talas-Fergana; (3) the Fergana depression and its mountain frame, in Western Uzbekistan - Western and Northwestern, in Southern Uzbekistan - Southern megazones. The directions of the seismically active zones cooincide with the strike of large geotectonic structures. Khasanov & Gulomov (2006) identify seven seismogen zones: (1) Chatkal-Fergana, (2) Eastern Fergana, (3) Southern Fergana, (4) Southern Tien-Shan (with its Western and Nothern subzones), (5) Central Kyzyl Kum, (6) Piskom-Karjantau, and (7) Southern Uzbekistan zones. The most devastating previous earthquakes are listed in Table 3.1.

A devastating earthquake occurred in Tashkent on April 26, 1966, when more than 300,000 citizens lost their

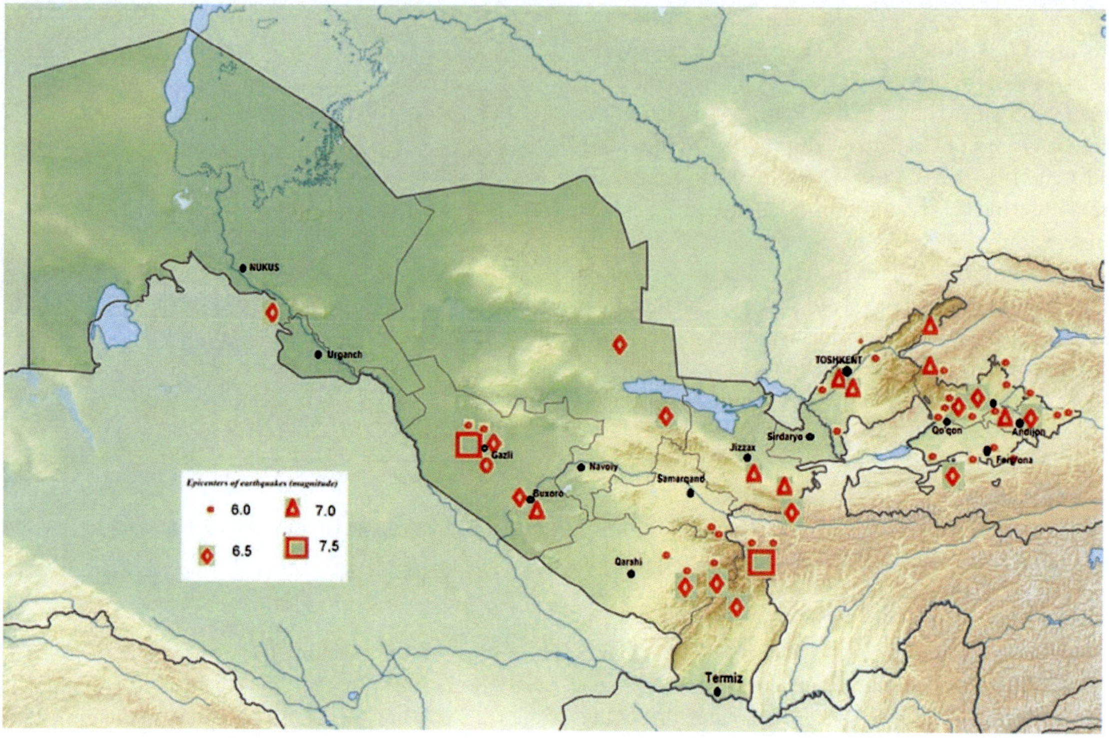

Fig. 3.11 Probability of earthquakes. (Compiled by the author, using data from Ulomov (2004) (public domain). File: Uzbekistan location map.svg: NordNordWest derivative work Виктор_В via Commons Wikimedia Relief Map of Uzbekistan; addition of objects with their names by Lola Gulyamova; https://creativecommons.org/licenses/by-sa/3.0/legalcode)

Fig. 3.12 Major seismic faults. (Compiled by the author, using data from Ulomov (2004))

Table 3.1 The most devastating previous earthquakes in Uzbekistan

Seismic zone	Seismic subzone	Earthquake Year	Richter scale magnitude	Location
Chatkal–Fergana	Piskom–Karjantau	1937	–	Piskom
		1946	–	Chatkal
		1959	–	Burchmulla
		1971	–	Abaybozor
		1972	–	Khalkabad
		1977	–	Tavaksay
		1980	–	Nazarbek
		1966	8	Tashkent
Southern Tien Shan	Southern Uzbekistan	1880	–	Urgut
		1984	–	Gallaaral
	Southern–Western Kyzylkum	1975	7–9	Gazli
Fergana	Southern Fergana	1888	8	Kostakuz
		1902	9	Andijon
		1903	9	Aim
		1923	8	Garden
		1924	7–8	Kurshab
	Northern Fergana	1927	8	Namangan
		1984	–	Pap
		1977	–	Shahimardan
		1982	–	Chimyon
		1984	–	Uygursay

homes (1966–2016 2016). The epicenter was located 5–10 km deep, and the total energy released was estimated to be 10^{14} J (Khasanov and Gulomov 2006). From 1915 to 1953, according to data from the Tashkent seismic station, earthquakes occurred in Tashkent 89 times: 22 earthquakes measuring 4 points on the Richter scale, 6 measuring 4–5 points, 15 measuring 5 points, 2 measuring 6 points, and 1 measuring 8 points.

The northern, southern, and western Fergana seismic subzones are the most active ones, with a high probability of earthquakes measuring 8 or more points on the Richter scale (Khasanov and Gulomov 2006).

In addition to natural causes, human activity can also cause earthquakes. Extraction of substances from the ground, such as natural gas, has increased the movement of blocks inside the Southern Kuljuktag fault and caused a sequence of severe earthquakes measuring between 7.3 and 9 points on the Richter scale in Kyzylkum, near the city of Gazli (Khasanov and Gulomov 2006; Wolchover 2011).

The probability of earthquakes in Uzbekistan is also higher because there is a vast zone of tectonic deformation, which may extend 1000 km from the border between the plates. This may cause earthquakes such as the one that occurred in the southern Ferghana Valley on July 19, 2011. This earthquake was an example of one with a combination of causes: a global process in the form of Eurasian and Indian Plate motion, and a regional reverse slip, most likely on a south-dipping thrust fault between the Pamir Mountains to the south and the Ferghana basin to the north (Vervaeck 2011).

3.2 Mineral Resources

The land and subsoil of Uzbekistan store untold wealth, and this country is one of the richest in minerals. Its reserves include 70 different types of mining for chemical raw materials, building materials, raw materials in the form of sorbents for cleaning of various industrial and food products, abrasive materials, and ornamental stones, also contain platinum metals, chromium ore, rare and other metals, perlite, caustic soda, mineral wool, basalt rocks, and undeveloped reserves of tungsten, silver, iron, tin, bismuth, lithium, manganese, mercury, strontium, fluorspar, graphite, limestone flux, high-alumina raw materials, mineral paints, and other types of minerals. New explorations are now discovering deposits of iron, manganese, coal, oil shale, nontraditional sources of gold, and uranium. According to the State Committee of the Republic of Uzbekistan, the monetary value of the minerals in the subsoils of Uzbekistan is US$7.9 trillion. Further details are available from https://map.uzgeolcom.uz/#. Several of the main deposits are shown in Fig. 3.13.

On the plains, there are deposits of oil, gas, self-seed salt, and building materials. In the mountain foothills, one can find deposits of coal, noble metals, nonferrous metals, fluorite, and building materials. According to 2017 data on the

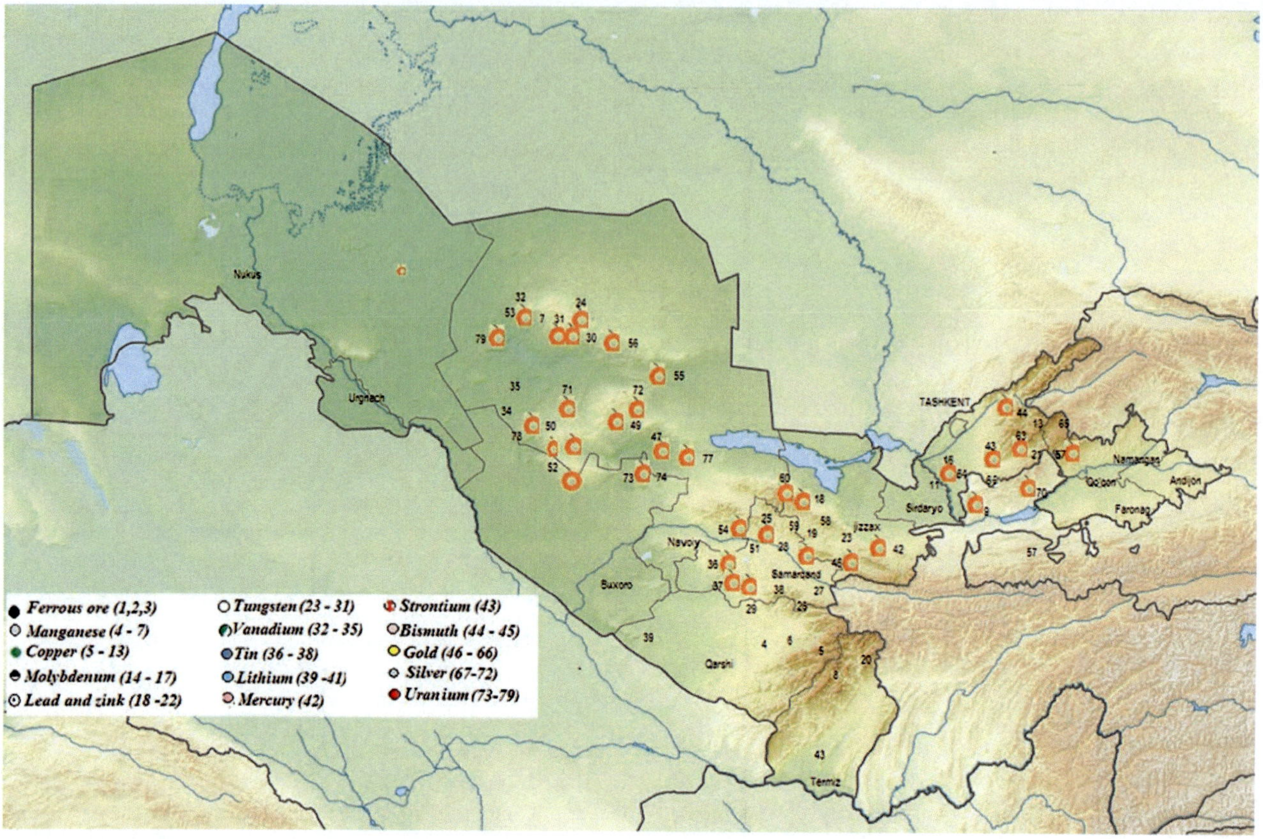

Fig. 3.13 The main deposits of mineral resources in Uzbekistan. (Compiled by the author, using data from Atlas 2017, courtesy of the State Committee for Land Resources, Geodesy, Cartography and State Cadastre of the Republic of Uzbekistan)

balance of mineral reserves in the Republic of Uzbekistan, 1931 deposits and plots of various fossils had been discovered in this country as of January 1, 2017 (Table 3.2; Fig. 3.14).

Turamuratov (2016) reports that the main gold deposits are in the Kyzylkum, Samarkand, and near-Tashkent geological–economic regions; the main copper reserves are located in the Almalyk (Olmaliq) region, where, in addition to the main resource, the extracted ore contains silver, cadmium, selenium, gold, and indium.

The mineral resources are very diverse. On the basis of previous estimates (Index Mundi 2015; Index Mundi data; RB Asia 2018; Uzbekistan Daily 2018; Toovey 2015), the main resources are listed in Table 3.3.

To date, 3100 deposits and occurrences of minerals have been discovered, including about 40 rare metals (niobium, tantalum, beryllium, lithium, rubidium, and cesium) in the main Tien Shan rare metal belt (Flynn 2016).

Twelve deposits of nonferrous and rare metals are located in different parts of the country. Uzbekistan has different kinds of industrial minerals and mines chemical raw materials, mineral fuels, gemstones, and refractory and ceramic raw and construction materials. Because of their unique geology, the mountains hold a wide range of the natural wealth of this

Table 3.2 Mineral deposits in Uzbekistan

Mineral type	Number of deposits
Construction	814
Fresh and mineral groundwater	624
Hydrocarbons	234
Metal ores	151
Mining	55
Chemical mining	32
Stone-colored raw materials	30
Coal and oil shale	7
Ferrous metals	5
Nonferrous metals	12
Precious metals	97
Radioactive metals	38
Total	**2099**

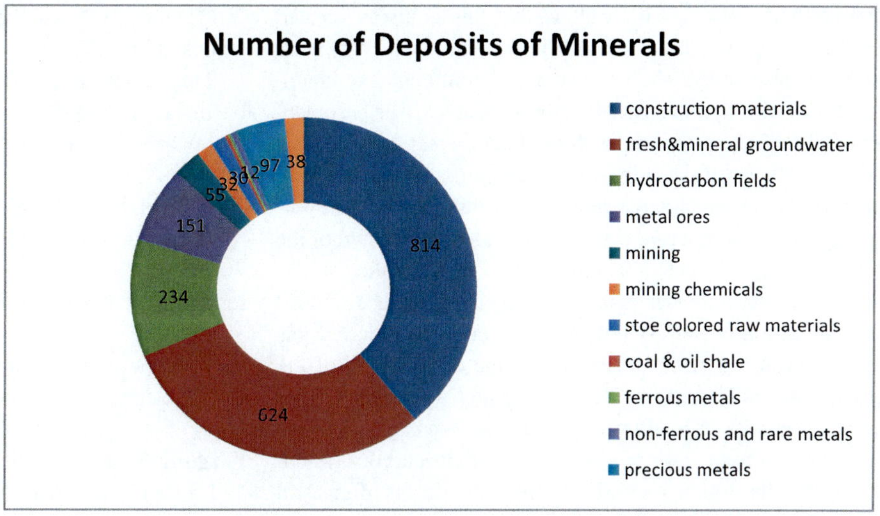

Fig. 3.14 The number relates to the number of deposits of the relevant materials. For example, number 814 means that 814 deposits of construction materials exist

Table 3.3 The main mineral resources of Uzbekistan

Mineral	Average annual production	Year
Natural gas[a]	$4,98 \times 10^{10}$ m^3	2020
Oil and gas condensate	2, 718.5 th t	2020
Coal	4,133 thousand tonn	2020
Gold[c]	93 t	2015
Silver	60 t	2015
Copper	100,000 t	2015
Lead	6000 t (lead content)	2015
Zinc	25,000 t (concentrate and direct shipping ore)	2015
Molybdenum	500 t (contained molybdenum)	2015
Uranium	2385 t (enriched uranium)	2015

Compiled by the author, using data from the State committee on Statistics of the Republic of Uzbekistan (2022) and Toovey (2015)
[a]Source: here and below State Committee on Statistics of the Repubic Uzbekistan (2022)
[bc]Source: Toovey (2015)

country. Most of these minerals are in high demand, and many deposits may be developed by open mining.

According to the State Committee of the Republic of Uzbekistan for Geology and Mineral Resources, Uzbekistan is among the top ten countries in the world for mineral resources (gold, uranium, gas, copper, potash, and phosphates).

Groundwater resources are also rich in predicted reserves of elements such as iodine, bromine, lithium, rubidium, cesium, strontium, boric oxide, and germanium. These resources are widespread all over the country and, according to Turamuratov (2016), they are located in Upper Jurassic, Coniacian, and Aptian aquifers in the Bukhara–Karshi (Qarshi) artesian basin; in Albian–Senoia–Turan, Senoia–Paleocene, and Alay aquifers in Surxondaryo province; in the Massaget aquifer in the Fergana artesian basin; and in Cretaceous aquifers in the southern near-Aral basin.

Recently discovered oil shale is estimated to lie at the depth 600 m and amount to 47 Gt in five deposits in central Kyzylkum, Sangruntau in Navoiy province, and Baisun in Surxondaryo province. Reports (Turamuratov 2016) reveal that the element content in the original material of the oil shale includes such raw materials as molybdenum, cadmium, selenium, nickel, vanadium, cobalt, copper, zinc, barium, aluminum, manganese, and phosphorus.

The subsoils in Uzbekistan also contain platinum metals, chromium ore, rare and other metals, abrasives, perlite, caustic soda, mineral wool, basalt rocks, and undeveloped reserves of tungsten, silver, iron, tin, bismuth, lithium, manganese, mercury, strontium, fluorspar, graphite, limestone flux, high-alumina raw materials, mineral paints, and other types of minerals.

Not all of the deposits are well studied and ready for industrial use. The known deposits of natural resources have been discovered in just 20% of the whole country.

3.2.1 Gas and Oil

Gas and oil are important natural resources as fossil fuels, which consist of hydrocarbons or molecules made of carbon and hydrogen atoms. Hydrocarbons are considered a type of organic chemical. In Uzbekistan, oil and gas are also sources for the chemical industry.

Since gas and oil do not occur in all rocks at all locations, the goal of controlling oil fields and regions with significant amounts of accessible oil in the groundwater is of political and economic importance. Geological conditions influence gas and oil development, storage, movement, and extraction. In Central Asia, a combination of favorable conditions started with Mesozoic events when the Cimmerian collision farther south split the old Altaids. Researchers (Hendrix and Davis 2000;

Torsvik and Cocks 2016) contend that this collision created extensional basins just north of the Aral Sea, which contain Jurassic and younger shallow water and continental sedimentary rocks with significant hydrocarbon reserves. The basement was uplifted along the major thrust faults, creating mountain ridges. Between these ridges, large compression basins were formed, within which large sedimentary thickness and important hydrocarbon reserves accumulated. The best known of the compression structures runs through the Fergana Valley.

A hydrocarbon reserve is a known supply of gas and oil held in groundwater. Whether the reserve is called a gas reserve or an oil reserve depends on the consistency of the gas and oil. Such reserves develop through specific associations of materials, conditions, and time—what geologists refer to as a hydrocarbon system. The hydrocarbon system consists of the source rock, the thermal conditions of gas and oil formation, the migratory pathway, and the trap.

Rock is called reservoir rock when its body has space in which the gas and oil can reside and it has channels through which the gas and oil can move. The reservoir rock is typically sandstone and carbonates. Because of the strength of its primary mineral (SiO_2 [quartz]) and its stability to reaction, a sandstone reservoir is considered high quality. The porosity, or the proportion of pore space in the rock, plays an important role. For comparison, shale has low porosity at about 10%, whereas poorly cemented sandstone has high porosity at 35%. This means that about one-tenth of a block of shale versus about one-third of a block of sandstone consists of open space. There are pores where the gas and oil occur and thus are distributed through the rock. The flow of gas and oil through the rock depends on its permeability, or the interconnection of cracks cutting through the rock. The greater the porosity is, the more capacity the reservoir rock has to hold gas and oil, and the greater the permeability is, the easier it is for the gas and oil to be extracted. The movement of gas and oil from the source rock into reservoir rock takes thousand to millions of years.

Gas and oil reservoirs exist if they are held within groundwater by means of a geological configuration called a trap. A trap usually consists of a seal rock—a relatively impermeable rock such as shale, salt, or unfractured limestone—which lies above the reservoir rock and prevents the hydrocarbons from rising further.

All of these conditions coincide in Central Asia in two main gas and oil basins: the Mesozoic Amu Darya basin and the Cenozoic–Mesozoic South Caspian basin. A third hydrocarbon-producing area is the Fergana basin.

In Uzbekistan, the gas reserves have been estimated (Khasanov and Gulomov 2006) to be 66 × 10^9 m^3 (1.3% of the world reserves). According to an estimate from the World Energy Council (2009), the proven recoverable natural gas reserves are 1.85 × 10^9 m^3. According to official data, 4,98 × 10^9 m^3 of gas and 806,000 t of oil were extracted in 2017 (State Committee on Statistics of the Republic of Uzbekistan 2022).

The major discovered gas reserves in the Amu Darya basin are in the Bukhara–Khiva and southwestern Gissar hydrocarbon zones.

The Bukhara–Khiva Hydrocarbon Zone The reservoir rocks in this zone are Mesozoic carbonates, containing about 70% of the oil–gas reserves of this country. The biggest gas reserves are at Shurtan, Mubarek, and Gazli.

The other reserves are in the Fergana Valley, Surxondaryo, southwestern Gissar, and Ustyurt.

Fergana Valley At the end of the nineteenth century, gas and oil were found in Mesozoic and Cenozoic areas in the Fergana Valley. In 1904, industrial extraction of oil was started by the Belgium Company in northern Soh, southern Olamushuk, Polvontosh, Chimyon, and Shorsuv (in what is now Fergana province). In 1992, oil reserves were found in Mingbulak in Namangan province, where an oil spill occurred on March 2, 1992. This spill was caused by a blowout that caught fire and burned for 2 months. About 2,110,000 barrels of oil were contained behind an emergency dam.

Surkhandarya In the Sherabad–Surxondaryo hydrocarbon zone, oil and oil–gas reserves are found in reservoir rocks in an intermountain depression of Paleocene limestone.

Southwestern Gissar In this hydrocarbon zone, oil has been found in reservoir rocks of Jurassic carbonates and in Mesozoic carbonates in Kashkadarya province.

Uzbekistan's demand for gas products is met by its own resources, and about 20% of its gas production is exported. The major importer of gas from Uzbekistan is China.

3.2.2 Coal

Coal has the most widely distributed reserves; it is mined in more than 100 countries. According to Schweinfurth (2016), coal is composed of a complex mixture of organic and inorganic compounds. This sedimentary rock is a fossil fuel because its composition includes primarily carbon and variable quantities of such elements as hydrogen, sulfur, oxygen, and nitrogen. The geological process of its formation takes millions of years. The ideal conditions for coal formation were provided by the wide, shallow seas of the Carboniferous Period. However, coal is known to have been created in most

Table 3.4 Proven recoverable coal reserves in Uzbekistan in 2017

Mineral type	Reserve
Anthracite and bituminous coal	47 Mt
Lignite	1853 Mt
Total reserves	1900 Mt
Proportion of world total reserves	0.2%

Compiled by the author, using data from O'zbekko'mir (2018)

geological periods, except during the Permian–Triassic extinction event.

According to data from the coal producer O'zbekko'mir, Uzbekistan's explored reserves amount to 1.853 Mt of brown coal and 47 Mt of bituminous coal. The total estimated coal resources are over 5.7 Gt (Table 3.4). Lignite, or brown coal (the lowest-quality coal), is used exclusively as a fuel for electric power generation. The reserves of bituminous coal, a dense sedimentary rock, are limited in quantity. This coal is also used primarily as fuel in steam–electric power generation.

Two coal fields are now being developed: the Angren brown coalfield in the Tashkent region and the Shargun anthracite deposit in the Surxondaryo region. Some bituminous coal is produced from the Baysun field, also in the southern region of Surxondaryo (Fig. 3.15).

O'zbekko'mir reports that at the Angren coal deposit, the Angrenskiy coal strip mine is extracting coal in an open-cut mining manner, and the open joint-stock company (OJSC) Erostigas produces gas from brown coal, using a method of groundwater gasification. At the Shargun and Baysun deposits, charcoal is extracted by open-cast mining and groundwater gasification methods.

All of these reserves are in Jurassic rocks. According to data from O'zbekko'mir, 3,993,419.5 t of coal are mined annually, including 70,000 t of charcoal. The main consumer of coal fuel is the power sector, whose share of the total coal consumption is over 85% (Coal Industry 2018). Coal fuel demand also comes from enterprises of industry, social and utility companies, and the population.

Six investment projects costing US$690.5 million were planned, including modernization of the joint-stock company (JSC) Shargunkumir (costing US$105.5 million) and construction of the Angren mine (costing US$165 million) and the Nishbash mine (costing US$175.2 million).

Coal is also an important resource in chemical production. Syngas is used for producing methanol, hydrogen, and carbon monoxide, which serve as the chemical building blocks in manufacturing of olefin, acetic acid, formaldehyde, ammonia, urea, and other compounds. By using inexpensive coal, it is possible to produce a wide range of valuable commodities. The interest in using coal has tended to increase in many countries with abundant coal resources. However, the coal-to-chemical process requires substantial quantities of water, and this is a challenge in settings where water resources are limited.

3.2.3 Gold

The main minerals produced in Uzbekistan are gold and uranium. In 2016, Uzbekistan was listed in tenth place among countries with the largest confirmed gold reserves, and it was listed in eleventh place with regard to extraction of gold. Uzbekistan is the world's ninth-largest gold producer. In 2015, Uzbekistan produced 103 t of gold (Focus Economics 2016). The main deposits are in the Kyzylkum, Samarkand, and Tashkent regions. The projected reserves are twice as large as the discovered ones.

Forty-six gold deposits have been discovered, of which 29 have been identified, and 12 of them are in use. The gold of Uzbekistan has been recognized as the purest gold of high quality, receiving international awards and prizes. In 1999, at the Tokyo Stock Exchange, the golden emblem of Uzbekistan, with a purity ratio of 99.99%, was recognized as meeting the highest purity criteria and accepted as a standard of gold quality worldwide.

Uzbekistan's gold reserves are found in Cambrian and Lower Paleozoic rocks in the Western Tien Shan and Nurota Mountains. The largest gold deposits are in central Kyzylkum; among them, the largest in the world is the Murumtau deposit. Others are the Ajibugut, Bulutkon, Balpentog, Aristotnog, Turbay, Marjonbulak, Olmaliq, and Kyzylolmasai deposits. Recently discovered deposits are the Yangi Davon, Birinchi Sentyabr, Jepsay, Roxat, Pistali, and Urtalik deposits. The known reserves and forecast resources of each of them reach tens or more tonnes of gold. About 86% of the gold currently produced comes from Kyzylkum, with the remainder coming from Olmaliq in the Tashkent region and from Chadak in the Namangan region.

The central Kyzylkum area is rich with deposits of "black shale" types of uranium and gold. The preliminary zone of the gold deposits follows the Tien Shan orogenic belt (Fig. 3.15).

The reserves of gold are estimated at 5300 t. Some of the main gold deposits are at Almalyk in Tashkent region in the Western Tien Shan (93 million ounces), at Zarmitan in the Nurota region of Samarkand province (11 million ounces), and at Muruntau, Kyzylkum, Navoiy province (175 million ounces).

Kempe et al. (2016) note that the Muruntau deposit is one of the largest single gold deposits worldwide, with resources in excess of 5000 t. The high quality of the ores makes the Muruntau and the prospective Tamdibulak deposits in the central Kyzylkum Desert unique. It is known (US Geological Survey 2015) that the Kyzylkum Desert is made of three main tectonic units, and the rocks are classified to be of Lower Paleozoic carbonaceous and sulfidic classic segments. The Muruntau deposit occurrence is explained (Kempe et al. 2016) by a multistage process of sedimentation, regional metamorphism including thrusting, and magmatism with

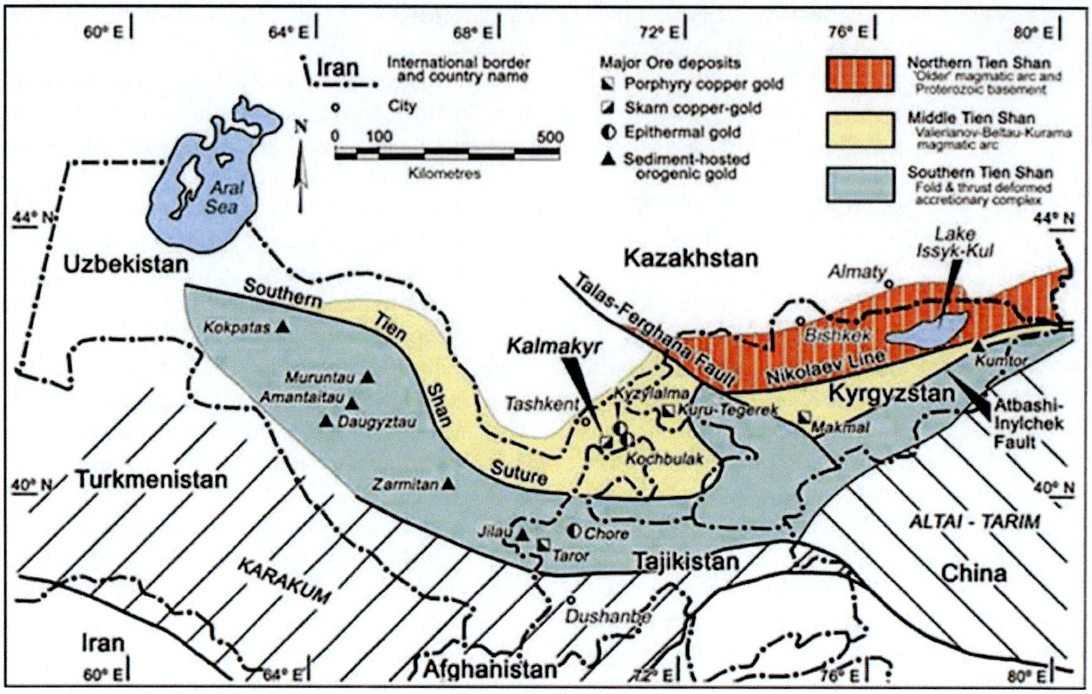

Fig. 3.15 The Tien Shan gold belt. (Reproduced from Natal'in and Celâl Şengör (1996) with permission from Elsevier through the Copyright Clearance Center's RightsLink® service order 4223730099284, November 7, 2017)

formation of hornfels aureoles and several stages of hydrothermal activity.

The Muruntau gold deposit was found in 1958 by Mordvintsev, Khamyshkin, and Lukýanov. The mine was opened a few years later. The history of this deposit is unique, since it was discovered in Soviet times and all information about it was classified. The discovery of this deposit contradicted the then prevailing theories regarding gold deposits, and it was unique in scientific research. Researchers (Kempe et al. 2016; Morelli et al. 2007) argue that the complicated genesis of gold mineralization at Muruntau is the reason why there is less agreement on current data interpretations and the understanding of the formation history (and this is also why it is necessary to re-examine genetic models of such giant orogenic gold deposits).

The Muruntau gold deposit is the largest single open-pit mine in the world. It is located within the Beltau–Kurama volcano–plutonic belt (US Geological Survey 2015; Menzie et al. 2013). This mining involves relatively low extraction costs. It is estimated that Uzbekistan may be on top of the world's largest gold reserves. Experts (Focus Economics 2016) suppose that 2500–5300 t of gold are yet to be mined. The reserves at Muruntau are expected to last until approximately 2032. Samples of polymetallic ore are shown in Figs. 3.16, 3.17, 3.18, and 3.19.

The southern Tien Shan gold belt includes large deposits such as Zarmitan, Jilau, Taror, and Savoyardy, with a similar genesis to that of the Muruntau and Tamdibulak deposits. It is assumed (Kempe et al. 2016) that gold ores from the Middle to Late Paleozoic Era were distributed and generally controlled by the final closure of the Turkestan Ocean. The Zarmitan gold deposit represents part of the Late Paleozoic Tien Shan orogenic belt. The strike length of this zone is estimated to be approximately

Fig. 3.16 Polymetallic ore from central Kyzylkum. (Photo by Lola Gulyamova, used with permission from the Geologic Museum of the National University of Uzbekistan)

Fig. 3.17 Polymetallic ore from central Kyzylkum. (Photo by Lola Gulyamova, used with permission from the Geologic Museum of the National University of Uzbekistan)

Fig. 3.18 Gold inside quartz–pyrite ore from the Kochbulak deposit, Kurama Ridge, Tashkent region. (Photo by Lola Gulyamova, used with permission from the Geologic Museum of the National University of Uzbekistan)

7 km, and the thickness varies from 200 to 1500 m. The total resource of the 84 major lodes is 32 Mt at 9.8 g/t Au and 14.6 g/t Ag.

The second largest mining company is the Olmaliq Mining and Metallurgical Combine, located in the Tashkent region. It primarily produces copper but also mines gold, silver, lead, and zinc as by-products. The Olmaliq porphyry Cu–Au system consists of several giant ore deposits and is associated with the Middle to Late Carboniferous Period, with a pulse of magmatic activity within the Devono-Carboniferous Valerianov–Bertau–Kurama volcano–plutonic belt. According to estimates (US Geological Survey 2015; Menzie et al. 2013), the Chadak gold mine in the

Fig. 3.19 Polymetallic ore from central Kyzylkum. (Photo by Lola Gulyamova, used with permission from the Geologic Museum of the National University of Uzbekistan)

Namangan region and the Marjanbulak gold mine in the Jizzakh region have total reserves of nearly 30 t of gold.

The mining operations for gold, iron, lead, silver, tin, and other minerals are 17 centuries old. About 20 ancient gold mines have been discovered in the Tien Shan Mountains and the Kyzylkum Desert; they have now been supplemented by newly opened ones.

3.2.4 Silver

Silver deposits are found in Navoiy and Namangan provinces. The Okjitpes, Kosmanchi, Viskovoltniy, and Aztepa deposits are of industrial importance. The silver reserves are estimated to be 30,000 t.

3.2.5 Uranium

Uranium does not occur in a free metallic state in nature. Because of its high reactivity and interactivity with nonmetals, it is an element in many intermetallic compounds. Deposits of uranium have been found in more than 25 locations in Uzbekistan. The main part of them is located in the central Kyzylkum Desert, stretching from the Uchkuduk deposit in the northwest to the Nurabad deposit in the southeast. Commercial mining started in 1958 at the Uchkuduk deposit, using open-pit and groundwater operations. Since 1965, when in situ leaching (ISL) recovery methods came into use, they have dominated the production scene. Extraction of uranium is carried out in an efficient and environmentally friendly way, using groundwater leaching. The sole producer is the state-owned Navoiy Mining and

Metallurgical Complex (NMMC). In 2005, according to a World Energy Council (2009) estimate, the uranium output was about 2300 t, corresponding to 5.5% of the global output, with cumulative production of 28,069 t.

In 2016, the Republic of Uzbekistan was among the top ten countries in terms of uranium reserves;.in terms of production, it occupied twelfth place. During uranium processing, rhenium, scandium, selenium, and rare earth elements are also extracted. The prepared uranium reserves and their growth prospects, as well as use of advanced technologies, will ensure stable operation of the uranium mining complex in the long term.

In addition to the existing known deposits, new ones have been discovered: the Istiqlal, Severniy Kanimeh, Aulek, Aktau, Kendiktube, Maylisay, and Janubiy Sugrali deposits. In further development, the aim is to carry out geological prospecting for nontraditional types of uranium mineralization (black shale, hydrothermal, etc.) at the Turtku–Yanbash, Ustuk–Fazilman, Dakan–Nurota, and Arnasay sites.

At present, the demand for uranium in the world market is increasing because its use for peaceful purposes is expanding. Today, Uzbekistan exports uranium to Russia and the USA. Renewal of interest in nuclear power has led to increases in construction in many countries. The project in Uzbekistan is being conducted jointly with Russia.

3.2.6 Ferrous Metals

So far, indicative deposits of more than 20 types of ferrous metals (including iron, titanium, manganese, and chrome) have been discovered. However, these deposits have not been identified for mining and industrial processing. The major deposits are at Temirkon (containing 32% iron ore) and Surenota (36%).

3.2.7 Nonferrous Metals

Uzbekistan has unspecified reserves of nonferrous metals: copper, lead, rare tungsten, molybdenum, tin, bismuth, mercury. Copper ores and more than 15 types of nonferrous metals are extracted. Reserves of 22Gt of nonferrous ores have been discovered at three deposits in the Almalyk Rude Field. These copper reserves can provide the Almalyk Mining and Metallurgical Combine with raw material for 200 years ahead. The raw materials include large deposits of phosphorites, graphite, asbestos, talc, chemical raw materials (salt, sulfur, and phosphorites [essential for agriculture]), bentonites, glauconites, potassium salts (for the glass industry), shale, quartz, quartz sand, kaolin, and others. Various decorative stones, from marble to color deposits, are widely found as well (Figs. 3.20, 3.21, 3.22, and 3.23).

Fig. 3.20 Antimony from the Handiza deposit, Surkhandarya region(Photo by Lola Gulyamova, used with permission from the Geologic Museum of the National University of Uzbekistan)

Fig. 3.21 Bornite, a mineral containing copper and iron sulfide (Cu_5FeS_4), from the Uchkulach deposit, Jizzakh region. (Photo by Lola Gulyamova, used with permission from the Geologic Museum of the National University of Uzbekistan)

3.2.8 Groundwater

In Uzbekistan, groundwater is used as an essential part of the water resources for the needs of the population, industry, and land irrigation. These resources include those that form naturally in catchment areas and those formed by filtration on irrigated land. About 10×10^6 m^3 of groundwater are used for irrigation, industry, and drinking water. The approved fresh groundwater resources are sufficient to provide half the population with drinking water.

According to the State Committee of the Republic of Uzbekistan for Geology and Mineral Resources, as of January 1, 2017, at 97 deposits of fresh and slightly saline groundwater, the reserves were 75.6 m^3/day (874.8 m^3/s),

Fig. 3.22 Copper from the Olmaliq deposit, Tashkent province. (Photo by Lola Gulyamova, used with permission from the Geologic Museum of the National University of Uzbekistan)

Fig. 3.23 Alunite, a hydrated aluminum–potassium sulfate mineral, from Aktash, Karjantau Ridge, Western Tien Shan, Tashkent region. (Photo by Lola Gulyamova, used with permission from the Geologic Museum of the National University of Uzbekistan)

and the approved reserves were 16.8×10^6 m^3/day (195 m^3/s), of which the total usage was 5.5 m^3/day (63.4 m^3/s). Ninety-five percent of the fresh groundwater is in the mountain hydrogeological regions. The groundwater resources are sufficient and, according to one estimation (Khasanov and Gulomov 2006), these resources total 31.5×10^9 m^3/year, with fresh and low-salinity water resources of 50.6×10^9 m^3/day. In Uzbekistan, there are nearly 100 groundwater fields, 77 of which hold fresh groundwater.

Groundwater resources are distributed unevenly across the country. This is dependent on the geology; the quality, amount, and flow module of the groundwater vary from place to place. Eighty percent of the usable groundwater resources are formed in the mountain foothills and mountains.

Table 3.5 Approximate distribution of groundwater in Uzbekistan

Hydrogeological region (artesian basin)	Groundwater volume (m^3 x 10^6/day)	Percentage
Tashkent	10.97	31.4
Surkhandarya	4.0	23.8
Southwestern Gissar		13.5
Zerafshan	6.03	11.6
Nurota–Turkistan	0.87	6.7
Fergana	15.39	5.9
Mirzacho'l		5.1
Central Kyzylkum Desert fault		2
Total	**50.57**[a]	**100**

Compiled by the author, using data from Khasanov and Gulomov (2006, p. 107)
[a]Source: Geology and Mineral Reserves of Uzbekistan

Groundwater is formed also by return waters discharged into rivers (51%) and into depressions (33%). A high level of pollution is the reason why only 16% of this water may be recycled for irrigation. Several hundred reservoirs are created by return drainage waters. Actually, they are stagnating with unstable water and salinity. The need to improve management of their use and environmental maintenance, with an aim to gain ecological balance is becoming a crucial issue.

On the basis of water quality, rock lithology, sources, and natural conditions (geology, climate, and topography), all groundwater deposits have been categorized by Khasanov and Gulomov (2006) into five major regions: mountains, 2 mountain foothills, intermountain depressions, and plains. According to Khasanov and Gulomov (2006), in the mountain zone, the groundwater resources are estimated to be 105 m^3/s in nine hydrogeological regions. Table 3.5 lists information about the approximate distribution of groundwater in some of these regions.

The mountain groundwater resources are about 650 m^3/s. The groundwater reserves are located mainly at an altitude of 1500–3500 m above sea level (Khasanov and Gulomov 2006). In the foothills, the water reserves are estimated to be 295 m^3/s; their salinity differs depending on the location and altitude.

In intermountain depressions, the quality and salinity of the groundwater vary and depend on its flow. Groundwater salinity is greater in such places as central Fergana, the steppes of Mirzachoul and Karshi, and the Lower Zerafshan River, where the natural conditions are worse (low flow, a dry climate, high summer temperatures, and high evaporation).

On the plains, the salinity of the groundwater resources is greater because chloride and sodium levels are higher, and this water is not suitable for drinking.

Global warming is resulting in shortages of surface water flow, causing more intensive use of groundwater as an alternative to irrigation; groundwater now accounts for about

6.4% of total irrigation water. Intensive agricultural development can contribute to deterioration of groundwater quality. One negative example is in the Aral Sea region, where water moves through the landscape and collects soluble salts. On irrigated land, poor drainage of soils and surface aquifers has brought water tables to the surface in low-lying areas, creating major land degradation problems and soil salinity. Researchers (Rakhmatullaev et al. 2012) have pointed out the looming challenge of strong disturbances of natural groundwater formation processes.

According to data from the United Nations Development Programme (UNDP) (2007), Uzbekistan's water deficit is estimated to be 2.5 km^3/year in the Amu Darya river basin and 1.5–3.0 km^3/year in the Syr Darya basin. According to the water security indicator model (WSIM) estimated by ISciences (ISciences 2022), during the past several years, this deficit has been moderate in winter and moderate to severe in summer.

Big reservoirs were constructed in the twentieth century for irrigation of cotton fields. In Uzbekistan, there are 55 man-made water reservoirs with a total gross volume capacity of 19 km^3 and a total surface area of 1450 km^2. Today, all possible sites for reservoir construction have already been used. Scientists note that the reservoirs have lost their dead storage capacities because of sedimentation, and in the near future, they may lose their functionality. In 1969, the Arnasai topographic depression became a huge permanent lake after a catastrophic flood. It comprises the Aidarkul, Tuzkan, and Upper Arnasai, with a total area of about 3491 km^2 (UNDP 2007).

About 80 dams have been created on rivers; the major ones are at Qorataka, Kukkul, and Badak in the watershed of the Chirchiq River. Only two dams are used for energy production (those at Andijon and Charvak); the others are used only for irrigation purposes. Dams such as the Ikhnoch, Shovurkul, Qorataka, and Katta dams pose a flood risk. Floods in the foothills occur mainly in deforested mountain areas during the rainy season in April, May, and June. It has been estimated (Khasanov and Gulomov 2006) that the flooding may cover a foothill area as big as 93,500 km^2.

Groundwater is a valuable natural resource in the arid climate of Uzbekistan. Groundwater is one of the two main sources of freshwater supply available for use within the hydrological cycle. As part of the continuous circulation of water between the atmosphere and the land, the amount and quality of groundwater depends on many factors, including geology. Groundwater resides and travels beneath the ground surface and occurs in a great variety of the rock formations and soil types that make up the Earth's crust. The most useful groundwater is associated with sedimentary rock types such as gravel or sand. To be an aquifer, saturated sediments must have underlying geology that contains sufficient quantities of groundwater. Like oil fields and mineral deposits, groundwater is not present at all locations. The status of an aquifer may change over time, depending on rainfall, climate change, and glacial melting. Gravels that have progressively become buried over geological time and that may once have formed high-yielding aquifers become less permeable and lose their ability to store water. This makes it difficult to manage groundwater resources. To become groundwater, rainfall must reach the water table when percolating downward from the surface and moving through the overlying soil. Aquifers store water from rainfall that fell months or years ago and convey it. The porosity and permeability of rocks determines their ability to store freshwater and to serve as an aquifer. The level of porosity of fine-grained clays or sands dermines how well they function as an aquifer. However, the high porosity of clay and its low permeability is the worst combination for making a good aquifer. Well sorted and rounded stones that have interconnected pore spaces without clays or sand to prevent water flow are the ideal material for hosting an aquifer. Sand and gravel deposits, fractured granite, and limestone with solution cavities make good aquifers (Table 3.6).

In the mountains, groundwater deposits are formed in faults in intermountain depressions in artesian basins.

Hydrogeological Regions in the Mountains

In the mountains, groundwater forms from rainfall and melting of snow and glaciers. The flow and formation of the groundwater depends on the lithology and precipitation. In these hydrogeological regions, the most favorable conditions are at an altitude of 1500–3000 m, where the precipitation is sufficient for groundwater formation. At an altitude lower than 1500 m, because of decreasing rainfall and increasing evaporation, the flow module dramatically decreases, dropping to between 0.10 and 1–3 g/s.

Depending on the porosity and permeability of the rocks, the flow module of the groundwater is 12 l/s within carbon dioxide rocks, 8–9 l/s in sedimentary–magmatic rocks,

Table 3.6 Porosity of sediments and rocks

Material	Porosity (%)
Unconsolidated sediment	
Soil	55
Gravel	20–40
Sand	25–50
Silt	35–50
Clay	50–70
Rocks	
Sandstone	5–30
Shale	0–10
Solution activity in limestone and dolostone	10–30
Fractured basalt	5–40
Fractured granite	10

Sources: Heath (2004) and Carey (2009)

Table 3.7 Groundwater and its use

	Water resources ($m^3 \times 10^6$/day)					
	Usable water		Approved water		Drilled groundwater	
Hydrogeological region	Total	Freshwater	Total	Fresh drinking water	Total	Water from approved sources
Fergana	15.39	9.19	9.90	3.31	12.06	2.09
Tashkent	10.97	2.60	3.68	2.27	4.07	1.48
Mirzachoul	3.53	1.39	1.37	1.30	3.90	0.21
Nurota–Turkestan	0.87	0.37	0.43	0.37	0.51	0.18
Zerafshan	6.03	3.56	1.11	1.11	2.70	0.49
Kashkadarya	1.24	1.05	0.59	0.59	0.99	0.33
Surkhandarya	4.0	2.26	0.50	0.50	0.92	0.14
Bukhara–Turtkul	1.65	0.23	0.50	0.17	1.28	0.05
Central Kyzylkum	0.89	–	0.54	–	0.26	0.11
Eastern Kyzylkum	0.55	–	–	–	0.01	–
Aral Sea	1.01	0.04	0.34	–	0.13	0.13
Lower Amu Darya river	4.42	0.06	0.39	–	0.18	–
Ustyurt	0.02	–	–	–	–	–
Total	**50.57**	**20.95**	**19.35**	**9.62**	**26.91**	**5.21**

Compiled by the author, using data from Khasanov and Gulomov (2006, pp. 107–108)

3–5 l/s in metamorphic sandstone–shale, and 1–3 l/s in Mesozoic and Cenozoic terrigenous rocks (Heath 2004; Carey 2009). Table 3.7 lists some characteristics of groundwater resources and its use.

Groundwater mineralization is higher on the plains, particularly in the Republic of Karakalpakstan and in Bukhara and Kashkadarya regions.

Groundwater formed in fault karst and under the beds of mountain rivers may be used for different purposes, but its overuse may have negative impacts on the amount of groundwater and the river water that flows to the foothills, intermountain depressions, and plains.

Hydrogeological Regions in the Mountain Foothills

Groundwater and interlayer water are mainly formed by water coming down from high mountains and rainfall. The groundwater resources of this region are submerged in the ground and merged with the water in artesian basins located below. The salinity of the groundwater changes from fresh to salty. Freshwater occurs in gravel-prone areas. It is used for drinking and cattle watering. The capacity of the dynamic reserve is estimated to be 295 m^3/s.

Hydrogeological Regions on the Plains and in Intermountain Depressions

In intermountain depressions, the quality, amounts, and flow modules of the groundwater and interlayer water differ from one place to another because they are formed from various sources—i.e., rainfall and groundwater flow from mountains, rivers, canals, and reservoirs. Most groundwater flows to the lower side of a depression. The salinity is dependent on the flow module; the higher the module is, the better the groundwater quality is. In the central Fergana, Mirzacho'l, Qarshi steppe, and Lower Zerafshan regions, the salinity of the groundwater is higher since the flow module is lower, the climate is dry, and the temperatures are higher. However, in intermountain depressions, the resources are sufficient and all demands for groundwater are fully met.

The Fergana depression is the richest in groundwater resources. Surrounded by mountains, it has favorable conditions for forming groundwater, with a volume of 289 m^3/s.

The main sources of groundwater on the plains are water that flows down from mountains and foothills, water that rises under pressure from lower layers, and (partly) surface water and rainfall. The aridity of the climate and high evaporation, together with low levels of precipitation, are the main reasons for the low flow module and high salinity. The amount of evaporation leads to deterioration in the groundwater quality, and the water is almost unsuitable for drinking, containing sodium and chloride salts.

The plains of Uzbekistan include the Amu Darya, Syr Darya, and Ustyurt artesian basins. Their natural groundwater resources are estimated to be 120 m^3/s. Most of it is directly related to the irrigated delta plains. In these basins, the freshwater constitutes approximately 10% of the total groundwater volume. The water enters the groundwater reservoir via several pathways: ground infiltration of precipitation and infiltration from streams, canals, artificial recharge ponds, and water treatment systems. Such fresh groundwater may be found on the edges of major canals on irrigated lands.

Artesian waters in Uzbekistan are located much deeper, in many places at depths of 100–400 m. The quality of the water is not the same at all locations. The water between the layers near the surface is fresh and slightly frozen; the deeper the layer is, the higher its temperature and mineralization are.

In the artesian basin in the desert region, interlayer waters are found in sandy rocks. The quality of this water is adequate for watering cattle, sustaining pastures, and irrigating small oases.

It is necessary to use special equipment for purification of groundwater in places with high salinity (1.5–5 g/l). Today, in the southern Aral Sea region, more than 100 special devices for water purification are functioning and provide potable water for drinking.

Changes in the water regime caused by irrigation of agriculture lead to higher salinity of the fresh groundwater in the basins of the Amu Darya river in Karakalpakstan and Xorazm, and the Zerafshan River in Bukhara province.

Mineral Groundwater

In Uzbekistan, because of the variety of geology and natural conditions, various groups and types of mineral waters are encountered and have been identified. They are formed in Jurassic, Cretaceous, and Quaternary rocks at different depths, with different mineralization levels and temperatures. The mineral waters contain carbon dioxide, radon, sulfur gases, iodine, bromine, lithium, barium, iron, and other chemical elements. Unique mineral waters are located around Tashkent, in Kyzylkum, in the Zarafshan Valley, and in the Ustyurt, Bukhara–Karshi, Surkhanndarya, and Fergana basins. More than 100 mineral deposits have been examined, and their waters have been used in dozens of sanatorium resorts, mineral water–bottling facilities, and iodine extract plants. Currently, 49 deposits and 30 sites are in operation, and 87 have been drilled. The number of wells currently operating on these mineral waters is 95, and they sustain more than 100 health resorts and more than 20 bottling facilities. Reserves of thermal water with temperatures exceeding 200 °C are used for treatment and as sources of heating.

In the Bukhara-Karshi–Surkhandarya, Ustyurt, and Fergana artesian basins, rubidium, cesium, chalk, germanium, iodine, and bromine have been identified. The groundwater in these basins is of industrial importance.

More than 30% of Uzbekistan's population is experiencing an acute shortage of good-quality water, and its provision remains inadequate. Only 58% of the 119 cities, 31% of the 1064 urban-type settlements, and 26% of the 11,088 rural settlements are provided with drinking water, at the expense of the explored reserves of groundwater. The remaining settlements depend on other sources of water supply (surface water, unapproved reserves, wells, waterways, etc.).

Researchers (Rahmatullaev et al. 2012) note that as a result of massive irrigation development during the Soviet era and intensive using of chemicals in agriculture, the surface runoff quality has been degraded and hydraulically related groundwater has also been affected. At present, the systems for groundwater use and management are underdeveloped and inadequate.

3.2.9 Loess

Loess (a loamy, unstratified deposit formed by wind or by glacial melt water deposition) is distributed in temperate and semiarid regions of the world. Ten percent of the surface soils on Earth are made up of loess (Muhs 2007). Carbonate is its significant characteristic component. The type and amount are dependent on the distance from the source, the process of formation, and the rainfall. According to Muhs (2007, 2018), loess is mostly made up of silt and loamy particles (up to 50%) and clay particles (up to 10%). Loess is not very strong, and water is its major enemy; when soaked with water, it can crumble and collapse.

Wind and water erosion may easily destroy loess. Water erosion causes raindrop, surface, rill, and gully formations. In places without protective vegetation, rainfall dissolves particles of exposed, degraded soil and carries them away. This results in landslides in the rainy season, and such hazardous events occur in many parts of Uzbekistan.

Wind contributes to loess formation by transporting fine dust and depositing it above ground. Silt can be moved by both water and wind, resulting in finer segregation of particles on the basis of their size and weight. Muhs (2018) argues that past glacier activity from 3 to 65 Ma contributed to loess formation too. A typical feature of loess is vertical overhanging walls, which form as a result of repeated deposition of dust. One of the attributes of loess is its uniform structure of the soil profile at depth and on wash slopes. Quartz, feldspar, and mica are some common minerals that form loess (Fig. 3.24).

The main driving forces for loess formation are (1) a dust source, (2) adequate wind energy for transportation of the dust, (3) a suitable dust accumulation area, and (4) a sufficient amount of time. In Uzbekistan, different natural condi-

Fig. 3.24 A loess wall in the Tashkent region

Fig. 3.25 Loess formation starts in the high mountains. The Chimgan creek carries rocks in Tashkent region

tions (including tectonic activity and climate humidity) determine the composition, properties, and distribution of the two major types of loess on the mountains and piedmont plains. The initial material of loess formation comes from the mountainous and alpine areas. Loess formation occurs when glaciers grind rocks to a fine powder followed by the process of carrying it by streams to the end of the glacier. This sediment becomes loess that by nature is a sedimentary deposit.

Typically, loess formation is a sequence of events occurring when crushed rock is delivered to flat land by water flow. Loess formation starts in the high mountains when water sorts rocks (Fig. 3.25) on the basis of their fragmentation size.

Sediments are then sorted by wind power and may be returned to the mountainside, forming a loess-like soil substrate in piedmont areas. Loess massifs provide sedimentary materials that are eroded and transferred by water flows during snow melting and during spring and autumn rainfalls. Water transfers the loess again from piedmont areas toward steppes and deserts (Fig. 3.26). As a result of this water activity, the redeposited loess forms on flat lands. Such soils are found in the Mirzacho'l region. Redeposited loess is different from genuine loess. It has specific stratification common to all sedimentary strata formed by water erosion.

The opinions of geologists differ regarding the genesis of loess, which is widespread in foothills and valleys. Some geologists suggest that the main driving force of loess formation is fluvial, whereas others consider aeolian erosion a main driving force. The dominant point of view of Uzbekistani researchers (Mavlyanova et al. 2006) is based on the hypothesis that loess creation results from a combination of events. This process includes deposition of sediment from which alluvial soil is formed by rivers coming from the highlands over a long period of time. The initial material of loess formation originates in mountainous and alpine areas. The primary source of the parent rocks are ancient rocks above the limit of loess occurrence.

On the basis of long-term and detailed studies of Quaternary fine-grained deposits in the central and southern parts of Central Asia, Mavlyanov proposed a genetic classification of loess rocks. At the same time, two groups of these deposits were identified in the territory of Uzbekistan: (1) a group of loess rocks—aeolian loess and proluvial[5] loess—and (2) a group of loess-like rocks.

According to this theory (Smalley et al. 2006), the silt-sized material produced in the Tien Shan may be used to form loess deposits. Fluvial processes deliver this material onto the plains, forming water-laid sediments. The Syr Darya and Amu Darya are great silt-carrying rivers, moving the material away from the piedmont regions and out into the dry desert. The Amu Darya is the biggest carrier of suspension materials in the region. The desert acts as a reservoir of silt particles, which, under suitable climatic conditions, may be released to form loess deposits.

Uzbekistan is one of the places on Earth with large areas of loess deposits and soils. On the Mirzacho'l steppe, the thickest loess deposits (up to 13–20 m thick) are found in places relatively near the mountains and the Syr Darya, which, by its transportation activity, distributes sediments

[5]A proluvial deposit is made by water on a plain; this term is used to describe silty deposits.

Fig. 3.26 The Chirchiq River transfers loess in Tashkent region

into the desert regions. In the Fergana depression, for example, the greatest thickness of loess deposits (15–25 m thick) is observed in its northeastern, eastern, and southern parts. The thickness of the loess can even reach 70 m. Tuychieva et al. (2016) suggest that this thickness changes from 15–20 m to 70–80 m within the city of Tashkent.

Within Uzbekistan, according to Khasanov and Gulomov (2006), loess-like rocks are widespread almost everywhere (except in certain areas of development of bedrock, pebbles, and aeolian sands) in different geomorphologic conditions and at different absolute heights:

1. In steppe and desert–steppe regions with the lowest absolute elevations, 20–40 m below sea level in some areas (within the territory between the Caspian and Aral Seas)
2. At oases, confined to river valleys, with absolute elevations mainly from 200 to 500 m
3. In the foothills of ridges and their spurs, at absolute elevations from 500 to 1500 m (in Central Asia, this is the main zone of distribution of loessial rocks)
4. In mountain valleys, usually at altitudes of 1500–3000 m
5. In high-altitude areas, particularly in the low parts of the crests of most ridges, mainly at absolute elevations of 3000–3700 m

Scientists (Toychiev and Stelmakh 2011a, b) agree that loess is an excellent archive of past climate and environment reconstruction. Muhs (2007, 2018) notes that loess constitutes one of the most important deposits that serve as terrestrial records of the Quaternary Period. Understanding the geochemistry of loess and paleosol can tell us much about past atmospheric circulation, past temperature and moisture regimes, and even tectonic processes.

Fig. 3.27 Loess was used as a main material in construction of the castle Itchan Kala in Khiva, Khorezm region

It is loess that has created good opportunities for intensive agriculture in the region and attracted people seeking a settled lifestyle. Its physical properties facilitate cultivation operations, ensuring proper soil aeration and root growth. Its porosity (ranging from 34% to 60%, depending on whether the loess is richer in clay or sand) makes it able to retain enough soil moisture and provide adequate amounts of minerals for good crop growth without use of chemical fertilizers.

Loess has been very popular for construction of cave dwellings in many Asian countries. Its ease of cleavage creates an opportunity for easy excavation and use in construction. Many ancient monuments are built of bricks made of loess (Fig. 3.27).

Fig. 3.28 Turquoise from central Kyzylkum. (Photo by Lola Gulyamova, used with permission from the Geologic Museum of the National University of Uzbekistan)

However, human activity (including land development and intensive irrigation) has led to decreases in areas of loess and loss of its properties, transforming it into loess-like rock. Loess is one of the main natural resources of Uzbekistan because of its ability to transform into very rich soil under appropriate climatic conditions.

3.2.10 Gemstones

The geology of Uzbekistan is manifested by many types of industrial stone gems: amethyst, rodingite, jade, amazonite, topaz, and petrified wood. Various minerals (beryls, amethyst, crystal, turquoise, malachite, lapis lazuli, azurite, calcite onyx, snake coils, fluorite, garnet, agate, tourmaline, rhodonite, cacholong, and opal) are found in the Piskom, Chatkal, Ugam, and Sultanizdag Mountains, as well as in the Kyzylkum Desert.

Turquoise

Geology determines the geographic isolation of the turquoise-bearing central Kyzylkum mineralogical province. The central Bukantau deposits in the Kyzylkum Desert and the deposits in the eastern part of this country in the Chatkal–Kurama district are examples of complicated mineralization. The central Bukantau region was historically famous for its deposits of turquoise, which are associated with areas of development of the carbonaceous–carbonate–siliceous formation in the Kyzylkum Desert. It is believed by Turk horsemen that turquoise protects them from damages and falls of any sort. Different kinds of jewelry reflect the beliefs of the local people and their respect for this stone (Figs. 3.28 and 3.29).

The appearance of high-quality pink, creamy to brick-red chalcedony is traced along the northern foothills of the Karatau among the rocks of the red carbonate–terrigenous[6] formation dating back to the Middle and Upper Paleozoic Era. Amethyst manifestations are localized in mineralized cracks of granitoid intrusions of the Northern Tien Shan and Chatkal–Kurama folded zones. Deposits of rhodonite are associated with spilite–keratophyre[7] manganese-containing formations in the Sultanuizdag (in Lower Amu Darya) and Nurota folded regions. In the weathering crusts of central Kyzylkum, manifestations of cacholong and opal are known (Figs. 3.30, 3.31, 3.32, 3.33, 3.34, and 3.35).

Marble

Uzbekistan is famous for Gazgan marble, which is used for facing external and internal walls, floors, steps, columns, windowsills, sun protection elements, dishes, vases, and other items. In particular, a lot of Gazgan marble has been used for decoration of theaters, museums, metro stations, hotels, and official buildings in Tashkent and other cities

[6]Terrigenous means formed in the sea from material derived from the land by erosion.

[7]A keratophyre is a volcanic rock of intermediate composition.

Fig. 3.29 Traditional jewelry of the Turkic people

Fig. 3.30 Carnelian from central Kyzylkum. (Photo by Lola Gulyamova, used with permission from the Geologic Museum of the National University of Uzbekistan)

Fig. 3.31 Garnet from Aksakata, Ugam Ridge, Western Tien Shan, Tashkent region. (Photo by Lola Gulyamova, used with permission from the Geologic Museum of the National University of Uzbekistan)

Fig. 3.32 Garnet from Aksakata, Ugam Ridge, Western Tien Shan, Tashkent region. (Photo by Lola Gulyamova, used with permission from the Geologic Museum of the National University of Uzbekistan)

(Samarkand, Navoiy, Karshi, Kokand, and Bukhara). In 1934–1935, during construction of the first stage of the Moscow Metro, warm-toned Gazgan marble was used for its decoration. Then, in the cities of Petersburg (St. Petersburg), Kiev (Kyiv), Tbilisi, Baku, Tashkent, Yerevan, Minsk, and Novosibirsk, metro stations were lined with multicolored Gazgan marble (Figs. 3.36 and 3.37). In many other cities of the former Soviet Union, many construction sites are lined with marble from this deposit; examples are the Ukraine Palace in Kiev, a monument in Baku, a circus in Novosibirsk, a drama theater in Bishkek, an administrative building in the Space Flight Control Center in Korolyov, and the Tretyakov Gallery in Moscow.

Marble blocks were delivered to Czech and Slovakia, Japan, Belgium, and Italy, and marble goods were delivered to Switzerland, Mongolia, and Italy.

Today, perhaps, there is not a single significant building or structure in our country without marble decoration.

The Nurota Mountains are the source of Gazgan marble. This deposit is unique in its reserves, its coloring, and its physicomechanical properties. The area of the deposit is a low ridge of hills with relative elevations of 10–30 m. The layered marble is crumpled into a sloping anticline fold on the northwest stretch. The thickness of the suite is 950 m. The productive horizon is 75 to 150 m thick, and the thickness of the individual layers ranges from 0.4 to 6 m. The

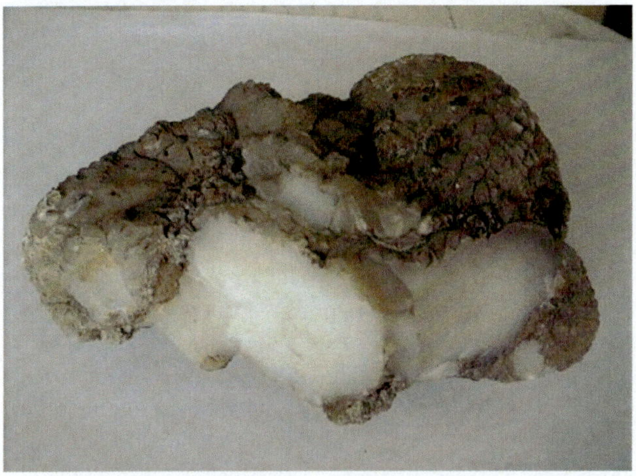

Fig. 3.33 Cacholong. (Photo by Lola Gulyamova, used with permission from the Geologic Museum of the National University of Uzbekistan)

Fig. 3.34 Fluorite, a mineral form of calcium fluoride. (Photo by Lola Gulyamova, used with permission from the Geologic Museum of the National University of Uzbekistan)

marble is fine grained and multicolored, with more than 35 colors (such as cream with gray, dark gray, or yellowish-pink). The practice of field exploitation shows that the different shades of the colored marble do not have sustained zones and are connected to each other by mutual transitions. In terms of color, Gazgan marble is divided into two types: warm colored and cold (gray) colored.

According to the available archival materials, the Gazgan marble deposit has been under development since 1929 (the Kash-Kash plot), but numerous artifacts on the Kanitahta site indicate that quarrying of marble for domestic needs (dishes and bowls) and for production of sculptural products was carried out many hundreds of years ago.

For further development of geological research and mining technology, Uzbekistan is establishing cooperation with

Fig. 3.35 Onyx from the Chatkal Ridge, Tashkent region. (Photo by Lola Gulyamova, used with permission from the Geologic Museum of the National University of Uzbekistan)

Fig. 3.36 A metro station in Tashkent, decorated with Gazgan marble

Fig. 3.37 Decoration on the facade of a theater named after Alisher Navoiy

foreign partners (Rio Tango, UK; Japan Oil, Gas and Metals National Corporation [JOGMEC], Japan; Polymetal, Russia; Korea Resources Corporation [KORES] and Korea Institute of Geoscience and Mineral Resources [KIGAM], South Korea; Realgold Resources, Canada; Ministry of Land and Natural Resources, China; Mine Reclamation Corporation [MIRECO], South Korea; and Institute for Remote Sensing, India).

3.2.11 Oil Shale

Oil shale, ranging from Cambrian to Tertiary in age, occurs in many parts of the world. Its organic matter is composed chiefly of carbon, hydrogen, oxygen, and small amounts of sulfur and nitrogen. Oil shale is deposited in a wide variety of environments. However, many of these deposits have not been explored and mined. The reason is that oil shale cannot compete with petroleum-based crude oil. Today, petroleum-based crude oil is cheaper to produce than shale because of the additional costs involved in mining oil shale and extracting energy from it. According to an estimate from the World Energy Council (2009), Uzbekistan's shale oil resources are 8386 million barrels.

3.3 Energy Resources

The remoteness of Uzbekistan from major international commodity markets and the high energy intensity of its economy are major challenges in the energy industry (UN 2004). Uzbekistan is self-sufficient in various energy resources. This gives an opportunity for further development and improvement of their usage. Self-sufficiency in energy resources is a major advantage for Uzbekistan. In this regard, it is the leading country in the region, with 31 hydropower plants (which generate 48×10^9 kWh of electrical power each year) and 100-million-gigacalorie thermal power plants (TPPs), with a total capacity of 10.6 GW, which generate 85% of the country's electric power (Aripdjanov and Adilov 2017).

Many challenges in today's regional energy generation and distribution are the consequences of unfair allocation of water and energy resources, as well as an inefficient system of water and electricity management (inherited from Soviet times) and a poor mechanism for Central Asian regional exchange of fuel and energy. Unevenly distributed water and energy resources in the region are causing challenges in their management, which requires a consensus between Central Asian states.

Another challenge is aging infrastructure, leading to concerns about the security and reliability of the power supply in Uzbekistan. One of the priorities of the current development of the energy industry is construction of additional facilities and upgrading of existing ones.

The government of Uzbekistan is the single owner of the country's natural resources, and its policy results in efficiency of water and energy resources management. Today, the government is initiating programs to attract foreign investment for technological innovations and joint projects in mining and exploitation. According to the World Bank Group (2013), the obsolete electricity infrastructure and poor management have led to loss of 20% of net generation. The World Bank Group (2013) suggests that one of the priorities is producing added-value petroleum products by using natural gas as a raw material and decreasing importation of crude oil, to encourage use of gas-powered engines. Significant modernization of gas-fired plants may increase efficiency by 40%.

It was previously estimated (UN 2004) that in 2000–2020, Uzbekistan's production of nonrenewable energy resources (natural gas, oil, coal) would be: coal 2 Gt, oil 350 Mt, gas 2×10^9 m^3, uranium 83,700 t, and hydropower 15×10^9 kWh/year. Thus, Uzbekistan is rich with nonrenewable energy resources, and it may become richer if it receives the benefits of all of its potential renewable energy sources, such as solar, wind, and biomass energy. The government is implementing policies to meet 21% of all energy needs from renewable sources by 2031, reducing energy intensity and implementing energy-saving technologies and systems (Aripdjanov and Adilov 2017; Aminjonov 2016; Eshchanov et al. 2013). From the global, regional, and local perspectives, one of the major issues is implementation of new energy, including "green" technologies. Researchers (Aminjonov 2016; Eshchanov et al. 2013) have expressed concern that this country does not pay enough attention to all potential renewable energy sources such as as solar, wind, and biomass energy. All programs on use of alternative energy sources are still at the initial stage of study and experimentation.

According to data from Uzbekenergo, in 2017, 85.7% of electricity was generated by TPPs of the state-owned electricity company Uzbekenergo and 13.3% by hydroelectric power plants (HPPs) of the state-owned electricity company Uzbekgidroenergo.

The World Bank's Ease of Doing Business indicator ranks Uzbekistan in 146th place (IFC and World Bank 2014) and 83rd out of 190 countries with regard to electricity access (UNDP 2017). Experts (Aminjonov 2016; Eshchanov et al. 2013) suggest that one of the priorities of Uzbek government policy, in addition to development of medium and large hydropower potential, should be promotion of small and minihydropower projects.

In 2014/2015, according to estimates (BP 2016), Uzbekistan's consumption of different energy sources, measured in million tonnes of oil equivalent, was as follows: oil 2.8/2.8, natural gas 43.9/45.3, coal 1.2/1.1, and hydroelectricity 2.3/2.5. The total energy consumption was 50.3/51.6. Natural gas was the main fossil fuel, with a trend toward an increase in its usage.

According to data from Uzbekenergo (2018, 2019), in 2017–2018, the major consumers of electricity were industry, the population, and agriculture, consumption by which decreased from 20.7% to 19.4% (Fig. 3.38).

Around half of Uzbekistan's population lives in rural areas and experience significant electricity shortages and power cuts due to high distribution losses, illegal energy tapping, and the poor condition of infrastructure in remote areas.

3.3.1 Coal Industry

According to data from the World Energy Council (2009), the proven recoverable reserves of coal in Uzbekistan are 3 Gt, with production of 3.2 Mt. Bituminous coal is mined on a small scale—about 70,000 t/year. Data (BP 2016) show that the total proven coal reserves are 1900 Mt of oil equivalent, production is 1100 t of oil equivalent, and consumption is 1100 t of oil equivalent.

According to data from the O'zbekko'mir coal producer, the volume of coal mining in 2017 amounted was between about 3Mt and 4.1Mt, including 70,000 t of charcoal. The main consumer of coal fuel is the power sector, whose share of total coal consumption is over 85%. Demands for coal fuel also come from industrial enterprises, social and utility companies, and the population.

McHutchon (2017) notes that a resolution issued by the president of the Republic of Uzbekistan (titled "On the Development and Modernization of the Coal Industry in 2017–2021") envisaged further development and modernization of the coal industry in 2017–2021. According to this document, it was planned to introduce modern coal mining at the Angren and Nishbash coal mines. The purpose was to increase coal production to 11.1 Mt by 2021 to meet the needs of the population and industry. Six investment projects worth US$690.5 million were to be realized, including modernization of JSC Shargunkumir (costing US$105.5 million) and construction of the Angren mine (costing US$165 million) and the Nishbash mine (costing US$175.2 million).

Coal is also an important resource in chemical production. Syngas is used for producing methanol, hydrogen, and carbon monoxide, which serve as chemical building blocks in manufacturing of olefin, acetic acid, formaldehyde, ammonia, urea, and other compounds. By using inexpensive coal, it is possible to produce a wide range of valuable commodities. Interest in using coal tends to increase in many countries with abundant coal resources. Coal-to-chemical processes require substantial quantities of water, which poses a challenge in a setting of limited water resources.

Power stations and boiler units account for 90% (by volume) of Uzbekistan's coal consumption according to a United Nations (2004) estimate and 85% according to a Uzbekenergo (2017) estimate. When coal is used for electricity generation, it is usually pulverized and then burned in a furnace with a boiler to convert boiler water into steam, which is then used to spin turbines and turn a generator. The efficiency of this multistage process is only about 25–45%. The efficiency may be improved if, instead of being pulverized and burned directly as fuel, the coal is gasified to create syngas, which is then burned in a gas turbine. Descamps et al. (2008) report that such integrated gasification com-

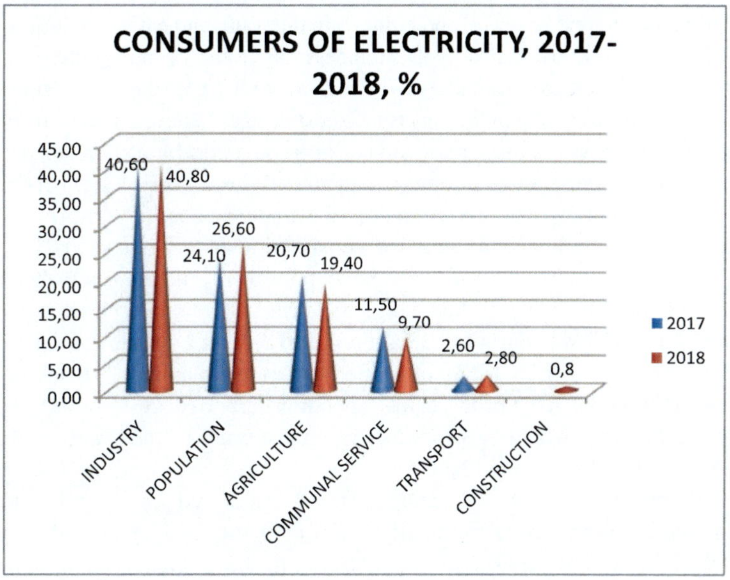

Fig. 3.38 Consumers of electricity in 2017–2018. (Compiled by the author, using data from Uzbekenergo (2018, 2019))

bined cycle (IGCC) power plants have higher efficiency of 42–45% when using bituminous coal.

A project to increase the efficiency of TPPs (Mitsubishi Corporation 2016) was implemented by Uzbekistan together with Mitsubishi Corporation (MC) and Mitsubishi Hitachi Power Systems Ltd (MHPS). MC and MHPS were awarded a full turnkey contract by JSC Uzbekenergo for construction of a combined cycle power plant at the Turakurgan Thermal Power Station (TTPS) in eastern Uzbekistan and the Navoiy Thermal Power Station (NPTS). After completion of construction in 2020, this TTPS plant had the capacity to supply 900 MW of power and the NPTS had the capacity to generate 450 MW of power and 200 Gcal/h of heat. These plants were expected to supply power amounting to about 10% of the country's overall energy in 2020.

3.3.2 Gas and Oil Industry

Uzbekistan is considered an oil- and gas-rich country, and it can satisfy all domestic needs from own sources. Its gas output is the highest in the region. The local resource base of the gas industry allows further growth (UN 2004). According to the national holding company (NHC) Uzbekneftegaz, the recoverable hydrocarbon reserves in Uzbekistan constitute over 2.5 Gt of conditional fuel, of which 65% pertains to shale gas reserves.

To attract investment in development and modernization of this industry, Uzbekistan has liberalized its legislation. As a result of this reform, the state oil and gas company NHC Uzbekneftegaz and the Russian oil company Lukoil started a joint project in the Gissar Group's gas condensate fields in southeastern Uzbekistan, where there are proven reserves of 241 million barrels of oil equivalent (Lukoil 2016). Uzbekneftegaz invested some US$200 million in geological prospecting to find heavy oil and bitumen in the south and east of Uzbekistan in the Korsagly–Dasmamagin area and in the Besharcha block in Surxondaryo district, as well as in the Fergana area in the east of the country. Uzbekneftegaz believed it would be possible to increase annual oil production by at least 100,000 t in those areas after completion of these works.

Lukoil commenced implementation of the active phase of the Kandym Early Gas Project for construction of a gas-processing plant (GPP) and arrangement of the Kandym group of deposits in the Bukhara region, worth US$2.66 billion. This GPP was put into operation in 2018, with a projected capacity of 8.1×10^9 m^3 of gas per year. It was planned that in the initial stage of the fields' operation, 2.2×10^9 m^3 of gas per year would be produced and transported to the Mubarek Gas Processing Plant.

Moreover, Lukoil started testing the operation of two preliminary gas-processing terminals (PGPTs) in the area of North Shady, and deposits in Kuvachi-Alat in the Bukhara region of Uzbekistan, as part of the Early Kandym Gas Project. The total capacity of the units is 2.2×10^9 m^3 of gas per year.

The gas pipeline, with a gas capacity of 20×10^9 m^3 of gas, was constructed by Uzbekneftegaz and the China Natural Petroleum Corporation. It is the fourth line of the Uzbek section of the Central Asia–China gas pipeline.

Uzbekenergo has also completed construction of the external power supply Ustyurt Gas Chemical Complex, which involved construction of a substation with a capacity of 220 kW. The capacity of this natural gas chemical complex allows processing of 4×10^9 m^3 of natural gas per year, along with production of 400,000 t of polyethylene.

Exploration and extraction of hydrocarbon deposits has also been started in the Uzbek part of the Aral Sea.

The major partner in Uzbekistan's oil and gas industry is Lukoil, which is planning to extract a total of about 18×10^9 m^3 of gas at all of its projects in Uzbekistan. Lukoil reports (Kandym-Khauzak-Shady 2017) that the biggest of these projects is the Kandym group of gas reservoirs, which includes six separate gas condensate plots in the Karakul district of Bukhara province, close to the border with Turkmenistan: Kandym, Kuvachi-Alat, Akkum, Parsankul, Xodji, and Western Xodji. This joint project by Lukoil, which has a 90% share in the profit, and Uzbekneftegaz, which has a 10% share in the profit, is intended to extract and process 8×10^9 m^3 of gas per year. Eighty percent of the extracted gas is for export to China, with the remaining 20% going to Kazakhstan and Russia. About 500,000 t of gas condensate is bought by Bukhara oil refinery, where the condensate is processed and goes to the domestic market for domestic purposes. The complex, with a capacity of 8×10^9 m^3 of gas per year, processes hydrogen sulfide gas to produce refined blue fuel, stable gas condensate, and commercial sulfur. The facility consists of two production lines, with terminals for shipment of commodity products, including gas supply for export through the main gas pipeline.

Uzbekistan's demand for gas products is being met by its own resources, and about 20% of the gas is exported. The major importer of gas from Uzbekistan is China. According to a report, in January–October 2017, 2.73 Mt of gas was exported by Uzbekistan to China, 18.32 Mt by Turkmenistan, and 229,140 t by Kazakhstan. According to the agreement between Uzbekistan and China, in 2018–2020, exports of gas to China were projected to reach 10×10^9 m^3, and, according to the contract between Gazprom (Russia) and Uzbekneftegaz (Uzbekistan), exports of gas to Russia were projected to increase by 4×10^9 m^3/year. In 2016, Gazprom imported 6.2×10^9 m^3 of gas from Uzbekistan.

The World Energy Council (2005) estimates that Uzbekistan's proven recoverable natural gas reserves are – 1850×10^9 m^3, with production of 59.7×10^9 m^3 and consumption of 47.2×10^9 m^3. The proven resources of shale oil are 8386 million barrels (or 1200 Mt).

3.3.3 Hydropower and Thermal Energy

Hydraulic energy is production of electrical energy by channeling of water that has accumulated naturally in a reservoir through water turbines. Of the various internationally agreed sustainable development goals, one is an increase in the share of renewable energy in the energy mix. In relation to this goal, hydroelectric potentials are to be explored and utilized by regional cooperation in the field of generation, transmission, and distribution of energy. Hydropower is the most advanced and flexible type of renewable energy and is the cleanest method of power generation. Its development has a long-term economic advantage. Use of reservoirs for hydropower storage is beneficial for water management, which is key for sustainable development especially in arid zones. The multipurpose functions of reservoirs can bring security of water supply and power.

In Uzbekistan, according to data from Uzbekenergo (2017), the total capacity of the 28 hydropower plants (HPPs) is 12.3×10^6 kV, and they generate more than 6.5×10^9 kV (totaling 60×10^9 kW) each year, representing 11.5% of the total power generation. Thermal power stations use gas to produce 86% of the country's electricity.

The Charvak HPP (total capacity 600 MW) and the Xojakent HPP (total capacity 165 MW) are the biggest HPPs. They generate 23% of the potential hydropower energy. These HPPs are located in the mountains of the Western Tien Shan in the Tashkent region. The new Piskom HPP will be constructed on the Piskom River in this region as well. Its capacity is 400 MW, with annual electricity generation of 900×10^6 kVh. All three HPPs will generate 27.4×10^9 kVh/year.

A challenge that Uzbekistan faces, as do other countries, is the need for continuous improvement in hydropower technology in terms of environmental performance, materials, efficiency, operation range, and costs. Priority is given to large schemes, which support grid systems and power urban and industrial centers. Less attention is paid to small-scale hydropower, which plays an important role in remote areas.

In 2016, UzDaily (2016) reported that modernization works were being performed at the Charvak HPP, involving replacement of the hydrogenerator starter winding and installation of new feed systems. The Russian company Power Machines was a partner in the modernization, improving the capacity of each hydraulic unit from 155 to 175 MW. This modernization provided an opportunity to generate an additional 120×10^6 kWh/year.

In Uzbekistan, there are more than 430 rivers and other waterways, totaling 12,700 km in length. Not all of them are used for electricity generation. For example, the average annual flow of the Zerafshan River is 5.27 km^3, whereas that of the Kashkadarya river, with a sink, is 1.34 km^3.

The major TPPs are located close to sources of energy in Talimarjan, Navoiy (1250 MV), Syrdarya (3000 MV), Tashkent (1860 MV), and Yangi Angren (2100 MV) in the Tashkent region (Cabinet…2017). The TPP in Sirdaryo is the largest in the region.

Electricity is generated primarily by use of gas, which constitutes 93.5% of the fuel used (Fig. 3.39). Some attempts have been made by the government to replace the power industry's use of natural gas and oil products with use of coal (Aripdjanov and Adilov 2017).

3.3.4 Alternative Energy Sources

The urgent need to fundamentally transform present energy systems into more sustainable systems is a major goal of global development. Many risks are connected to security of supply in energy systems because of the potential damage they can do in nature. The experience of countries with advanced alternative energy systems (such as the Netherlands and the USA) shows that this transformation requires not only technological but also political measures to protect the natural life support systems on which humanity depends. Countries whose governments have adopted specific policy mechanisms (such as the feed-in laws in the European Union) are achieving good results. In the Netherlands, local trains use electricity generated by wind energy, thus reducing the cost of transportation. In many countries, governments have adopted policies and financing mechanisms to support solar energy applications. Appropriate policy measures stimulate not only solar applications but also creation of new industries, new jobs, and new economic opportunities, as well as protecting the environment. Researchers (Aminjonov 2016; Abdullaev and Isaev 2005; Aripdjanov and Adilov 2017) agree that Uzbekistan has a lot of opportunities for development of alternative energy sources.

Alternative technologies have become cost effective for some applications, but they are still expensive for other applications such as grid electricity. To be implemented in Uzbekistan, these technologies must be promoted by government policies, such as policies that grant financial support and create a favorable business environment for manufactur-

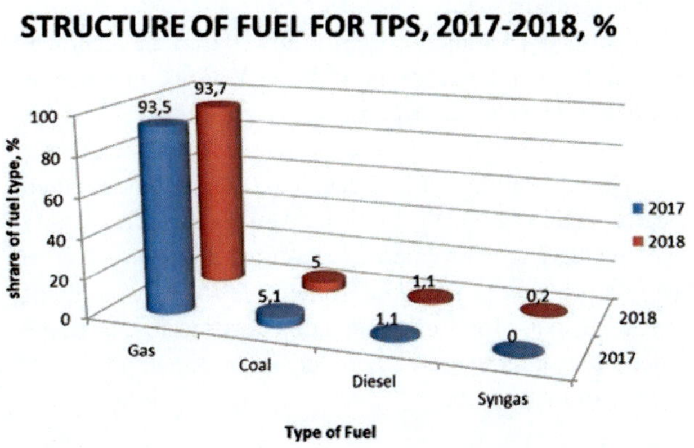

Fig. 3.39 The fuel source for electric power generation at thermal power stations in 2017. (Compiled by the author, using data from Uzbekenergo (2018, 2019))

ers and producers. Implementation of most of these technologies is dependent on importation of the technology itself, as well as appliances, materials, and services. It makes these technologies more expensive and makes their implementation challenging.

According to UNDP (2012) data, the total installed electricity-generating capacity of Uzbekistan in 2012 was 12,517 MW, with only 397 MW of installed renewable energy.

Solar Energy

Solar energy is the most abundant permanent energy source, and it is available for use in its direct form of solar radiation. The solar resource is almost uniform throughout all regions of the world. It creates opportunities for economically attractive applications of solar energy not just in the sunniest regions.

In Uzbekistan, the price of solar panels is still high. Some households have installed them, but they are still not widespread (Fig. 3.40).

Fig. 3.40 A solar energy panel in Samarkand province

The presence of long periods of sunny days each year is a strong argument for development of solar energy in Uzbekistan (Abdullaev 2005). This energy source represents about 99% of the total renewable energy potential (Aminjonov 2016), but this potential is not yet well used. The government (Times of Central Asia 2014; Aripdjanov and Adilov 2017) initiated several programs for development of this industry by establishing the first solar plant in Samarkand region to produce 100 TW of electricity each year, and another three solar stations were planned to be operational by 2020 in the Namangan and Surkhandarya regions (Aripdjanov and Adilov 2017).

Wind Energy

Complex mechanisms involving the rotation of the Earth, heat energy from the Sun, the cooling effects of the oceans, the temperature gradients between land and sea, and the physical effects of mountains generate winds. However, only small turbines are being tested in the windy mountains of the Western Tien Shan (Fig. 3.41).

Fig. 3.41 A wind turbine on the Chatkal Ridge, Tashkent region

An estimate (Aminjonov 2016) showed that the technical potential for wind energy is 1600 MW. This industry is extremely undeveloped. A conservative preliminary analysis (Jegelevicius 2015; REVE 2015) estimated that Uzbekistan's wind energy potential is more than 520,000 MW of installed capacity and more than 1×10^9 MWh/year. This study (Jegelevicius 2015) revealed two promising zones—one in Navoiy province and the other in the south of Karakalpakstan—for further detailed measurements to be conducted in order to more accurately define the wind energy potential of these areas.

Biomass Energy

Bioenergy is generated through use of wood, agricultural residues, or vegetables as sources of energy. This industry could be developed in Uzbekistan, which produces 2–3 Mt of cotton per year. Some estimates (United Nations 2013) show that potential sources of bioenergy are cotton stalks and bulrush plants (available amount 10–15 Mt/year), solid communal waste (available amount 30×10^6 m^3/year), and agricultural waste (available amount 100×10^6 m^3/year). Cotton plant stems are the main source of biomass energy for 65% of the country's rural population, which mostly uses cotton stalks for cooking and space heating.

To sum up, taking into account growing energy demands in Uzbekistan, development of alternative energy sources can improve the level of energy security in the country. Aminjonov (2016) argues that there is substantial potential to develop different types of energy sources, which, in combination, will complement each other, thus establishing a self-sustaining energy system.

3.4 Land Resources

These vital resources, together with water, are most important for human wellbeing. Land degradation is becoming the most dangerous global threat, which has already affected more than 3.2 billion people in the world (United Nations 2018). As reported by Watts (2018), the Intergovernmental Science-Policy Platform on Biodiversity and Ecosystem Services (IPBES)—which aims to provide a "global knowledge base about a threat that is less well-known than climate change and biodiversity loss, but closely connected to both and already having a major economic and social impact"—has revealed that "land degradation costs more than 10% of annual global GDP [gross domestic product] in lost ecosystem services such as carbon sequestration and agricultural productivity" and can cause vegetation loss and increase the risks of flooding and landslides.

3.4.1 Soils

The nature of Uzbekistan and its physical geography vary significantly between the northern and southern parts of the country, and between the plains and the mountains. These differences are manifested by a variety of natural resources, including soil. Soil is formed under such driving forces as topography, geology, hydrology, and hydrogeology. The complexity of the combination of lithology, structure, hydrogeological conditions, and climate contributes to formation of different types and subtypes of soil. Likewise, mosaic geology and geomorphology, climate, and soil drainage are extremely complicated and characterized by variations in their properties from one place to another. Sandy and rocky desert soils occur on the plains, whereas brown shade soil is found in the foothills and mountains. Irrigated soil replaces other types at oases. Combinations of zonal and intrazonal soils[8] are common in many places.

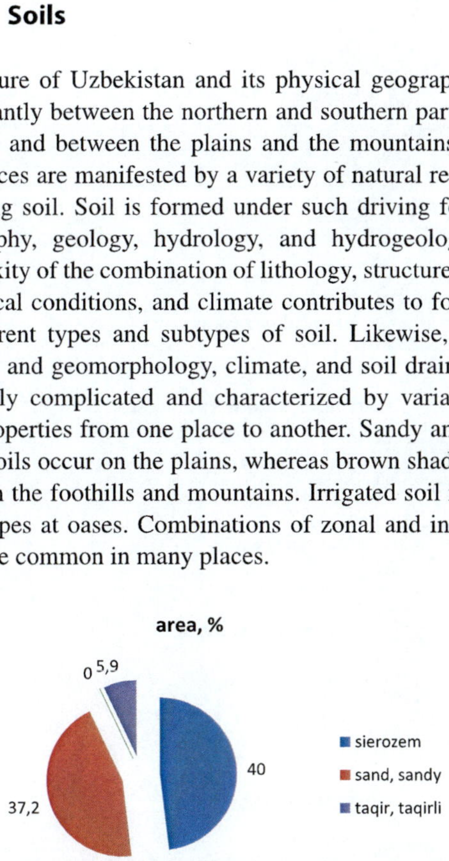

Fig. 3.42 Composition of soils in the desert zone. (Compiled by the author, using data from Khasanov and Gulomov (2006))

Fig. 3.43 Takir soils in the Kyzylkum Desert, Bukhara region. (Source: Uploaded by Ivtorov (CC-BY-SA 3.0), via Wikimedia Commons)

Soils on Plains and in Foothills

On the plains, desert soils are formed under an extra-arid dry climate. This formation is extremely dependent on the hydrological and biological conditions, resulting in creation of a mosaic of different types of sierozem,[9] takir,[10] pasture, and swampy saline soils. Soil formation is dependent not only on the climate but also on the properties of the underlying rocks and the depth of the groundwater. This is the reason for co-occurrence of various azonal[11] soils with different levels of humidity and salinity in the same zone.

The desert zone constitutes 61.16% of the country's area at altitudes of 0–400 m above sea level in the central and northern parts of Uzbekistan. This zone is 27,004,000 ha in size, and it is estimated (Khasanov and Gulomov 2006) that sierozem soil constitutes 40% of the desert area, while sand and sandy soils account for 39% and takir-type soil for 5.9% (Figs. 3.42 and 3.43). The northern part is moderate, while the southern part is in the dry subtropics.

[8]According to the *Merriam-Webster Dictionary* (https://www.merriam-webster.com/dictionary/intrazonal_soil), intrazonal soil is "a major soil group, classified as a category of the highest rank and including soils with more or less well-developed soil characteristics determined by relatively local factors (as the nature of the parent material) that prevail over the normal soil-forming factors of climate and living organisms."

[9]According to *Encyclopaedia Britannica* (https://www.britannica.com/science/sierozem), sierozem is alkaline soil in semidesert and desert environments where the supply of organic substances and the humus content are extremely low. Gray-brown soils form in the temperate zones, and gray desert soils (sierozems) develop in the arid subtropics.

[10]A takir is a clay soil area with a flat surface in desert and semidesert environments, marked by polygonal cracks in the dry season.

[11]According to the *Merriam-Webster Dictionary* (https://www.merriam-webster.com/dictionary/azonal_soil), azonal soil is "a major soil group often classified as a category of the highest rank and embracing soils that lack well-developed horizons because of immaturity or other factors that have prevented their development."

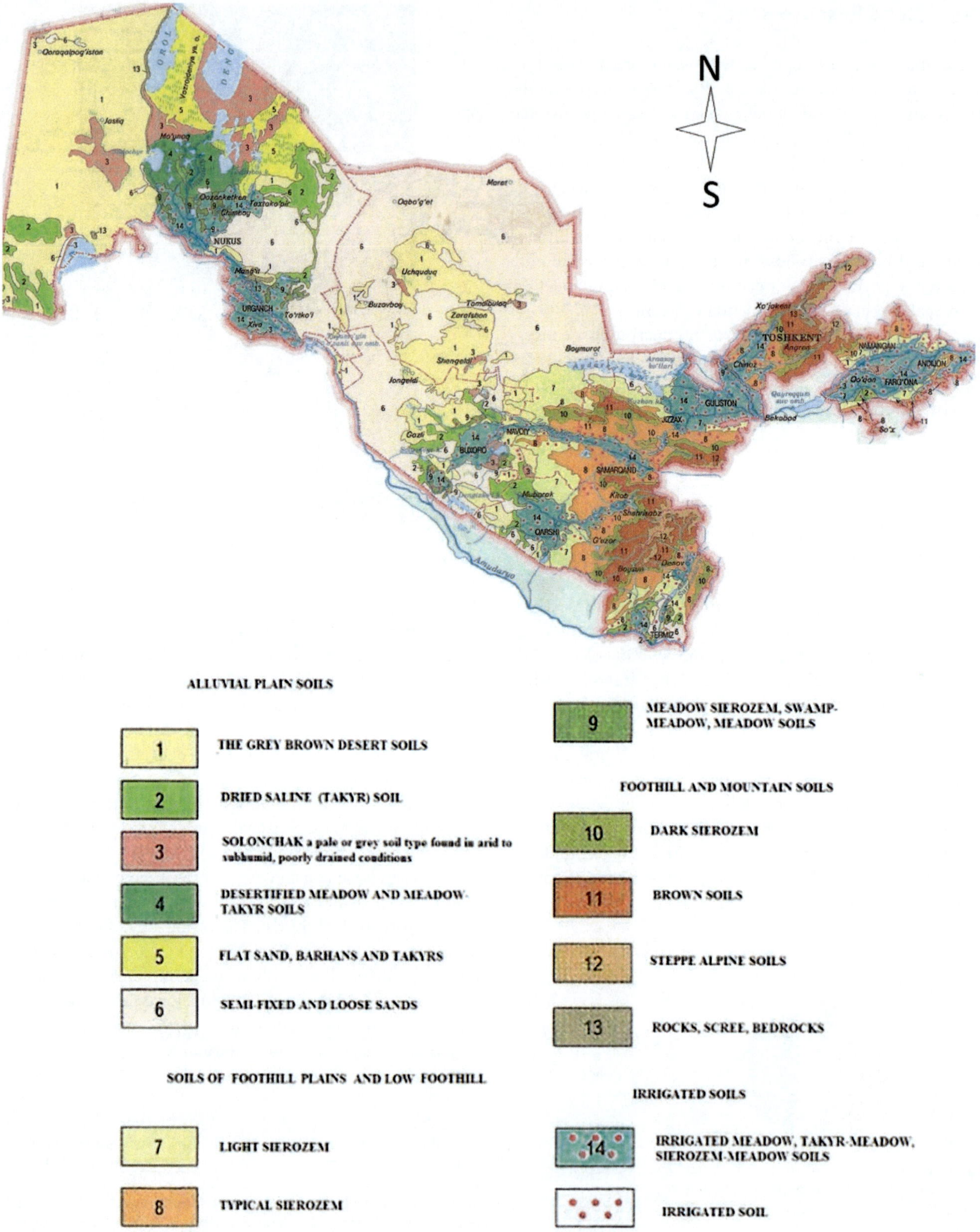

Fig. 3.44 Soil map. (Image courtesy of the State Committee for Land Resources, Geodesy, Cartography and State Cadastre of the Republic of Uzbekistan)

The border between the northern and the southern parts is in southern Ustyurt, the Lower Amu Darya river, and the northern Kyzylkum Desert. Because of the difference in climate, the ecology is not the same in both parts of the desert (Fig. 3.44).

In the northern subzone of the Kyzylkum Desert, the climate is very dry and continental, and precipitation is scarce but evenly distributed over the seasons. The prevailing landscape is a rocky desert with rare vegetation.

From the perspective of the climate, the genesis of the soils, and the content of vegetation and animals, the southern subzone has many similarities with the desert of Middle East and may be considered an extension of it. In contrast to the northern subzone, the precipitation is not evenly distributed between the seasons. It is more frequent in the spring and winter than in the summer dry season, when no rain may fall at all. This creates different conditions for soil formation at altitudes of less than 250 m and at altitudes of 250–400 m above sea level.

Researchers (Gintzburger et al. 2003; Khasanov and Gulomov 2006) note that the desert soils have low organic matter content, high levels of calcium and associated gypsum, and a high probability of salinity. All soils can be grouped into 19 types (Khasanov and Gulomov 2006), which may be aggregated into five major types, based on their location, water supply, groundwater depth, and salinity (Table 3.8).

In the desert zone of Uzbekistan, a mixture of zonal and intrazonal soils is widespread. Such intrazonal soils as solonchak,[12] takir, swamp, and pasture soils may be found everywhere.

The main zonal soils are sierozem, takir, and sand in combination with swamp, saline, and pasture soils (Table 3.9). These have been identified in proluvial foothills and in residual mountains (Ustyurt, Devxona in the Bukhara region, and in the central Kyzyl Kum). They are poor in humus, and only 30,000 ha out of an area of 11,025,000 ha is cultivated (Khasanov and Gulomov 2006).

Use of places with takir soil for agriculture requires advanced agrotechnical technology to improve the quality of the soil.

Hydromorphic[13] soils, which are more valuable for agriculture, are represented by pasture, pasture–swamp, swamp, and saline soil types in many valleys and deltas. These soils are formed in the foothills by water that has low mineral content and is rich in carbonates. Meadow soil contains 1–4% humus.

Swamp soils occur in depressions with high humidity and poor drainage in deltas, close to rivers, and even in the desert. These soils are not used for agriculture.

The origin of salinity in soils can differ; it may be due to high salinity levels in the underlying rocks or/and groundwater close to the surface, accumulation of salt spread by the wind, poor drainage, and human activity. Salinity of soils is a major issue in irrigated lands all over the country, particularly in Jizzakh, Kashkadarya, and Sirdaryo provinces.

Irrigated soil is the result of many years of recultivation of pasture, swamp pasture, and saline soils.

Table 3.8 Grouping of soils

Soil group		Soil type	Groundwater depth (m)
I	Automorphic soil[a]	Takir (barren)	>5
		Desert sand	
		Brown shade	
		Light brown shade	
		Dark brown shade	
		Brownish	
II	Semihydromorhic soil	Pasture, swamp pasture	2–5
		Swamp and pasture–sandy, pasture–takir	
III	Hydromorphic soil[b]	Swamp	<2
		Swamp pasture	
		Swamp pasture and swamp	
!V	Different types of saline soil, depending on the depth of the salinity level[c]	Saline takir	
		Typical saline	
		Saline pasture	
		Saline swamp pasture	
		Saline pasture–swamp	
V	Irrigated soil	New irrigated soil	
		Irrigated soil	
		Old irrigated soil	

Compiled by the author, using data from Khasanov and Gulomov (2006, p. 114)

[a]Automorphic soils are dry, excessively drained, mostly coarse-textured, shallow soils. The plant-available water supply in the 75-cm soil layer is <50 mm, and there may be a risk of soil drought during the vegetation period

[b]According to the *Merriam-Webster Dictionary* (https://www.merriam-webster.com/dictionary/hydromorphic; accessed 12 Jun 2022), hydromorphic soil is "developed in the presence of an excess of moisture which tends to suppress aerobic factors in soil-building"

[c]If the soil surface is more than 3% contaminated with salt, it is considered saline

[12]A saline horizon occurs on arid land, and, in semiarid climatic conditions combined with strong evaporation, there are intrazonal soils, which have been termed *white alkali soils* (Matthews 2014).

[13]Hydromorphic soils include swamps (gleysols, black soils) and climo-hydromorphic soils (meadow soils).

Table 3.9 Composition of desert soils

Soil type	Area (km^2)	Environments	Locations
Sand and sandy soil		Deserts	Kyzylkum Desert, Lower Zerafshan, Lower Amu Darya
		Steppes	Kashkadaryo, Central Fergana, Mirzacho'l
Takir	17.8	Foothills, valleys, river deltas, intermountain depressions, alluvial plains	Amu Darya, Kashkadarya, and Syr Darya rivers
Saline	12.7	Desert, deltas, terraces, dried meanders, lakes, dried Aral Sea bed	

Compiled by the author, using data from Khasanov and Gulomov (2006, p. 115)

Table 3.10 Composition of mountain and foothill soils

Soil type	Area (km^2)	Altitude (m)	Humus content (%)	Locations
Light sierozem	25.90	250–400	2	Foothills of the Mirzacho'l steppe; alluvial fans of the Fergana Valley; upper terraces of the Zerafshan, Kashkadaryo, and Surxondaryo rivers
Ordinary sierozem	30.50	300–600	1.5–2.5	Foothills, plains, lowlands
Dark sierozem	10.50	600–1400	3–4	Mountain slopes
Brown and gray in mountain forests	16.60	1000–3000	8–10	Pasture zones
Light gray	5.40	2500–2600		Summer pasture zones

Compiled by the author, using data from Khasanov and Gulomov (2006, pp. 117–125)

Soils in Mountains and in Foothills

In the mountains and foothills, the soil distribution is strongly dependent on the altitude. It results in formation of not just different types of soil but also different soil composition. The natural conditions (precipitation, vegetation, temperature, evaporation) change at different elevations, and these changes are demonstrated by changes in humus in different parts of the mountains and foothills. The higher the mountains are, the more precipitation and vegetation contribute to formation of soil with higher levels of humus.

Sierozem soils are identified on the loess plains in the foothills. This type of soil is considered a soil of the mountain zone. This soil changes in accordance with changes in the natural conditions in vertical belts. Sierozem soil has been identified in the Chirchiq–Ahangaran basin at an altitude of 250–400 m and in the southern mountains at an altitude of 1500–1600 m. This type of soil is rich in carbonates, which increases its value for agriculture. Its humus may reach a depth of 15–18 cm. Since the groundwater is quite deep, these soils are not saline (Table 3.10).

3.4.2 Agricultural Land Resources and Food

Uneven distribution of water resources throughout the country and between seasons limits the use of land for irrigated agriculture. The permanent pasture area is 235,000 km^2, or over 80% of all agricultural land. Only about 10% (4300 km^2) of the total area (448,978 km^2) is irrigated, and the arable land area is 3300 km^2. According to data from the Food and Agriculture Organization of the United Nations (FAO) (2017), the composition of the soils in the irrigated agricultural areas depends on the locations of the major feeding rivers. In regions in the lower reaches, takir and gray soil constitute between 7% and 18% of the total area, but they are absent in the upper reaches. Typically, sierozem accounts for 30% of the soils in the middle and upper reaches but is almost absent in the lower reaches.

According to Khasanov and Gulomov (2006) and the FAO (2003), Uzbekistan's soils in the irrigated areas are spread over three climatic zones, each characterized by different temperatures, groundwater, and salinity levels. Such soils have hard and loamy textures in the northernmost of these climatic zones, found in Karakalpakstan and the Xorazm region. But irrigated soils in the Tashkent foothills and in the Samarkand region belong to the sierozem and sierozem–pasture types, and they display low salinity content.

Clays and pasture soils in the central climatic zone have low salinity too; moreover, they have favorable hydrophysical properties. These occur in Fergana Valley and in the Tashkent and Samarkand regions, where the temperature sum represents accumulated heat energy over a year that exceeds 4000 °C. In Uzbekistan, according to Babushkin L.N. (2004), 10 agroclimatic zones are defined depending on the annual sum of temperatures which varies from 3,500 °C to 5,000 °C depending on geographic location. In a subzone of this central climatic zone, with 240–270 mm of precipitation per year, the irrigated soils consist of nonsaline clays, clay–pasture, and meadow types. These occur in the Sirdaryo, Jizzakh, and Samarkand regions. In the latter two regions and in the Kashkadarya region, with even less (220–240 mm) precipitation and a higher sum of annual temperatures (4200–4400 °C), clays and nonsaline or low-salinity pasture soils occur as well.

In the even hotter (average annual sum of temperatures 4600–5000 °C) and dryer (annual precipitation 70–180 mm) southern climatic zone, mostly (light) clay and desert soils occur in the Kashkadarya and south Bukhara regions, and local clay–pasture is found in old irrigated, slightly alkaline soils with less annual precipitation (80–120 mm), and in irrigated soils where extremely dry conditions prevail, with annual precipitation rates of 70–100 mm (FAO 2003).

The prevailing soil is sierozem in rain-fed agricultural areas (3415 km^2), although old irrigated meadow (427 km^2), marsh alluvium (461 km^2), light sierozem (275 km^2), and gray meadow (208 km^2) also occur in irrigated agricultural areas (1526 km^2).

These soils constitute 90% of the total area in the lower reaches (in the Karakalpakstan Republic, Xorazm, and parts of the Bukhara and Navoiy regions) and the central reaches (in the Samarkand, Surxondaryo, Kashkadarya, and Jizzakh regions).

In the upper reaches, the irrigated agricultural areas cover 12,020 km^2, and sierozem and pasture–marsh–marsh alluvium comprise 97% of the total area (Khasanov and Gulomov 2006; FAO 2003, 2017).

According to an estimate by the FAO (2017), the arable land resources available for further cultivation are limited. During 1988–2017, they decreased from 48,540 to 47,700 km^2, or from 10.85% to 10.66%. The agricultural land area decreased from 27,724 km^2 in 1992 to 26,770 km^2 in 2014 (FAO 2017).

The rapidly growing population and limited land and water resources available for intensive use pose problems for food security. According to FAO (2017) data, the average protein supply has not changed; it was 80 g per capita per day in 1990 and in 2015, and 1.6 million people (5%) are undernourished. The cost of food is rising. The average value of food production (in constant 2011 dollars per person) increased from 228 to 358, the depth of the food deficit increased from 20 to 29 kCal per capita per day, and the GDP per capita (at purchasing power parity [PPP] in constant 2011 international dollars) increased from $3031 to $5316.90.

The value added to the GDP by agriculture is decreasing: for comparison, in 1988–1992 it was 34.83%, and in 2013–2017 it was 18.26%.

Researchers (Schroeder, Meyers 2016) note that Uzbekistan does not import agricultural products but has a low ranking in regard to the average dietary energy supply adequacy indicator and the percentage of the population with access to improved water sources.

3.5 Water Resources

Water is the major issue for discussion, tensions, and cooperation in the region. The transformation of the Central Asian states after the collapse of the Soviet Union created challenges regarding water access and water resource management. In the Soviet era, the state control system maintained the distribution of water resources between states. The political decentralization impacted water management negatively. The difference in the priorities of the upstream and downstream states added more tensions over water resources. Access to water has become crucial, especially for irrigated agriculture in Uzbekistan, where the demand for water is huge.

In 2017, the government of Uzbekistan proposed joint activities for building hydropower plants in neighboring Kyrgyzstan and Tajikistan to fulfill their interests in increasing access to electricity. This initiative for joint development of Eurasian hydroelectric resources, alongside a policy of increasing water efficiency, may be as helpful for regional stability as it is for development of the regional mechanism of water management.

In spite of the dry climate, a lot of rivers and lakes feed the lands of Uzbekistan with water. However, they are unevenly distributed over the country, with low volumes in the main part of western Uzbekistan (Fig. 3.45). This difference in distribution is caused by differences in climate, hydrology, and topography between eastern and western Uzbekistan. Precipitation is unevenly distributed over the country and over the seasons as well. The rain off is higher in the winter and the spring. Summer is the driest season, when many rivers dry up.

The majority of the rivers (about 90%) start in the high mountains of the neighboring countries Kyrgyzstan and Tajikistan. The volumes of ice and snow in these mountains play a big role in the water balance too. The tributaries located in Uzbekistan contribute only 6% to the volume of the Amu Darya river and 15% to the Syr Darya (Khasanov and Gulomov 2006).

The continental dry arid climate and the orography of the place create challenges for preserving water in sandy deserts and on plains.

3.5.1 Rivers

This country possesses a variety of rivers, creeks, and streams in the eastern mountainous area. In Uzbekistan, most rivers start in glaciers in the western and the southern–western upper slopes of the Piskom and Gissar Mountains at an altitude of 3500 m or greater. These rivers flow down the slopes and bring valuable loess to the plains. The loess stays suspended in the water for a long time, making the water muddy and brown. The water resources are complemented by rivers in the intermountain depressions and plains in the foothills of the surrounding mountains. These are intensively used for irrigation in the fertile intermountain depressions via a dense network of canals. The high level of solar radiation, dry and hot air in summer, little precipitation, and the flat surface and

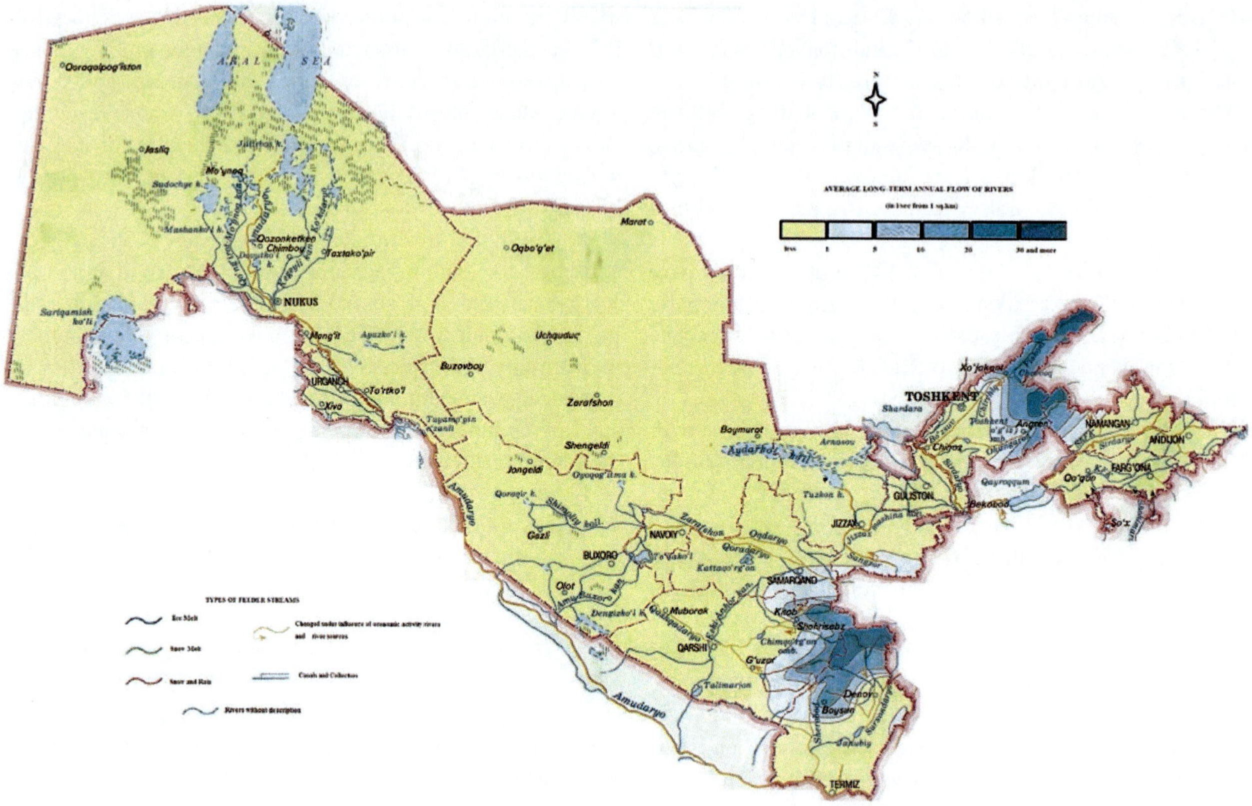

Fig. 3.45 Map of the main rivers in Uzbekistan. (Source: Atlas 2017, courtesy of the State Committee for Land Resources, Geodesy, Cartography and State Cadastre of the Republic of Uzbekistan)

sandy soils of the plains do not create natural conditions for flowing water. All rivers have their own life cycles, depending on their sources. Large rivers such as the Amu Darya and Syr Darya have their sources in glaciers and do not dry up in the hot summer. Because of high evaporation, the majority of the secondary rivers lose water and even dry up or disappear in the sandy desert; thus, they do not reach the bigger rivers. In intermountain clay alluvial areas and on alluvial delta plains, some temporary rivers do flow in the rainy season, but they dry up in the summer.

In accordance with the classification suggested by the hydrologist V. L. Shults (1949), all of the rivers can be grouped into four main groups based on the locations of their headwaters and the sources that feed them: (1) glaciers and snow, (2) snow and glaciers, (3) snow, and (4) snow and rain (Table 3.11).

The rivers in the first group have their sources in mountains outside Uzbekistan. This group includes the main rivers of Uzbekistan: the Amu Darya, Zerafshan, Isfayramsoy, Oqsuv, Xojabaqirgan, Sox, and Isfara Rivers. The rivers of this type have a stable water balance; the highest volume (30–38% of the annual volume) is in July–August during the most intensive snow- and ice-melting period. All of them have sources in the high mountains of Pamir and Alay (Table 3.11).

Table 3.11 Main rivers in Uzbekistan

River group and main rivers	Sources	Altitude (m)	Period of high volume
Group 1 rivers			
Amu Darya	Glaciers	4500	July–August
Zerafshan			
Isfayramsoy			
Oqsuv			
Xojabaqirgan			
Sox			
Isfara			
Group 2 rivers			
Syr Darya	Seasonal snow and rain	–	May–June
Zerafshan			
Narin			
Qurshob			
Qoradarya			
Chirchiq			
Surxondaryo			
Group 3 rivers			
Kashkadarya	Seasonal snow	<3400	April–May
G'uzordarya			
Sangzor			
G'ovasoy			
Group 4 rivers			
Kalas	Seasonal snow and rain	<2000	March–April
Zaminsuv			
Sheroboddarya			

Compiled by the author, using data from Khasanov and Gulomov (2006, pp. 92–93)

The second group of rivers includes the Syr Darya, Narin, Qurshob, Qoradarya, Chirchiq, and Surxondaryo rivers, which are mainly formed from seasonal snow and rain in the mountains of Kyrgyzstan and Tajikistan. The main volume of 30–40% is seen in May–June (Table 3.11).

The third group of rivers includes the Kashkadarya, G'uzordarya, Sangzor, and G'ovasoy Rivers, which are formed by seasonal snow at altitudes of less than 3400 m in the Gissar Mountains. Sixty percent of their volume is seen in April–May (Table 3.11).

The fourth group of rivers includes the Kalas River, the Zaminsuv River, and the Sheroboddarya, which are formed by seasonal snow and rain at altitudes of less than 2000 m. Eighty percent of their volume is seen in March–April, and in summer, some of them dry up (Table 3.11).

In addition to the main rivers, a lot of smaller streams flow in the mountains. More than 6500 such streams are found in the Fergana Valley and more than 120 in the watershed of the Zerafshan River. These are formed by snow–rain and groundwater at altitudes of 1000–2000 m. The main volume of 50–100% is seen in February–June; later in the year, they have less volume or may even dry up.

The average annual flow of the major part of the rivers on the plains is less than 1 l/s from an area of 1 km^2. It is only in small areas in the high mountains of the Western Tien Shan and Gissar that the flow may be more than 30 l/s from 1 km^2. Climate change is causing intensive ice and snow melting and decreasing the debit of the main rivers. According to estimates (Semakova et al. 2015), the areas of glaciation at selected sites in the basins of the Kashkadarya, the Surxondaryo, and the Piskom River decreased by 20–50% during the past 50 years. Shrinking of glaciers results in significant decreases in water flow. Researchers (Yunusov et al. 2017) note that between 1980 and 2016, the water intake from the Amu Darya and the Zerafshan River for irrigation of lands in the Kashkadarya oasis decreased by 36×10^6 m^3/year.

The Amu Darya River

In Central Asia, the most famous and from the perspective of abounding in water the biggest river is Amu Darya. This river is 2,540 km long, including 1,000 km within Uzbekistan. It is formed by the junction of the Vakhsh and Pyanj Rivers in the Pamir Mountains. From this junction, the Amu Darya flows to the northwest into the southern remnants of the Aral Sea. At the start, it is named the Vahjir; after the conjunction with the Pamir River, it changes its name to the Pyanj, and after the conjunction with the Vahsh, it is named the Amu Darya. Several tributaries join the Amu Darya in its upper flow: the major ones are the Kafirnigan in Tajikistan and the Surxondaryo in Uzbekistan, from the north, and the Kunduz in Afghanistan, from the south. On the plain, no tributary joins the Amu Darya. In ancient times, the Amu Darya was named the O'kuz, Chorjo'y, Omuya, Oxus, O'x, Jayhun, and Omul. The flow is strongest in summer and weakest in January and February. Evaporation, filtering, and diversion of water for irrigation cause significant loss of its flow (Fig. 3.46).

The drainage basin of the Amu Darya river is 534,739 km^2 in area. The mean discharge is about 97.4 km^3 of water annu-

Fig. 3.46 The Amu Darya river in its lower flow near Urgench, Khorezm region

Fig. 3.47 The Syr Darya river at the border between Uzbekistan and Kazakhstan

Fig. 3.48 The glaciers of the Piskom Ridge in the Western Tien Shan. (Source: Public domain under Creative Commons CC0 via StocksSnap)

ally. The existence of the Amu Darya is dependent on glaciers in the Pamir and Tien Shan Mountains. The river flows mostly in lowlands where it rarely rains. Only 200,000 km^2 of the watershed contributes to the river. The high evaporation, plain surface, and low level of precipitation on the steppe and in the desert cause loss of water.

The Syr Darya River

The longest river in Central Asia, the Syr Darya (2800 km long, including 2000 km in Uzbekistan) starts in a glacier named after Petrov on Oqshiroq Mountain in the Central Tien Shan. At the beginning, it is named the Narin, and after conjunction with the Qoradaryo, its name changes to the Syr Darya. In the Fergana Valley, many tributaries are used for irrigation and thus do not reach this river. The Syr Darya has tributaries after leaving the Fergana Valley from the east side of the flow, and they are the Ahangaran, Chirchiq, Kalas, and Aris Rivers. It does not have any tributary from the south of its flow. Its fullest flow is in spring and summer from April onward, and the highest flow is in June (Fig. 3.47).

3.5.2 Glaciers

It was estimated by Khasanov and Gulomov (2006) that the annual precipitation on the slopes of the high mountains is about 1000 mm, and they are the main sources of glacier formation at altitudes of 3000–3500 m above sea level. In Uzbekistan, the rivers start at 547 small glaciers with a total area of 332.2 km^2. The biggest of them—the Ayutur-3, measuring 3.8 km^2 in area—is in the Tashkent region. Glaciers contribute up to 70% of the water flow in the major river systems of the Central Asian region (Fig. 3.48).

Climate change and air pollution have resulted in decreases in the water reserves held in glaciers. According to estimates (Kutuzov 2009), the total surface area of the glaciers in the region decreased by 19% between 1990 and 2003. One study (Hagg and Mayer 2016) estimated that by 2025, small glaciers may disappear and the total surface area of the glaciers may shrink by 25%. This study also revealed that as a result of global warming, the glacial resources in the region declined by almost 30% in the twentieth century, with high potential for a further decrease in the annual stream flow. Abdullaev and Akbarzadeh (2010) noted that the Fedchenko Glacier, which is the longest in the region (measuring 70 km in length in the twentieth century), Since 1933 retreated by almost 1 km, with its surface area decreasing by 11 km^2, losing a total of about 2 km^3 of ice. Water–energy forecast scenarios point to further melting of glaciers and consequent changes in runoff, with small (5–10%) to very substantial (10–40%) effects in the long run. Experts (EC IFAS 2011) predict that by 2050, the runoff of the Amu Darya river will decline by 10–15% and that of the Syr Darya will decline by 6–10%. Glacial melting is resulting in serious social, environmental, and economic challenges for the whole region by reducing water resources and posing the threat of desertification.

3.5.3 Lakes and Reservoirs

The tradition of construction of reservoirs (in the form of a *sardoba*—a small water reservoir for gathering winter and spring rainfall and runoff water [*hauz*]) has existed for many generations. Major reservoirs were constructed in the twentieth century for irrigation of cotton fields. In Uzbekistan, there are 55 man-made water reservoirs with a total gross volume capacity of 19 km^3 and a surface area of 1450 km^2. Today, all possible sites for reservoir construction have already been used. Scientists note that reservoirs have lost their dead storage capacities because of sedimentation, and in the near future, they may lose their functionality. In 1969, the Arnasai topographic depression became a huge permanent lake after a catastrophic flood. It comprises the Aidarkul, Tuzkan, and Upper Arnasai, with a total area of about 3491 km^2 (UNDP 2007).

About 80 dams have been created on the rivers; the major ones are at Qorataka, Kukkul, and Badak in the watershed of the Chirchiq River. Only two dams are used for energy production (those at Andijon and Charvak); the others are used only for irrigation purposes. Dams such as the Ikhnoch, Shovurkul, Qorataka, and Katta dams pose a flood risk. Floods in the foothills occur mainly in deforested mountain areas during the rainy season in April, May, and June. It has been estimated (Khasanov and Gulomov 2006) that the flooding may cover a foothill area as big as 93,500 km^2.

The existing lakes are unevenly distributed, and they are primarily of small size in valleys, deltas, and intermountain depressions. In the mountains, they are located mainly at altitudes of 2000–3000 m above sea level. There are 21 lakes in the mountains of the Fergana Valley, 20 in the watershed of the Chirchiq River, and 16 in the watershed of the Zerafshan River.

Several lakes larger than 200 km^2 are located on the plains: (1) the Aral Sea, (2) Arnasay (2000 km^2 in size), (3) Aydar (1309 km^2), (4) Tuzkon (515.8 km^2), (5) Sariqamish (28,500 km^2), (6) Sudochye (337 km^2), and (7) Dengizkul (267 km^2).

Arnasoy, Aydar, and Tuzkon are in the watershed of the Syr Darya river; Sarikamish and Sudochye are in the watershed of the Amu Darya river; and Denizkul is in the watershed of the Zerafshan River. The genesis of these lakes (Khasanov and Gulomov 2006) is due mainly to water erosion and accumulation processes. Aydarkul and Sudochye are artificial reservoirs, huge drainage basins. In the mountains, lakes originate from tectonic and glacial processes as a result of movement of rocks down the slopes. Man-made mountain reservoirs and natural lakes can pose serious threats.

According to the FAO (2016), there are at least 50 reservoirs in Uzbekistan with a total capacity of over 22 km^3. In the Syr Darya basin, the largest reservoirs (Charvak and Andijon) are multipurpose dams used for irrigation, flood control, and hydropower production. The Charvak reservoir, which hosts one of the largest hydropower plants in Central Asia, is on the Chirchiq River and has a storage capacity of 1.99 km^3 and power generation capacity of 600 MW. The Andijan reservoir on the Karadaryo river in the Fergana Valley has a storage capacity of 1.9 km^3. In the Amu Darya basin, the largest reservoir is the Tuaymuyun in Xorazm province, with a storage capacity of 7.8 km^3, comprising four separate reservoirs. One reservoir in this system (Kaparas) provides drinking water for the Karakalpakstan area, which is experiencing severe environmental problems as a result of the shrinkage of the Aral Sea. Most of the reservoirs were built more than 25 years ago. Since then, almost all of them have been exposed to siltation and lost up to 20–25% of their useful capacity.

Aral Sea

One of the four largest lakes in the world, this sea has been shrinking since the 1960s. Now it is just 10% of its original size. Today, instead of one lake with an area of 68,000 km^2, there are several lakes of smaller sizes. The eastern basin of the Aral Sea has dried up completely and is now named the Aralkum Desert.

It is not entirely clear how this sea was originally formed, and many different theories have been proposed to explain the time of its creation and the history of its life. A relic of the Turkestan Ocean, the Aral Sea depression was formed at the end of the Neogene Period. During this period, which lasted from 23 to 2.6 Ma, this hollow was filled in with water brought by the Syr Darya. Formation of the Aral Sea began about 5.5 Ma. In the late Pleistocene Period, (1–2 Mya) and the early Holocene Period, the Amu Darya changed its course from the Caspian Sea to the Aral Sea. Before that, the Amu Darya flowed through central Karakum; its flow through Uzboi to the Caspian Sea lasted from the tenth to the second millennia BCE (Zonn and Kostianoy 2013). Some researchers have suggested that a change in the flow of the main Amu Darya feeder into the shallow depression occurred after the beginning of the Holocene Period (Middleton 2002). Not all scientists agree with the assumption that a "Great Aral Sea" covered the whole area of the Sarygamysh depression and the Aral Sea in the fifth and sixth millennia BCE. Some researchers (Zonn and Kostianoy 2013) have assumed that the Amu Darya flow went into the Sarygamysh until the middle of the first millennium BCE.

Over time, the whole regional hydrographical system changed, and the Aral Sea has risen and fallen several times (Fig. 3.49). This process may be compared with the geological evolution of the Caucasus when the rhythmic fluctuation of the Black Sea and the Caspian Sea changed the landscape and nature many times (Bondyrev et al. 2015). The maximum regression of the Black Sea 1.7 Ma and the separation of the Caspian and Black Seas occurred during the Early Pleistocene Period, while in the Middle Pleistocene Period, they were connected to each other. Similar rhythmic fluctuations in the Aral Sea basin have been described in many works (Zonn and Kostianoy 2013; Hikmatov and Aytbaev 2002). Hydrologists (Hikmatov and Aytbaev 2002) note that in 1785, 1835–1850, and 1885, the sea level rose several times, and in 1825 and 1862, the sea level fell. Unfortunately, no comparative data regarding the evolutions of these three basins are available.

The Aral Sea is located in a dry continental climate zone, which features cold winters, hot summers, high levels of evaporation, and sparse rainfall. The precipitation (totaling

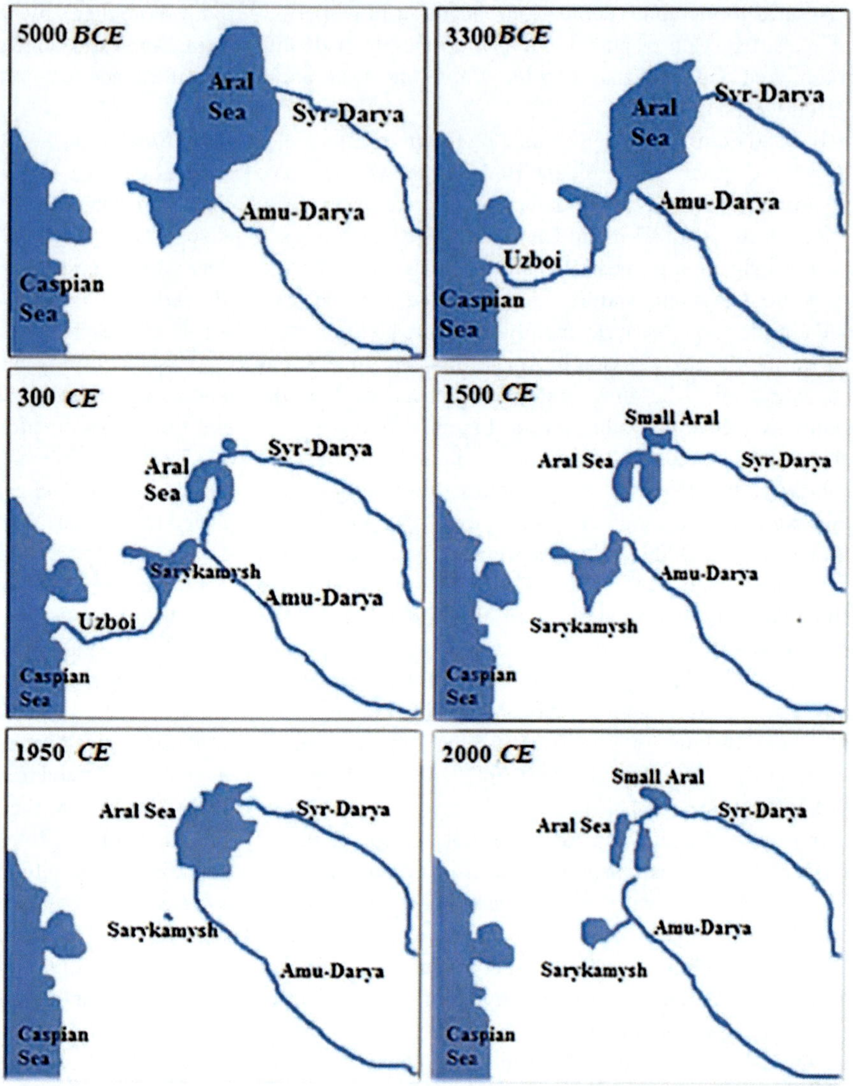

Fig. 3.49 Development of the hydrographical systems of Central Asia. (Reproduced from Zonn and Kostianoy 2013)

about 100 mm) in the spring and autumn is insufficient to replace the amount of water that evaporates, and this leads to a negative water balance in the sea and sensitivity to the amount of water that the Amu Darya and the Syr Darya bring in. Over the centuries, annual water level variations (of about 6 m) and seasonal water level variations (of 0.5–3 m) have been recorded (Hikmatov and Aytbaev 2002). The bed of the sea is inclined from east to west. The depth in the eastern part was previously 16 m, descending to a maximum of 69 m off the western shore.

The amount of water delivered by the rivers that reached the Aral Sea was dropping in the 1960s (Hikmatov and Aytbaev 2002). By the 1980s, the two rivers had dried up before they reached the Aral Sea. Within one decade, the sea receded to form two separate parts containing highly saline water. In 1992, the total area of the Greater Aral (the southern part) and the Small Aral (the northern part) had been reduced to 33,800 km^2 (Walters 2010). In spite of measures undertaken by the governments of the states surrounding the sea, to minimize use of water from the Amu Darya and Syr Darya rivers and to stabilize the water level of the Aral Sea, the results were not promising. The reduction in water use did not have a significant impact on the amount of water reaching the sea. At the beginning of the twenty-first century, the Aral Sea had receded into three separate lakes. The Greater Aral had divided into a long, narrow western lake and a larger, broader eastern lake. The water level had dropped to 36 m above sea level, and the water volume was reduced by three-fourths of what it had been in 1960. The largest retreat of the Southern Aral Sea occurred between 2005 and 2009. The eastern portion of the sea suffered the most drastic and immediate decline, diminishing by about

Fig. 3.50 The Aral Sea in 1998 (*left*) and in 2014 (*right*). (Source: NASA Earth Observatory (public domain))

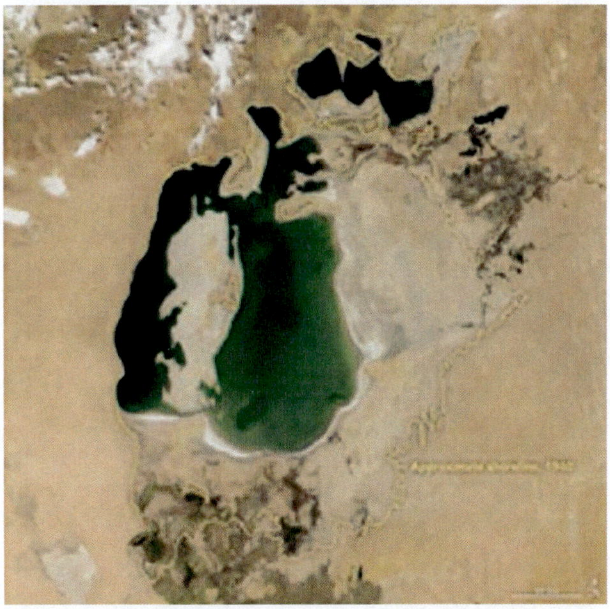

Fig. 3.51 The Aral Sea on August 25, 2000. (Source: NASA Earth Observatory (public domain))

four-fifths of its size. The water levels fluctuated annually between 2009 and 2016 in alternately dry and wet years (Micklin 2007; Micklin et al. 2014). In 2014, the eastern lobe of the Southern Aral Sea disappeared completely (Figs. 3.50, 3.51, 3.52, and 3.53).

Fig. 3.52 The Aral Sea on August 21, 2016. (Source: NASA Earth Observatory (public domain))

Fig. 3.53 Satellite image of the Aral Sea in 2010. (Source: NASA Earth Observatory (public domain))

Not all secrets of the Aral Sea have been fully revealed. The tragedy of the Aral Sea is the result of combined natural and human-induced processes. The main reason for the drying of the Aral Sea after the 1960s has been the change in river directions for water-intensive cotton agriculture and irrigation purposes during Soviet times. Construction of poorly built canals for irrigation led to water leaks and evaporation in the dry arid climate. It is estimated (UNECE 2005) that in some places of the sandy Karakum Desert, 30–75% of the water went to waste. One researcher (Thomas 2003) noted that in 1960–2000, the amount of water taken for cultivation of water-consuming cotton doubled. The nearly fivefold increase in the salinity of the water killed the natural flora and fauna, and this is the saddest part of the history of the Aral Sea. One study (Thomas 2011) revealed that by 2007, the salinity of the South Aral (the part of the sea located in Uzbekistan) reached 100 g/l. According to one study (Micklin 2007; Micklin et al. 2014) between 1960 and 2006, the area of the Greater Aral (located in Uzbekistan) and the Small Aral (located in Kazakhstan) diminished by 74% and the volume by 90%. Since 1960, when the commercial fishing catch exceeded 43,000 t, the Aral Sea has lost as much as 88% of its surface area and 92% of its volume. In 1996, only 547 t of fish were caught, much of it contaminated with pesticides. Meanwhile, other researchers (White 2014; Luxner 2015) found that the salinity had jumped from 10 parts per thousand (essentially freshwater) in 1960 to 92 parts per thousand (some three times the salinity of most oceans) in 2004.

Several remedial projects have now been implemented in the Northern Aral Sea. The Syr Darya Control and Northern Aral Sea Project (SYNAS-1), which aims to reduce water losses in the Kazakh part of the Aral Sea, has helped increase its volume by 68%. One report (World Bank 2014) points out that this change has halved the water salinity and allowed more freshwater to flow there, tripling fish production and improving the flora and fauna conditions, as well as the economy and the environment.

Construction of the 13-km Kok-Aral Dam on the Kazakhstan side of the sea was finished in 2005. This project has helped save the North Aral, enlarging its area. Luxner (2015) wrote that in 2016, the waters were coming closer to the city of Aralsk in Kazakhstan; the distance to the shoreline had decreased from 100 km (in 2000) to 20 km (in 2016).

Images released by the US National Aeronautics and Space Administration (NASA) in October 2014 showed that the eastern basin of the South Aral Sea was dry. Many experts (Micklin et al. 2014; World Bank 2014) are unanimous that attempts to save the entire Aral Sea are unrealistic because funding for such large investments are not available and closing irrigation systems would create even bigger economic and social hardships than the Aral Sea crisis itself ever did.

Some ancient irrigation systems and mausoleums surrounded by settlements were found on the bed of the Aral Sea, providing evidence that during the thirteenth and fourteenth centuries, the sea level might have been even lower than it is today. Scientists (Mackay 2014) suggest that by the late sixteenth century, it started to fill up again. The impacts of irrigation are compounded by climate change and an unfavorable water balance, making the Aral Sea very sensitive to all changes caused by nature and by humans.

3.5.4 Environmental Hazards

Environmental Consequences of the Drying of the Aral Sea The drying of the Aral Sea has resulted in damage to the environment and caused social, ecological, and economic challenges. According to research (Lindsey 2016), the salinity of the soil and water, in combination with the blowing of dust from the exposed seabed, is a major public health hazard. The high salinity and mineralization of the water has made it unfit for drinking purposes and killed off the once abundant supplies of sturgeon, carp, barbell, roach, and other fish. People who previously worked in fishing have lost their jobs, since the seaports are now far from the shore. The climate has become noticeably harsher, with extreme temperatures in summer and in winter. Winds blow across the exposed seabed and spread toxic dust contaminated with salt, fertilizer, and pesticides (Fig. 3.54).

Fig. 3.54 Satellite image of a plume of dust from the dried seabed of the Aral Sea. (Source: NASA Earth Observatory (public domain))

Fig. 3.55 The Aral Kum Desert, formerly the river delta of the Amu Darya

Fig. 3.56 Saline soil in the Amu Darya delta, Khorezm region

Fig. 3.57 Saline soil in the Amu Darya delta, Khorezm region

According to one study (UN 2004), inefficient management and intensive use of water and energy resources have impacted nature, resulting in pollution, degradation of soils, deterioration of biodiversity, and diminishing productivity of all natural resources (land, water, air, and mineral reserves). The nature of Uzbekistan is extremely sensitive to changes at the global, regional, and local levels. Efficient management of natural resources is a crucial issue as global warming, in combination with geological processes and human activities, creates new challenges. The interconnections between human activity and nature are not yet well understood by decision-makers and the public. The risk of water shortages as a result of melting of glaciers and shrinkage of snowfields and alpine woodlands adds such challenges as mud flows, floods, and land erosion. Today, we are witnessing increases in salinity, desertification of land (Fig. 3.55), increased mineralization, and toxic pollution of land and water. The most dangerous prospect is an imbalance of the whole environment, causing threats to humans and to flora and fauna, and bringing about irreversible adverse consequences.

The growth in the population and the increasing demand for water raise concerns regarding water quality. Increased diversion of water and discharges of polluted return flow into rivers and reservoirs create new challenge in water quality balance. Poor and obsolete sewerage systems in households and at municipal and industrial waste sites are sources for increased contamination of groundwater everywhere in Uzbekistan.

Large-scale development of irrigation and other types of use have changed regional hydrological cycles. The results are the reduction of the Aral Sea, destruction of its ecosystems, loss of fish production, and desertification of estuaries and the seabed, with negative impacts such as diminished water quality, public health effects of salt and dust transportation, and local climate changes. One report noted that the degradation of river and groundwater quality, high salinity and waterlogging, desertification of irrigated lands, instability of water levels, mineral content in water bodies, and return flow from irrigation are causing declines in bioproductivity and biodiversity, and threatening the sustainability of the ecosystem.

Salinity of Lands

One of the major challenges that decreases the productivity of irrigated land is salinity of the land, resulting from reclamation of areas naturally exposed to salinization and increased alkalinity of soil water. According to one estimate (Khamraev 2011), the saline irrigated land area is 21,709 km^2, or 50.7% of all irrigated land in the country. These lands are located in the lower reaches of the Amu Darya and Syr Darya basin. From 1995 to 2004, in Jizzakh, Kashkadarya, Namangan, Sirdaryo, Fergana, and Xorazm provinces, the proportion of moderately and highly saline lands increased by 14.4%. One researcher (Khamraev 2011) has expressed concern that in Xorazm province, because of a high groundwater level, the total area of saline land exceeds 80% of the total land area with low salinity (less than 3 g/l) (Figs. 3.56 and 3.57). The main reasons for the increase in the ground-

Fig. 3.58 Damage to construction is caused by a rise in saline groundwater in the city of Khiva, Khorezm region

water level are large water losses from canals and fields, caused by poor land planning, irrational use (losses) of water in the fields, and inadequate water diversion from irrigated lands. The high level of salinity and waterlogging are leading to productivity loss and an ecologically dangerous situation (Fig. 3.58).

Scientists (Madramootoo and Dukhovny 2008) have pointed out the need for development of a mechanism of investment or subsidized government loans (or loans from other sponsors)—as well as legal and financial mechanisms and technical support—to implement water-saving irrigation technologies.

References[14]

Abdullaev K, Akbarzadeh Sh (2010) Historical Dictionary of Tajikistan Publisher Scarecrow Press ISBN10:0810860619. Retrieved March 12 2018 from https://books.google.co.uz/books?id=OslIDwAAQBAJ&pg=PA145&lpg=PA145&dq=The+biggest+Fedchenko+losing+a+total+of+about+2+cubic+km+of+ice&source=bl&ots=vljVjiJxt&sig=ACfU3U1zvNM6wKc8dwvwifujIQN8etdX5A&hl=en&sa=X&ved=2ahUKEwjWwr_o9ZXhAhVDi8MKHTpZDGQQ6AEwAHoECAgQAQ#v=onepage&q=The/biggest/Fedchenko/losing/a/total/of/about/2/cubic/km/of/ice&f=false

Abdullaev DA, Isaev RI (2005) Resources of solar radiation and wind energies in Uzbekistan and system of their combined utilization. ISESCO Sci Technol Vision 1:76–82. http://www.isesco.org.ma/ISESCO_Technology_Vision/NUM01/doc/D.A.Abdullaev.pdf

Abzalov M (2007) Zarmitan Granitoid-Hosted Gold Deposit Tien Shan Belt, Uzbekistan. Econ Geol 102(3):519–532. https://doi.org/10.2113/gsecongeo.102.3.519. Derived from https://www.researchgate.net/publication/247864305_Zarmitan_granitoid-hosted_gold_deposit_Tian_Shan_Belt_Uzbekistan

Aminjonov F (2016) Alternative energy sources of Uzbekistan. Eurasian Research Institute. Akhmet Yassawi University. Accessible from May 30, 2016 at http://eurasian-research.org/en/research/comments/energy/alternative-energy-sources-uzbekistan

Aripdjanov U, Adilov R (2017) Uzbekistan – Energy. J Global Legal Insights. Accessed on July 12, 2019 from https://www.globallegalinsights.com/practice-areas/energy/global-legal-insights%2D%2D-energy-5th-ed./Uzbekistan

Artikov TA, Ibragimov RS, Ibragimova TL, Mirzaev MA (2020) Complex of general seismic zoning maps OSR-2017 of Uzbekistan in J Geodesy Geodynamics V 11(I4):273–292. DOI: https://doi.org/10.1016/j.geog.2020.03.004

Atlas The Physical Geography of Central Asia and Uzbekistan (2017) Yergeodezkadastr state committee Tashkent (In Uzbek O'rta Osiyo va O'zbekiston tabiiy geografiyasi Yergeodezkadastr Davlat qomitasi Toshkent 2017)

Bazhenov ML, Chauvin A, Audibert M, Levashova NM (1993) Permian and Triassic paleo magnetism of the southwestern Tien Shan; timing and mode of tectonic rotation. Earth Planet Sci Lett 118:195–212

Berzin NA, Coleman RG, Dobretsov NL, Zonenshain LP, Xuchan X, Chang EZ (1994) Geodynamic map of the western Paleoasian Ocean. Geologiya I Geofizika 35:8–29. (in Russian)

Biske YS, Seltman R (2010) Paleozoic Tian-Shan as a transitional Region between the Rheic and Urals-Turkestan Oceans. Gondwana Res. https://doi.org/10.1016/j.gr.2009.11.014. https://www.academia.edu/9647516/Paleozoic_Tian-Shan_as_a_transitional_region_between_the_Rheic_and_Urals-Turkestan_oceans

Bolt BA (2016) Earthquake. Geology. In: Encyclopedia Britannica https://www.britannica.com/science/earthquake-geology#ref837428

Bondyrev IV, Davitashvili ZV, Singh VP (2015) The geography of Georgia. Problems and perspectives, World regional geography book series. Springer

BP Statistical Review of World Energy (2016) Accessible June 2016 from https://www.bp.com/content/dam/bp/pdf/energy-economics/statistical-review-2016/bp-statistical-review-of-world-energy-2016-full-report.pdf

Buslov MM, Safonova IY, Watanabe T, Obut OT, Fujiwara Y, Iwata K, Semakov NN, Sugai Y, Smirnova LV, Kazansky YA, Itaya T (2001) Evolution of the Paleo-Asian Ocean (Altai-Sayan region, Central Asia) and collision of possible Gondwana-derived terrains with the southern marginal part of the Siberian continent. J Geosci 5:203–224

Cabinet of Ministers of the Republic of Uzbekistan. The State Inspection on Supervision for Electric power Industry "O'ZDAVENERGONAZORAT" portal. Current state and development prospects for the energy sector of Uzbekistan. Accessible from http://www.energonazorat.uz/en/energy-saving/ozbekiston-energetikasining-zamonaviy-ahvoli-va-taraqqiyotining-istiqbollari.html

Carey K (2009) Groundwater. Retrieved August 12, 2018 from https://www.lakeheadu.ca/sites/default/files/uploads/54/Geology/3015/Week/8b.pdf

Chaterjee ShP, Bishop B (2015) Himalayas (mountains, Asia). In: Encyclopaedia Britannica. https://www.britannica.com/place/Himalayas

Coal Industry (2018) In JSC Uzbekenergo. Retrieved August 20, 2018 from http://www.uzbekenergo.uz/en/activities/coal-industry/

Cole A, Seltmann R (2000) The role of granitoidsduring Variscan orogenic gold mineralization in the Tien Shan and Ural mountain belts of Central Eurasia. In: Bouchot V, Moritz R (eds) Orogenic gold deposits in Europe, editions BRGM, vol 297, pp 110–111

Descamps C, Kanniche M, Bouallou C (2008) Efficiency of an integrated gasification combined cycle (IGCC) power plant including

[14](1966–2016 2016)' with a citation for the following new reference, to be added to the reference list: 'UNDP [United Nations Development Programme] (2016) 1966–2016: 50 years since the Tashkent earthquake. United Nations Office for the Coordination of Humanitarian Affairs. https://reliefweb.int/report/uzbekistan/1966-2016-50-years-tashkent-earthquake. Accessed 8 Jun 2022.

CO_2 removal. Energy 33(6):874–881. Retrieved April 5, 2017 from https://www.researchgate.net/publication/223525623_Efficiency_of_an_Integrated_Gasification_Combined_Cycle_IGCC_power_plant_including_CO2_removal

Dolgopolova A et al (2017) Geodynamic evolution of the western Tien Shan, Uzbekistan: insights from U-Pb SHRIMP geochronology and Sr-Nd-Pb-Hf isotope mapping of granitoids. Gondwana Res 47:76–109. https://doi.org/10.1016/j.gr.2016.10.022

Earthquake Hazards Program (2016) US Geological Survey. Retrieved from http://earthquaketrack.com/p/uzbekistan/recent?mag_filter=5, http://earthquaketrack.com/p/uzbekistan/recent?mag_filter=7

Eshchanov BR, Plaat SMG, Eshchanov RA, Salaev SK (2013) Prospects of renewable energy penetration in Uzbekistan–perception of the Khorezmian people. Renew Sustain Energy Rev 21:789–797

FAO (2003) Food and Agriculture Organization of the United Nations Fertilizer use by crop in Uzbekistan. Rome. Corporate Document Repository Produced by Natural Resources Management and Environment Department. Annex. Soil Types in Uzbekistan. Accessible from http://www.fao.org/docrep/006/Y4711E/y4711e0a.htm

FAO (2016) AQUASTAT website. Food and Agriculture Organization of the United Nations (FAO). Website accessed on March 12, 2017. http://www.fao.org/nr/water/aquastat/countries_regions/Profile_segments/UZB-WR_eng.stm

FAO.AQUASTAT (2017) Retrieved March 20, 2017 from http://www.fao.org/nr/water/aquastat/data/query/results.html

Ferland R (2002) Activities of the international GPS service (IGS) reference frame working group. In: Schwarz KP (ed) Vistas for geodesy in the new millennium. International Association of Geodesy Simposia, vol 125. Springer, Berlin/Heidelberg

Flynn F (2016) Mineral Resources of Uzbekistan and Opportunities for Project Development State Committee of the Republic of Uzbekistan on Geology and Mineral Resources (GoscomGeology) Mineral Resources of Uzbekistan. Retrieved February 04, 2017 from https://slideplayer.com/slide/12741345/

Focus Economics (2016) U.S. Geological Survey and Focus Economics Consensus Forecast. http://www.focuseconomics.com/sites/default/files/wysiwyg_images/focuseconomicscommodities_may2016_0.png

Gansser A, Gruschke A, Olschak BC (1987) Himalayas. Growing mountains, living myths, migrating peoples. Facts on File, Oxford/New York/New Delhi

Gintzburger G, Toderich KN, Mardonov BK, Mahmudov MM (eds) (2003) Rangelands of the Arid and Semi-Arid zones in Uzbekistan. CIRAD-ICARDA, p 432. Accessible from May 13, 2016 https://www.researchgate.net/publication/303014592_Rangelands_of_the_Arid_and_Semi-Arid_zones_in_Uzbekistan

Golonka J (2009) Phanerozoic Paleoenvironment and Paleolithofacies maps: Cenozoic. Geologia 35(4):507–587

Grechko M (2015) Cause of Afghan Quake is a deep mystery. http://news.nationalgeographic.com/2015/10/151026-afghanistan-pakistan-deep-earthquake-faults-geology/#close. Accessed 26 Oct 2015

Hagg W, Mayer C (2016) Water of the Pamir-Potential and Constraints. In: Mapping transition in the pamirs, pp 69–79. https://doi.org/10.1007/978-3-319-23198-3_1

Heath RC (2004) Report. Basic ground-water hydrology. In: Water supply paper series 2020. U.S. Geological Survey. Retrieved June 15 2018 from https://pubs.usgs.gov/wsp/2220/report.pdf, https://doi.org/10.3133/wsp2220

Hendrix MS, Davis G (eds) (2000) Paleozoic and Mesozoic tectonic evolution of Central Asia: from continental assembly to intracontinental deformation. Geological Society of America, Boulder

Heubeck C (2001) Assembly of central Asia during the middle and late Paleozoic. In: Hendrix MC, Davis GA (eds) Paleozoic and Mezozoic tectonic evolution of central Asia: From continental assembly to intra continental deformation, vol 194. Geological Society of America Memoir, Boulder, pp 1–22. https://books.google.co.uz/books/about/Paleozoic_and_Mesozoic_tectonic_evolutio.html?id=OdFfiwV51zUC&redir_esc=y

Hikmatov FH, Aytbaev DP (2002) The Science of Lakes Tashkent (In Uzbek Ҳикматов Ф.Ҳ., Айтбаев Д.П. Кўлшунослик Тошкент:Университет)

Huang H, Zhang Z, Santosh M, Zhang D, Zhao Z, Liu J (2013) Early Paleozoic tectonic evolution of the South Tienshan Collisional Belt: evidence from Geochemistry and Zircon U-Pb geochronology of the Tie'reke Monzonite Pluton, Northwest China. J Geol 121(4):401–424. https://www.jstor.org/stable/pdf/10.1086/670653.pdf?seq=1#page_scan_tab_contents. Accessable 4 July 2013

Index Mundi data portal (2015) Uzbekistan natural resources. Accessed on August 12, 2019. http://www.indexmundi.com/uzbekistan/natural_resources.html

ISciences (2022) Water security indicator model. https://www.isciences.com/water-security-indicator-model. Accessed on July 15, 2022

Jegelevicius L (2015) Uzbekistan's ambitious wind power target signals new energy politics. In: Renewable Energy World, June 11. http://www.renewableenergyworld.com/articles/2015/06/uzbekistan-s-ambitious-wind-power-target-signals-new-energy-politics.html)

Kandym-Khauzak-Shady (2017) In Lukoil portal. Retrieved September 2, 2017 from http://www.lukoil.ru/Business/Upstream/KeyProjects/Kandym-Khauzak-Shady

Kempe U, Graupner T, Seltmann R, de Boorder H, Dolgopolova A, van Emmichhoven MZ (2016) The Muruntau gold deposit (Uzbekistan) – a unique ancient hydrothermal system in Southern Tien Shan. Geosci Front 7(3):495–528. https://doi.org/10.1016/j.gsf.2015.09.005. Accesseable May 2016 from http://www.sciencedirect.com/science/article/pii/S1674987115001139

Khamraev SR (2011) The role of land reclamation in the socio-economic progress in Uzbekistan and government support policies of water management. In: Madramootoo C, Dukhovny V (eds) Water and food security in Central Asia. NATO Science for Peace and Security Series C: Environmental Security. Springer, Dordrecht. https://doi.org/10.1007/978-90-481-9974-7_7

Khasanov IA & Gulomov PN (2006) The physical geography of Uzbekistan Tashkent p.p.297 (In Uzbek Ўзбекистон табиий географияси)

Kious JW, Tilling RI (1996) Understanding Plate Motions: Indian plate. In: This dynamic Earth: The story of plate tectonics. Geological Society of America Memoir, Boulder. http://eurasiatectonics.weebly.com/indian-plate.html

Kious WJ, Tilling RI (2015) This dynamic earth: the story of plate tectonics. General Interest Publication. https://doi.org/10.3133/7000097. Accessed from https://pubs.usgs.gov/gip/dynamic/himalaya.html on November 1, 2018

Lindsey R (2016) Shrinking Aral Sea. Accessible from August 21, 2016 http://earthobservatory.nasa.gov/Features/WorldOfChange/aral_sea.php

Lukoil secures $500-million loan for Uzbekistan Development (2016) Oil & Gas Journal. Accessible from November 16, 2016 from http://www.ogj.com/articles/2016/11/lukoil-secures-500-million-loan-for-uzbekistan-development.html

Luxner L (2015) Reviving the North Aral Sea. Aramco World. Accessible from September/October, 2015 http://www.aramcoworld.com/en-US/Articles/September-2015/Reviving-the-North-Aral-Sea

Mackay AW (2014) Humans drained the Aral Sea once before – but there no Free Refills this Time Around. http://www.rawstory.com/2014/10/humans-drained-the-aral-sea-once-before-but-there-are-no-free-refills-this-time-around/http://theconversation.com/humans-drained-the-aral-sea-once-before-but-there-are-no-free-refills-this-time-round-32513

Madramootoo C, Dukhovny V (eds) (2008) Water and Food Security in Central Asia, NATO science for peace and security series c: environmental security. Springer, Dordrecht

Main Indicators of Coal Industry of O'zbekko'mir JSC (2018) In Open Data Portal of the Republic of Uzbekistan. (In Uzbek:

"O'zbekko'mir" AJ ning 2018-yil uchun ko'mir sanoatining asosiy ko'rsatkichlari (dekabr oyi holatiga) O'zbekiston Respublikasi ochiq ma'lumotlar portal). Retrieved March 15, 2019 from https://data.gov.uz/uz/datasets/1523

Mao J, Konopelko D, Seltmann R, Lehmann B, WenChen, Yitian Wang, Eklund O and Usubaliev T (2004) Postcollisional age of the Kumtor golddeposit and timing of Hercynian events in the TienShan, Kyrgyzstan; Economic Geology v 99, pp 1771–1780

Mavlyanova NG, Machalet B, Takhmatullaev HLO, Hara-Dhand K, Jefferson IF (2006) The formation of loess deposits in the Tashkent region and parts of Central Asia; and problems with irrigation, hydrocollapse and soil erosion. Quat Int 152:59–69. https://doi.org/10.1016/j.quaint.2005.12.002

McHutchon (2017) Plans for moderniztion of the Uzbekistan in 2017–2021 In Eurasian Mining Information portal – February 27, 2017 Retrieved March 12, 2017 from https://www.minexforum.com/en/plans-for-modernization-of-the-uzbekistan-s-coal-industry-in-2017-2021/

Menzie WD, Soto-Viruet Y, Bermúdez-Lugo O, Mobbs PM, Perez AA, Taib M, Wacaster S, Staff (2013) Review of selected global mineral industries in 2011 and an Outlook to 2017. U.S. Geological Survey, Reston. https://pubs.usgs.gov/of/2013/1091/OFR2013-1091.pdf

Micklin P, Aladin NV, Plotnikov I (eds) (2014) The Aral Sea. The devastation and partial rehabilitation of a Great Lake. Springer

Middleton N (2002) The Aral Sea. In: Shahgedanova ML (ed) The physical geography of Northern Eurasia. Oxford University Press, Oxford, pp 3–498

MIT (2018) Geodesy and geodynamics. Global Time Series. Retrieved from www.gpsg.mit.edu/-tah/MIT_IGS_AAC?index2.html

Morelli, Creaser RA, Seltmann R, Stuart MA, Finlay M, Selby D, Graupner T (2007) Age and source constraints for the giant Muruntau gold deposit, Uzbekistan, from coupled Re-Os-He isotopes in arsenopyrite. J Geol 35(9):795–798. https://doi.org/10.1130/G23521A

Mossakovsky AA, Ruzhentsev SV, Samygin SG, Kheratovska TN (1993) The central Asian fold belt: geodynamic evolution and formation history. Geotektonika 6:3–32. (in Russian)

Muhs DR (2007) Loess deposits, origins and properties. In: Encyclopedia of quaternary sciences. Elsevier, pp 1405–1418. https://doi.org/10.1016/B0-44-452747-8/00158-7. Retrieved March 22, 2018 from https://www.sciencedirect.com/science/article/pii/B0444527478001587?via%3Dihub, https://www.researchgate.net/publication/316784380_LOESS_DEPOSITS_ORIGINS_AND_PROPERTIES

Muhs DR (2018) The geochemistry of loess: Asian and North American deposits compared. J Asian Earth Sci 155:81–115. https://doi.org/10.1016/j.jseaes.2017.10.032

Mukhin PA (1998) Uzbekistan. In: Moores EM, Fairbridge RW (eds) Encyclopedia of European and Asian regional geology. Springer/Chapman and Hall, pp 766–773

Mukhin PA, Abdullayev KA, Minayev VY, Khristov SY, Egamberdiyev SA (1989) The Paleozoic geodynamics of Central Asia. Int Geol Rev 30:1073–1083

Natal'in BA, Sengor C (2005) Late Paleozoic to Triassic evolution of the Turan and Scythian platforms: the pre-history of the Palaeo-Tethyan closure. Tectonophysics 404(3–4):175–202. https://doi.org/10.1016/j.tecto.2005.04.011. http://www.sciencedirect.com/science/article/pii/S0040195105001836. Accessable 1 August 2005

Nurtaev B, Kharin V, McCann T, Manchego MV (2013) The North Nurata Fault Zone, Uzbekistan—Structure and evolution of a Palaeozoic Suture Zone. J Geodyn 64:3–14. https://www.researchgate.net/publication/257097347_The_North_Nurata_Fault_Zone_Uzbekistan-Structure_and_evolution_of_a_Palaeozoic_Suture_Zone. Accessable March 2013

Rahmatullaev S, Huneau F, Kazbekov J, Djumanov J (2012) Groundwater resources of Uzbekistan: an environmental and operational overview. J Geosci 4(11):67–80. https://doi.org/10.2478/s13533-011-0062-y. This work is licensed under the Creative Commons Attribution-NonCommercial-NoDerivatives 3.0 License. BY-NC-ND 3.0

Redman CM, Leighton LR (2009) Multivariate faunal analysis of the Turonian Bissekty formation: variation in the degree of marine influence in temporally and spatially averaged fossil assemblages. PALAIOS 24(1):18–26. https://doi.org/10.2110/palo.2007.p07-072r

Ryabchikov AM et al (2018) Asia. In: Encyclopaedia Britannica. Accessible from https://www.britannica.com/place/Asia

Safonova I, Biske G, Romer RL, Seltmann R, Simonov V, Maruyama S (2016) Middle Paleozoic mafic magmatism and ocean plate stratigraphy of the South Tianshan, Kyrgyzstan. Gondwana Res 30:236–256

Schroeder, Meyers (2016) https://books.google.nl/books?id=XmBoCwAAQBAJ&pg=PA93&lpg=PA93&dq=average+protein+supply+in+eu+and+uzbekistan&source=bl&ots=u35TnEFgcT&sig=LEOg7EZawJo1G-QAHnslfa4fHLo&hl=ru&sa=X&ved=0ahUKEwi969LPq-rSAhVKDxoKHdmLC2sQ6AEIGjAA#v=onepage&q=average/protein/supply/in/eu/and/uzbekistan&f=false

Schweinfurth SP (2016) Coal – a complex natural resource. In: USGS. Science for a Changing World. – November 23, 2016. Retrieved August 12, 2018 from https://pubs.usgs.gov/circ/c1143/html/text.html

Semakova E, Gunasekara K, Semakov D (2015) Identification of the glaciers and mountain naturally dammed lakes in the Pskem, the Kashkadarya and Surkhandarya River basins, using ALOS satellite data. Geomat Nat Haz Risk 7:1081–1098. https://doi.org/10.1080/19475705.2015.1023852

Şengör A, Natalin B (1996) Paleotectonics of Asia: fragments of a synthesis. In: MRA T, Reinemund JA (eds) The tectonic evolution of Asia. Cambridge University Press, pp 486–640

Şengör A, Natal'in B, Burtman V (1993) Evolution of the Altaid tectonic collage and Palaeozoic crustal growth in Eurasia. Nature 364:299–307. https://doi.org/10.1038/364299a0

Shults VL (1949) Central Asian Rivers. Nauka, Moscow (in Russian)

State Committee on Statistics of the Republic of Uzbekistan. Industry (2022) Accessed https://stat.uz/en/official-statistics/industry on July 15, 2022

Status of the State Balance of Mineral Reserves of the Republic of Uzbekistan as of 01.01.2017 (2017) The State Balance of Resources The State Committee of the Republic of Uzbekistan for Geology and Mineral Resources. Retrieved August 20 2018 from http://www.uzgeolcom.uz/en/about/state_balance/14159

Sues HD, Averianov A (2016) Ornithomimidae (Dinosauria: Theropoda) from the Bissekty Formation (Upper Cretaceous: Turonian) of Uzbekistan. Cretaceous Res 57:90–110. https://doi.org/10.1016/j.cretres.2015.07.012

Thomas DS (2003) Aral Sea. https://en.wikipedia.org/wiki/Aral_Sea#cite_ref-physical1_11-1Weather statistics. Tashkent, Uzbekistan. http://www.geographic.org/global_weather/weather_data.php?month=01&year=1941&id=UZM00038457&path=weather_stations/ut000000000_zz999999999/UZM00038457.dly&name=Tashkent&country=Uzbekistan, https://www.yr.no/place/Uzbekistan/Tasjkent/Tashkent/statistics.html

Toovey LM (2015) Gold Mining in the Republic of Uzbekistan in Gold Investing News The Investing News Network® (INN) October 3, 2015. Retrieved February 22, 2017 from http://investingnews.com/daily/resource-investing/precious-metals-investing/gold-investing/gold-mining-in-the-republic-of-uzbekistan/

Torsvik TH, Cocks LRM (2016) Earth history and paleogeography. Cambridge University Press. https://doi.org/10.1017/9781316225523, https://www.cambridge.org/core/books/earth-history-and-palaeogeography/076D4FCFCEE4CAE6F8EF3B449594DD50

Toychiev HA, Stelmakh AG (2011a) Chronology of Quaternary loess deposits of Uzbekistan on the basis of paleo magnetic data. American Geophysical Union, Fall Meeting. SAO/NASA ADS Physics Abstract Service. http://adsabs.harvard.edu/abs/2011AGUFMPP23B1853T

Toychiev HA, Stelmakh AG (2011b) Chronology of Quaternary loess deposits of Uzbekistan on the basis of paleo magnetic data. American Geophysical Union, Fall Meeting. SAO/NASA ADS Physics Abstract Service. http://adsabs.harvard.edu/abs/2011AGUFMPP23B1853T

Turamuratov I (2016) 1895 Mineral Reserves are Discovered in the Lands of Uzbekistan In 1News. Economy – August 05, 2016 (In Russian Турамуратов И На территории Узбекистана окрыто 1895 месторождений. 1news.uz Экономикаю 05.08.2016) Retrieved August 20, 2018 from http://1news.uz/na-territorii-uzbekistana-otkryto-1895-mestorojdeniy/

Tuychieva MA et al (2016) Expert evaluation of modern state of complex geologic risk in the urbanized places of the Eastern Uzbekistan. Geol Miner Resour 5:67–75. (In Russian: Экспертная оценка современного состояния комплексного геологического риска на урбанизированных территориях восточного Узбекистана. // Геология и минеральные ресурсы 5:67–75 стр)

Ulomov VI (2004) On the role of horizontal tectonic movements in seismic geodynamics and seismic hazard prediction. J Earth Phys 9:14–30. (In Russian Уломов В.И. О роли горизонтальных тектонических движений в сейсмогеодинамике и прогнозе сейсмической опасности. Физика Земли № 9, с 14-30) Retrieved from http://seismos-u.ifz.ru/personal/centrasia2.htm

UN. United Nations Economic Commission for Europe (UNECE) (2013) Kouzmitch VV Strengthening cooperation of Central-Asian countries in using advanced technologies in energy efficiency and renewable energy sources. Accessible from https://www.unece.org/fileadmin/DAM/energy/se/pdfs/gee21/projects/AdvTech_IncreasingCooperation.pdf

UNDP (United Nations Development Program) (2007) Water, critical resources for Uzbekistan's future. UNDP, Tashkent

UNDP (2012) Renewable energy snapshot: Uzbekistan. Retrieved on March 16, 2017 from www.eurasia.undp.org/content/dam/rbec/…/Uzbekistan.pdf?

UNDP [United Nations Development Programme] (2016) 1966–2016: 50 years since the Tashkent earthquake. United Nations Office for the Coordination of Humanitarian Affairs. https://reliefweb.int/report/uzbekistan/1966-2016-50-years-tashkent-earthquake. Accessed 8 June 2022

United Nations Economic Commission for Europe (UNECE) (2005) Drainage Basin of the Aral Sea and other trans boundary Surface Waters in Central Asia. Retrieved 4 Feb 2016 https://www.unece.org/fileadmin/DAM/env/water/…/aral.pdf

US Geological Survey (USGS) (2009) This dynamic earth: the story of plate tectonics. Accessed from https://pubs.usgs.gov/gip/dynamics on October 18, 2018

USGS (2015, March 09) The Himalayas: Two continents collide. Accessed October 16, 2018 from https://pubs.usgs.gov/gip/dynamic/himalaya.html

Uzbekenergo (2018) Alternative sources of energy. In: Uzbekenergo portal. Retrieved from http://www.uzbekenergo.uz/ru/activities/alternative-energy-sources/

Uzbekenergo (2019) Structure of consumption of electricity in 2018 In Uzbekenergo. Activity: Technic-economic characteristics. (In Russian Деятельность. Технико-экономические показатели в 2018 году. Портал Акционерного общества «Узбекэнерго») Retrieved March 20, 2019 from http://www.uzbekenergo.uz/ru/activities/technical-and-economic-indicators/

UzDaily (2016) Uzbekistan Modernizes Charvak Hydro Power Station –January 18, 2016. Retrieved from https://www.uzdaily.uz/en/post/34864

Vervaeck A (2011) Understanding the deadly Uzbekistan/Kyrgyzstan earthquake. http://earthquake-report.com/2011/07/23/understanding-the-deadly-uzbekistan-kyrgyzstan-earthquake-july-19-2011/. Accessable 19 July 2011

Walters P (2010) Aral Sea recovery. In: National Geographic – April 22, 2010 Retrieved from https://news.nationalgeographic.com/news/2010/04/100402-aral-sea-story/

Watts J (2018) Land degradation threatens human Wellbeing, major report warns. In: Guardian. Retrieved March 26, 2018 from https://www.theguardian.com/environment/2018/mar/26/land-degradation-is-undermining-human-wellbeing-un-report-warns

White KD (2014) Chapter 12: Nature and economy in the Aral Sea Basin. In: Micklin Ph, Aladin NV, Plotnikov I (eds) The Aral Sea: the devastation and partial rehabilitation of a Great Lake, Springer Earth system sciences series. Springer-Verlag, Berlin/Heidelberg, pp 301–335

Windley BF (2015) Geologic history of Earth. https://www.britannica.com/science/geologic-history-of-Earth

Windley BF, Alexeiev D, Xiao W, Kröner A, Badarch G (2007) Tectonic models for accretion of the Central Asian Orogenic Belt. J Geol Soc 164:31–47. https://doi.org/10.1144/0016-76492006-022

Wolchover N (2011) Can humans cause Earthquakes? J Live Sci. http://www.livescience.com/32932-can-humans-cause-earthquakes.html. Accessible 1 January 2011

World Bank and Kazakhstan Plan Further Improvements in the Northern Aral Sea Area (2014) Press Release. Acceptable from July 10, 2014 http://www.worldbank.org/en/news/press-release/2014/07/10/world-bank-and-kazakhstan-plan-further-improvements-in-northern-aral-sea-area

World Bank Group (2013) Kochnakyan A et al., Uzbekistan Energy/Power Sector Issue Notes. –June 2013. Retrieved November 13, 2018 from http://documents.worldbank.org/curated/en/810761468318884305/pdf/ACS41460WP0Box0Issues0Note00PUBLIC0.pdf

World Bank Group the Global Facility for Disaster Reduction and Recovery (GFDRR) (2018) Uzbekistan. https://thinkhazard.org/en/report/261- uzbekistan/EQ#:~:text=Earthquakes%20can%20cause%20secondary%20hazards,and%20tsunami%20in%20coastal%20areas

World Energy Council (2009) Survey of energy resources interim update 2009. Retrieved February 2017 from https://www.worldenergy.org/wp-content/uploads/2012/10/PUB_Survey-of-Energy-Resources_Interim_update_2009_WEC.pdf

Yakubchuk A (2002) The Baikalide - Altaid, Transbaikal-Mongolian and North Pacific orogenic collages: similarity and diversity of structural patterns and metallogenic zoning. Geological Society, London, Special Publications. https://doi.org/10.1144/GSL.SP.2002.204.01.16

Yakubchuk A (2004) Architecture and mineral deposit settings of the Altaid orogenic collage: a revised model. J Asian Earth Sciences 23. https://doi.org/10.1016/j.jseaes.2004.01.006

Yakubchuk AS, Cole A, Seltmann R, Shatov V (2002) Tectonic setting, characteristics and regional exploration criteria for gold mineralization in Central Eurasia: the southern Tien Shan province as a key example. In: Goldfarb R, Nielsen R (eds) Integrated methods for discovery: global exploration in twenty-first century. Econ. Geol. Special Publication No. 9, pp 177–201

Yunusov G, Quvvatov D, Khyjamatov I (2017) Dynamics of irrigation water in Kashkadarya oasis. In: Materials of International Conference "Uzbekistan in the Central Asia Region: Geography, Geoeconomics, Geoecology" (In Uzbek Қашқадарё воҳасида суғоришга олинган сув миқдори динамикаси)

Zonenshain LP, Kuzmin MI, Natapov LM, Page BM (1990) Tectonic framework. In: Geology of the USSR: a plate-tectonic synthesis. American Geophysical Union, Washington, DC. https://doi.org/10.1029/GD021p0001

Zonn IS, Kostianoy AG (2013) The Turkmen Lake Altyn Asyr and water resources in Turkmenistan. In: The Handbook of environmental chemistry, vol 28. Springer-Verlag, Berlin/Heidelberg, pp 1–8. https://doi.org/10.1007/698_2012_218. http://www.kaspinfo.com/sites/default/files/atoms/files/The/Turkmen/Lake/Altyh/Asyl/and/water/resources.pdf. Accessible 22 March 2013

Economic Geography

4

Abstract

This chapter discusses the impacts of regional development on the economy of Uzbekistan. The main indicators of economic development are reviewed from a historical perspective. It is shown how the location influences the spatial variation in the development of manufacturing, services, trade, and finance. Brief information is given about industry, agriculture, transportation, and foreign economic relations.

Keywords
Economic development · Income distribution · Employment · Industry · Agriculture · Trade · Transportation

4.1 Regional Developments and the Economy

Historically, Uzbekistan served as an economic hub on the Great Silk Road and connected the Asian and European economies. Today, its development is closely related to human productive activity at the regional geographic scale and its openness to innovations from the East and the West.

The Asian Region
According to an Asian Development Bank (ADB) forecast, the Asian region was expected to expand by 5.2% in 2017 and in 2018, and Central Asia by 4.0–4.2% (Asian Development Bank 2017a).

By definition of IMF, 'Emerging Asia" refers to China, India, Indonesia, Malaysia, the Philippines, Thailand, and Vietnam. "Frontier and Developing Asia" refers to Bangladesh, Cambodia, Lao People's Democratic Republic, Mongolia, Myanmar, Nepal, and Sri Lanka.

By definition of ADB, Central Asia and Caucasus is one of the five developing Asia regions (https://www.adb.org/news/developing-asia-economies-set-grow-5-2-year-amid-global-uncertainty) (Figs. 4.1 and 4.2).

Further transition of the Asian region is dependent on greater improvements in productivity. The experience of the Asian economy is confirmation of the truth that innovation, human capital, and infrastructure are the pillars of growth, strengthened by institutions and policies and underpinned by microeconomic stability (Asian Development Bank 2017b, 2019). To achieve higher incomes, the Asian Development Bank recommends pursuit of new business opportunities and production of more complex goods and services that are novel to the market. These require further development of research-oriented education, with more investment in infrastructure (Asian Development Bank 2017b).

The Central Asian Region is rich in natural and agricultural resources, attractive mineral reserves, and fuels. Central Asia is a part of developing Asia, which is transforming from a low-income region into a middle-income region. After dissolution of the Soviet Union, all Central Asian countries, in their attempts to move toward a market economy, have suffered transitional problems, resulting in increased unemployment and inflation.

A weakening economy in neighboring partner countries due to a decline in oil revenues and a recession in the Russian Federation have caused economic problems in several countries in the Caucasus and Central Asian regions. The new tension is directing the Central Asian region's foreign relations eastward. Actually, these relations are mutually beneficial, since East Asia is looking for resources and markets (Contessi 2016). China is the trading partner for Uzbekistan (International Trade Center 2017).

In general, the orientation of the Central Asian region to the West and the East has led to development of transcontinental Eurasia as a catalyst for international change (Contessi 2016) and, to some extent, to a balance of driving economic forces.

L. Gulyamova, *The Geography of Uzbekistan*, World Regional Geography Book Series,
https://doi.org/10.1007/978-3-031-07873-6_4

Fig. 4.1 Growth in gross domestic product (GDP) in developing Asian regions and Central Asia in 2015–2019. The 2018–2019 figures are forecasts. (Source: Asian Development Bank (2019), CC-BY-3.0 IGO License (public domain))

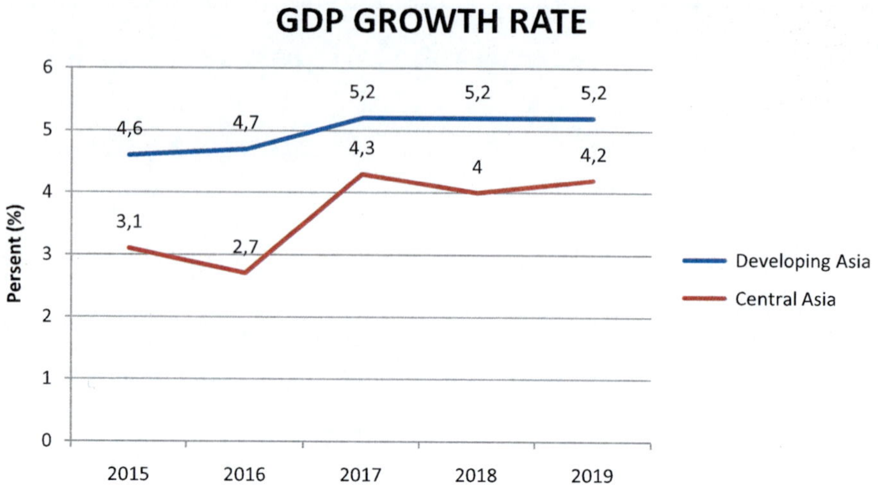

Fig. 4.2 Inflation rate in Asian regions in 2015–2019. The 2018–2019 figures are forecasts. (Source: Asian Development Bank (2019), CC-BY-3.0 IGO License (public domain))

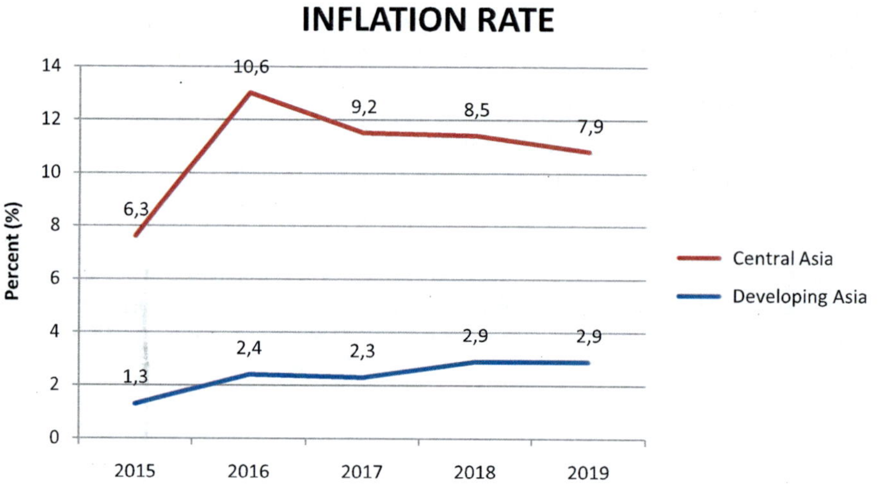

Table 4.1 Economic forecasts of gross domestic product (GDP) growth in Central Asian countries in 2017–2018

	GDP growth (%)							
Year	Armenia	Azerbaijan	Georgia	Kazakhstan	Kyrgyz Republic	Tajikistan	Turkmenistan	Uzbekistan
2017	2.2	−1.1	3.8	2.4	3.0	4.8	6.5	7.0
2018	2.5	1.2	4.5	2.2	3.5	5.5	7.0	7.3

Source: Asian Development Bank (2017a) (CC-BY-3.0 IGO)

Uzbekistan is in a more favorable state (Table 4.1) than its neighbors. From the regional perspective, Uzbekistan has good opportunities for further economic development. Uzbekistan's gross domestic product (GDP) changed from 14.31 $US billion in 2005 to $US billion in 2016 (which was about 4.7 times the 2005 value; Table 4.2). These changes are a result of the search for a proper economic strategy and attempts to find a niche in the world economy as well. Uzbekistan is still facing challenges caused by the heritage of its agriculture-oriented economy with a cotton monopoly. The GDP[1] of Uzbekistan grew at an average annual rate of 7.67%, increasing from US$14.95 billion in 1998 to US$49.677 billion in 2017. In 2017, however, the GDP dropped by 28.2% in comparison with 2016 (Fig. 4.3).

Rich natural resources (gold, oil and gas, and cotton production) provide opportunities for growth, with the highest rates in the region. However, experts from the International Trade Center (2017) point out the threat of dependence on natural resources and a few select markets, which makes Uzbekistan vulnerable to volatile fluctuations in prices and demand. Export continue to consist of natural resources and goods with low added value (cotton, copper, petroleum gases, vehicles, uranium, fertilizers, fruit, and vegetables).

[1] Source: World Data Atlas 2018 (https://knoema.com/atlas/Uzbekistan/GDP).

Table 4.2 Changes in Uzbekistan's gross domestic product (GDP) in 2006–2017

Year	Value (US$ billions)	Change (%)
2017	49.677	−28.20
2016	66.446	0.31
2015	66.904	5.35
2014	63.067	9.38
2013	57.690	11.34
2012	51.822	12.81
2011	45.915	18.90
2010	39.333	14.69
2009	33.689	14.07
2008	29.549	32.37
2007	22.311	28.77
2006	17.331	18.99

Source: Uzbekistan—gross domestic product in current prices (CC-BY-SA 4.0) via World Bank Data (2018)

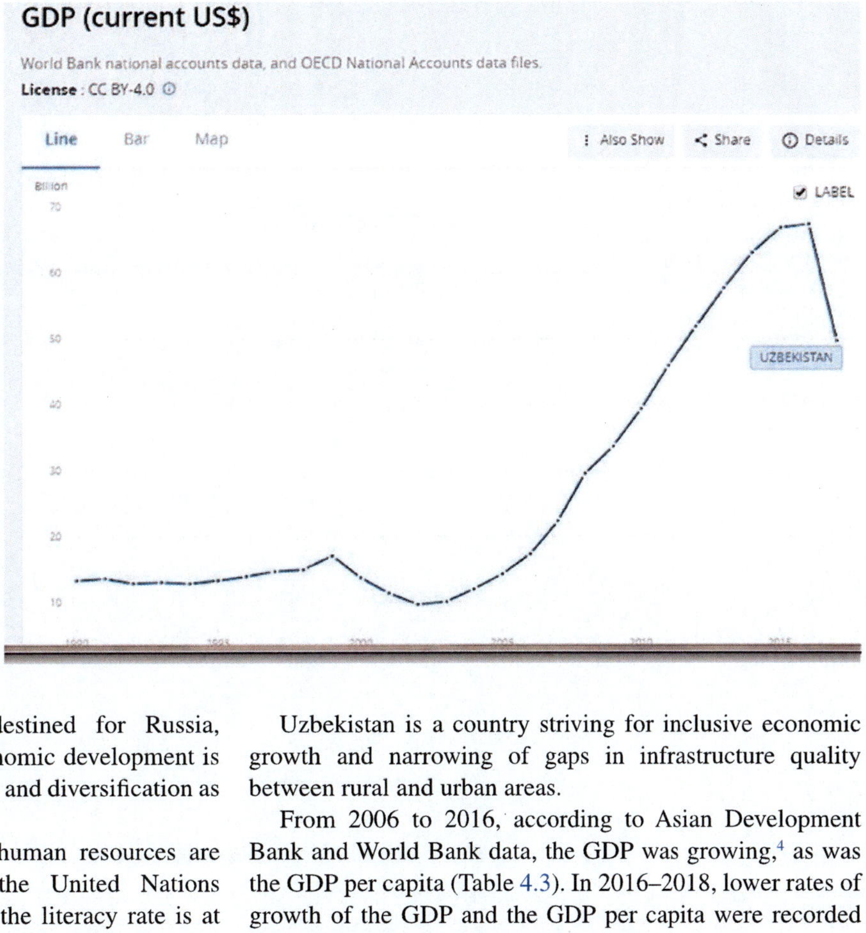

Fig. 4.3 Gross domestic product (GDP) in current US dollars. (Source: World Bank Data (n.d.) (CC-BY-4.0))

All of these goods are mainly destined for Russia, Kazakhstan, China, and Türkiye. Economic development is hampered by inadequate market access and diversification as well.

In addition to natural resources, human resources are attractive as well. According to the United Nations Development Programme (UNDP),[2] "the literacy rate is at 99.9%." The literacy rate of Uzbekistan's population[3] is one of the highest in the world.

Uzbekistan is a country striving for inclusive economic growth and narrowing of gaps in infrastructure quality between rural and urban areas.

From 2006 to 2016, according to Asian Development Bank and World Bank data, the GDP was growing,[4] as was the GDP per capita (Table 4.3). In 2016–2018, lower rates of growth of the GDP and the GDP per capita were recorded (Figs. 4.4 and 4.5).

According to International Monetary Fund (IMF) (2019) analytical data, "GDP per capita in Uzbekistan is 1520 US dollars in 2017, 1480 US dollars in 2018." This indicator was

[2]Source: United Nations Development Programme (UNDP) Human Development Reports (Creative Commons Attribution 3.0 IGO via UNDP).

[3]The estimation data for the section on Uzbekistan's population literacy are based on the latest data published by the United Nations Educational, Scientific and Cultural Organization (UNESCO) Institute for Statistics (accessed March 13, 2016).

[4]Uzbekistan's GDP per capita purchasing power parity (PPP) forecasts are projected using an autoregressive integrated moving average (Commons Attribution 3.0 IGO license; http://www.tradingeconomics.com/uzbekistan/gdp-per-capita-ppp).

Table 4.3 Gross domestic product (GDP) per capita in current US dollars

Year	2005	2006	2007	2008	2009	2010	2011	2012	2013	2014	2015	2017
GDP (US$)	546	654	830	1082	1213	1377	1565	1740.5	1907	2036	2132	2130

Source: Data Bank World Development Indicators © World Bank. http://hdl.handle.net/123123123123/123 License: Creative Commons Attribution license (CC-BY-IGO 3.0) via World Bank Group

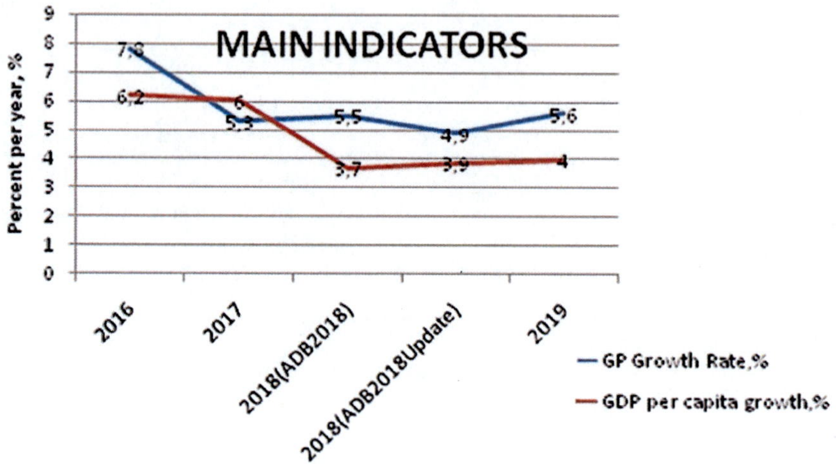

Fig. 4.4 Growth in gross domestic product (GDP) and GDP per capita in 2016–2018. The 2018–2019 figures are forecasts. (Compiled by the author, using data from Asian Development Bank (2019) (CC-BY-3.0 IGO) License (public domain))

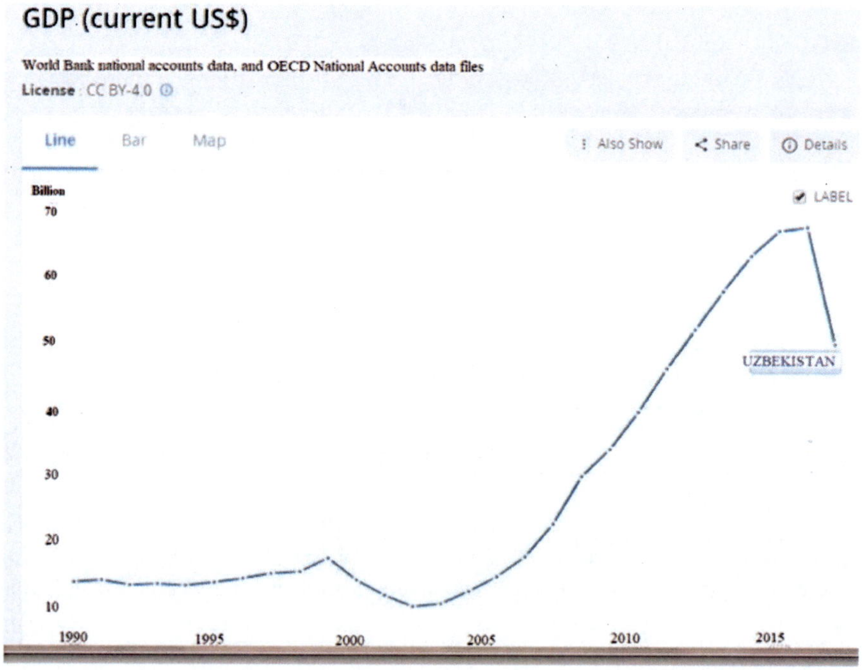

Fig. 4.5 Gross domestic product (GDP) per capita in current US dollars. (Source: (CC-BY-4.0) via World Bank Data (2019))

lower than that of emerging market and developing economies (US$5030) and the world average (US$10,860). According to the World Bank Group (2018a, b, c), the GDP per capita was US$2130 in 2017 (Table 4.3).

The reported growth slowed to 7.8% in 2016 from 8.0% the previous year. The GDP per capita annual growth also declined from 5.942% in 2016 to 3.547% in 2017 (Fig. 4.6).

Inflation was marginally lower at 8.4%, and the current account recorded a small surplus. According to a World Bank projection, the GDP was expected to grow by 5.0% in 2018 and by 5.1% in 2019. Earlier, the bank forecast that the growth of Uzbekistan's economy would be 7.6% in 2017 and 7.8% in 2018. According to "Global Economic Prospects" from the World Bank (January 2019, p. 6), the GDP growth at market prices would be 5.1% in 2019, 5.5% in 2020, and 6.0% in 2021.

The Asian Development Bank's experts (2019) also cut the estimated GDP growth; according to their estimates,

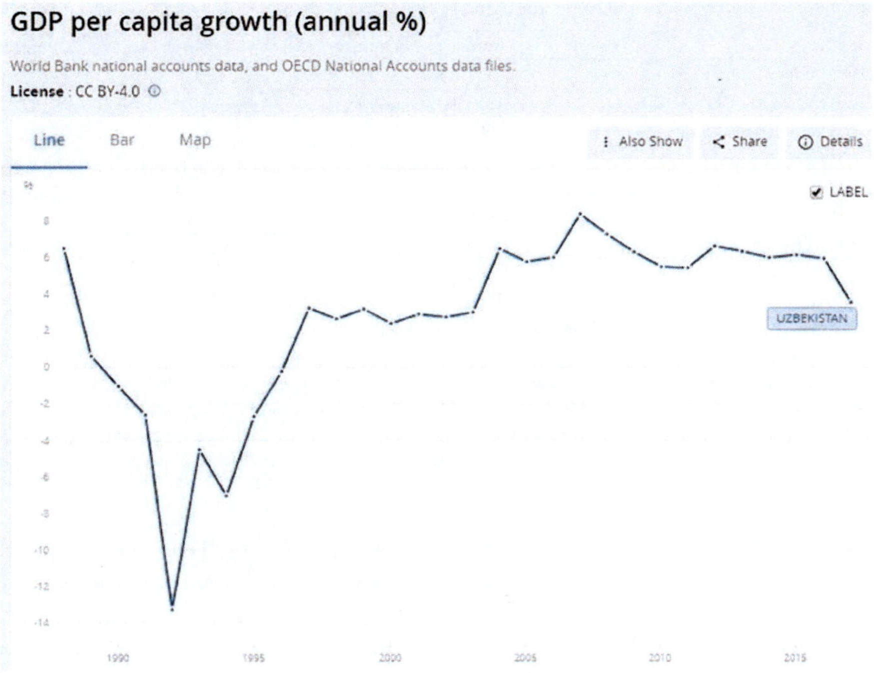

Fig. 4.6 Annual percentage growth in gross domestic product (GDP) per capita. (Source: (CC-BY-4.0) via World Bank Data (2019))

Table 4.4 Gross domestic product (GDP) growth in Uzbekistan from 2012 to 2018

	GDP growth (%)							
Data source	2012	2013	2014	2015	2016	2017	2018	2019
Asian Development Bank[a]	8.2	8	8.1	8	7.8	7.0	4.9	5.0
World Bank[b]	–	–	–	–	7.8 (estimated)	7.6 (forecast)	5.6 (forecast)	6.5

[a]Source: Asian Development Bank (2017b)
[b]Source: World Bank (2018a, b, c). [License: Creative Commons Attribution license (CC-BY-IGO 4.0) via World Bank Group]

Uzbekistan's GDP was expected to grow by 4.9% in 2018 and 5.0% in 2019. The per capita GDP growth for Uzbekistan was expected to be 3.9% in 2018 and 4.0% in 2019. Earlier, the Asian Development Bank forecast that currency depreciation and government spending would raise inflation to 9.5% in 2017 and 10.0% in 2018. The Asian Development Bank changed the inflation rates to 16.0% in 2018 and 14.0 in 2019 (Table 4.4).

In 2020, the forecasts for GDP growth were revised, considering the serious economic challenge posed by the COVID-19 pandemic. According to the World Bank, the decline was a result of both a significant decrease in foreign trade and domestic economic shocks. In 2020, the World Bank predicted that the growth in Uzbekistan's GDP would be 1.6%, while at the beginning of the year, by the estimates of the World Bank, the projected GDP growth of Uzbekistan was 5.7%—the highest indicator in the region. The GDP growth was projected to be 6.5% in 2021 and 5.5% in 2022 (Fig. 4.7). World Bank experts predicted that reduced tax revenues and additional costs to smooth out the crisis would increase the budget deficit to 5.6% of GDP in 2020 and 4.7% of GDP in 2021.

Fig. 4.7 Forecast annual percentage growth in gross domestic product (GDP). (Source: (CC-BY-4.0) via World Bank Data (2020))

Table 4.5 Macroeconomic indicators for Uzbekistan

Forecast	Actual[a]	Quarter 32,018	Quarter 42,018	Quarter 12,019	Quarter 22,019	2020
GDP per capita (US$)[b]	2031.05	2126	2126	2232	2232	2288
GDP per capita PPP (US$)	6253.10	6499	6499	6662	6662	6900
GDP annual growth rate (%)	4.9	5.2	5.5	5.5	4.8	4.8
GDP (US$ billions)	48.72	50	50	58.8	58.5	60
GDP from construction (US$ billions)	21,588	26,485	35,934	3450	22,625	39,768
GDP from agriculture (US$ billions)	35,210	48,596	73,327	5825	36,901	81,150
GDP from services (US$ billions)	66,562	81,268	123,219	33,665	69,758	136,366

Source: © World Bank. http://hdl.handle.net/123123123123/123 License: Creative Commons Attribution license (CC-BY-IGO 3.0) via World Bank Group

GDP gross domestic product, *PPP* purchasing power parity

[a]As of September 22, 2018

[b]Uzbekistan GDP per capita forecasts are projected using an autoregressive integrated moving average (ARIMA) model calibrated using analysts' expectations. The past behavior of Uzbekistan's GDP per capita is modeled using vast amounts of historical data, and the coefficients of the econometric model are adjusted by taking into account the analysts' assessments and future expectations. The forecast for Uzbekistan GDP per capita was predicted on April 26, 2017 (http://www.tradingeconomics.com/uzbekistan/gdp-per-capita/forecast%20%20uzbekistan)

In 2017, the GDP per capita was US$6514; when adjusted for purchasing power parity[5] (PPP), this was equivalent to 32% of the world average.

Asian Development Bank[6] experts suggested that devaluation of the national currency (soums [currency code: UZS]) and liberalization of monetary policy in 2017 would lead to better performance of the national economy in 2018. The Asian Development Outlook 2017a, b report underlined the public debt of Uzbekistan at 15.1% of GDP. The World Bank forecast steady economic growth in Uzbekistan in 2018–2020 (Table 4.5).

In 2017, one of the main drivers was industry, which (according to the State Committee of the Republic of Uzbekistan on Statistics) grew by 7.6% (compared with 4.7% in 2016). Production constitutes 80% of Uzbekistan's industry. Machinery and equipment played the leading role; production rose by 33% because of the growing demand for this type of product in the Russian Federation. Construction and services grew more slowly as a result of inflation. Agriculture was declining; it increased by 5.8% (compared with 6.8% in 2016), with slower growth in grain and livestock production.

The economy of Uzbekistan is still influenced by the principles of a planned economy. State-owned companies dominate the strategic sectors of the economy.

In the Doing Business indicator rankings in 2017, Uzbekistan ranked 87th[7] out of 190 countries and 24th out of 25 developing economies in Europe and Central Asia (World Bank Group 2016a, b).

Uzbekistan is a lower-middle-income economy on a per capita basis, with a gross national income (GNI) per capita of US$2160 (ranking it 163rd out of 217 countries) in 2015 and US$2000 (ranking it 142nd out of 190 countries) in 2017.[8]

This is reflected in the population's poverty. The population poverty rate[9] is estimated by the Asian Development Bank to be 62.1% (as of the year 2003).

Uzbekistan scored 23 points[10] out of 100 on the 2018 corruption perceptions index reported by Transparency International. According to this index, Uzbekistan ranked 158th out of 180 countries. The index uses a scale of 0 to 100, where 0 is highly corrupt.

[5]Macroeconomic analysis relies on several different metrics to compare economic productivity and standards of living between countries and across time. Purchasing power parity (PPP) is an economic theory that compares different countries' currencies through a market "basket of goods" approach. According to this concept, two currencies are in equilibrium or at par when a market basket of goods (taking into account the exchange rate) is priced the same in both countries. Uzbekistan's GDP per capita PPP forecasts are projected using an autoregressive integrated moving average (ARIMA) model calibrated using analysts' expectations. (From: Investopedia. What is Purchasing Power Parity? http://www.investopedia.com/updates/purchasing-power-parity-ppp).

[6]From Asian Development Outlook 2017a, b Update (CC-BY-3.0 IGO); http://www.adb.org/publications/Asian-Development-Outlook-2017-Update; https://www.adb.org/countries/uzbekistan/economy

[7]The data were current as of June 1, 2016 (except for the taxpaying indicators, which covered the period of January–December 2015). These indicators are used for analyzing outcomes and comparing conditions for a local entrepreneur to open and run a small or medium-sized business when complying with relevant regulations.

[8]Source: Creative Commons Attribution license (CC-BY-IGO 4.0) via World Bank Data.

[9]The population below $1.90 per day is the percentage of the population living on less than $1.90 per day at 2011 international prices (CC-BY-3.0 IGO; source: Asian Development Bank. Key indicators for Asia and the Pacific 2018; https://www.adb.org/publications/key-indicators-asia-and-pacific-2018).

[10]From *Tashkent Times*: "Uzbekistan ranks 158 out of 180 countries in corruption perceptions index 2018." January 29, 2019 (http://tashkent-times.uz/national/3459-uzbekistan-ranks-158-out-of-180-countries-in-corruption-perceptions-index-2018).

4.2 Inequality of Income Distribution

The Gini index[11] measures the extent to which distribution of income among individuals and households within an economy deviates from perfectly equal distribution.[12] This indicator measures relative wealth, not absolute wealth. A Gini index of 0 represents perfectly equal distribution. This index is convenient for measuring the degree of inequality. In Uzbekistan, the Gini index changed from 24.9 in 1988 to 35.3 in 2017 (Table 4.6).

At the same time, this index has limitations, and these coefficients are not unique. It is worth noting that the data are not strictly comparable across countries and even across years within a country. When measuring inequality with this index, it is recommended to consider such limitations as differences among countries that may bias comparisons of distribution.

According to the 2018 UNDP Human Development Report, Uzbekistan's human development index (HDI)[13] of 0.710 gives it a high human development rank (105th of 189 countries); for comparison, Norway has the top HDI of 0.949, giving it a very high human development rank. This indicator changed in Uzbekistan in 2010–2015, with average annual growth of 107% (Table 4.7).

The gender development index (GDI; Table 4.7) measures the ratio of female to male HDI. According to the United Nations (UN) classification, Uzbekistan is in the third group of countries, which comprises countries with medium equality in HDI achievements between women and men (the absolute deviation in the case of Uzbekistan is 7%).[14]

The inequality-adjusted HDI (IHDI) measures the loss in human development due to inequality. If the IHDI equals the HDI, it means there is equality in income distribution; the more difference there is between these two indexes, the more inequality there is and the IHDI falls below the HDI. Uzbekistan's IHDI was 0.489 in 2015 (Table 4.7).

Measured by the multidimensional poverty index[15] (MPI), the poverty rate is 0.013. This MPI estimation is based on data that refer to 2006, and it identifies the deprivation intensity in households at 36.6%. In Uzbekistan, 3.5% of the population is multidimensionally poor and 6.2% are near multidimensional poverty.[16]

Table 4.6 The Gini index in 1988–2017

Year	1988	1998	2000	2003	2017
Gini index	24.95	44.67	36.10	35.27	35.3

Source: World Bank Group Gini index (World Bank estimate) © World Bank. http://hdl.handle.net/123123123123/123 License: Creative Commons Attribution license (CC-BY-4.0) via World Bank Group (public domain)

4.3 Employment

In 2017, the employment rate was 94.2% (13,520,000 people)[17] among people of working age and the unemployment rate was 5.8% in the economically active population.[18]

The highest rates of employment are in the city of Tashkent and in Navoi and Tashkent regions (Table 4.8).

The youth unemployment rate (19.8%) is about twice the overall unemployment rate (10.1%).[19]

In January–September 2017, 82.2% of total employment was in the nonstate economy; 77.8% of total employment was in small and medium-sized enterprises[20] (Table 4.9).

When analyzed by types of economic activity, the shares of employment are 27.7% in agriculture, forestry, and fisheries; 13.6% in industry; 10.9% in trade; 9.5% in construction; and 8.3% in education[21] (Fig. 4.8).

Between 2008 and 2017, the numbers of people in employment increased from 11.035 million to 13.520 million.

4.4 Labor Migration

Uzbekistan contributes to the following types of global migration: (1) labor; (2) permanent residence; (3) the "brain drain"; and (4) human trafficking.

[11]Source: World Bank Development Research Group; the data are based on primary household survey data obtained from government statistical agencies and World Bank departments (https://data.worldbank.org/indicator/SI.POV.GINI).

[12]From https://data.worldbank.org/indicator/SI.POV.GINI

[13]A composite index measuring average achievement in three basic dimensions of human development: a long and healthy life, knowledge, and a decent standard of living (source: UNDP Human Development Report 2016; http://hdr.undp.org/en/countries/profiles/UZB).

[14]Source: UNDP Human Development Report 2016; http://hdr.undp.org/en/composite/GDI

[15]The multidimensional poverty index (MPI) identifies multiple overlapping deprivations suffered by households in three dimensions: education, health, and living standards. The education and health dimensions are each based on two indicators, while the standard of living is based on six indicators (source: http://hdr.undp.org/en/content/multidimensional-poverty-index-mpi).

[16]Source: UNDP Human Development Report 2016 [Creative Commons Attribution 3.0 IGO license].

[17]Data from the State Committee of the Republic of Uzbekistan on Statistics (https://www.gazeta.uz/ru/economy/), accessed January 16, 2017.

[18]Source: State Committee of the Republic of Uzbekistan on Statistics (2018).

[19]Source: UNDP Human Development Report 2016.

[20]Source: State Committee of the Republic of Uzbekistan on Statistics (2018).

[21]Source: State Committee of the Republic of Uzbekistan on Statistics (2018).

Table 4.7 Indexes for measuring average achievements in basic dimensions of human development

	Armenia	Azerbaijan	Georgia	Kazakhstan	Kyrgyz Republic	Tajikistan	Turkmenistan	Uzbekistan
HDI	0.743	0.759	0.769	0.794	0.664	0.627	0.691	0.675
IHDI[a]	0.674	0.659	0.672	0.714	0.582	0.532	–	0.590
Gini index: World Bank, 2003	32.4	16.64	38.5	26.35	27.37	30.77	40.77	35.27
Gini index: UN, 2015	31.5	31.8	40.1	26.3	26.8	30.8	–	–
Palma ratio[b]	1.2	1.2	1.9	0.9	1.0	1.2	–	1.5[c]
Average HDI annual growth (2010–2015): UN, 2015	0.39	0.48	0.72	0.49	0.98	0.64	0.78	1.07
GDI[d]	0.993	0.940	0.970	1.006	0.967	0.930	–	0.946

Gini index source: World Bank estimate, 2003, Asia; http://www.indexmundi.com/facts/indicators/SI.POV.GINI/map/asia (CC-BY-4.0)

HDI, IHDI, GDI source: United Nations Development Programme (UNDP) Human Development Report 2016; http://hdr.undp.org/en/composite/IHDI [Creative Commons Attribution 3.0 IGO license]

GDI gender development index, *HDI* human development index, *IHDI* inequality-adjusted human development index, *UN* United Nations

[a]The IHDI is the income index value adjusted for inequality in income distribution on the basis of data from household surveys (source: UNDP Human Development Report 2016; http://hdr.undp.org/en/composite/IHDI)

[b]The Palma ratio is a measure of inequality, calculated as the ratio of the richest 10% of the population's share of gross national income (GNI) divided by the poorest 4% share. It is based on the work of Palma (2011); see http://hdr.undp.org/en/composite/IHDI

[c]Source: UNDP Human Development Report, 2015

[d]The GDI is the ratio of female to male HDI values; see https://reliefweb.int/sites/reliefweb.int/files/resources/2016_human_development_report.pdf

Table 4.8 Employment in different regions of Uzbekistan in 2017

Country/region	Workforce (thousands)	Economically active population (thousands)	People employed in the economy (thousands)	People employed in the economy (%)
Republic of Uzbekistan	18,666.3	14,357.3	13,520.3	94.2
Republic of Karakalpakstan	1054.6	688.8	647.2	94.0
Andijan	1722.2	1404.1	1319.2	94.0
Bukhara	1073.1	874.6	826.5	94.5
Jizzakh	756.2	492.6	468.2	95.0
Kashkadarya	1784.6	1296.6	1218	93.9
Navoiy	547.2	445.7	422.4	94.8
Namangan	1549.5	1099	1034.9	94.2
Samarkand	2079.6	1629.8	1523.1	93.4
Surkhandarya	1423.9	1062.7	991.7	93.3
Syrdarya	475.2	371.9	353.1	94.9
Tashkent	1623.9	1360.9	1289.6	94.8
Fergana	2031.1	1630.5	1525.7	93.6
Kxorezm	1000.6	773.3	729.3	94.3
City of Tashkent	1544.6	1226.8	1171.4	95.5

Source: State Committee of the Republic of Uzbekistan on Statistics (2018) (public domain)

Table 4.9 Employment in the state and nonstate economies in January–September 2017

Country/region	People employed (thousands)		
	Number (thousands)	State employment (%)	Nonstate employment (%)
Republic of Uzbekistan	13,518.5	17.8	82.2
Republic of Karakalpakstan	651.2	22.0	78.8
Andijan	1313.7	13.4	86.6
Bukhara	825.4	17.8	82.2
Jizzakh	461.7	18.3	81.7
Kashkadarya	1205.6	16.4	83.6
Navoiy	423.9	35.6	64.4
Namangan	1022.0	15.6	84.4
Samarkand	1525.1	14.3	85.7
Surkhandarya	1000.7	16.8	83.2
Syrdarya	347.5	19.1	80.9
Tashkent	1309.3	15.6	84.4
Fergana	1520.1	15.4	84.6
Khorezm	730.3	16.1	83.9
City of Tashkent	1182.0	28.4	71.6

Source: State Committee of the Republic of Uzbekistan on Statistics (2018)

Labor migration is increasing, and the main destinations are Russia, Kazakhstan, the United Arab Emirates, Türkiye, the Republic of Korea, and Europe.[22]

The State Committee of the Republic of Uzbekistan on Statistics reports that in 2017, the main destinations were Russia and Kazakhstan (Fig. 4.9).

[22]Source: International Organization for Migration (2018).

According to World Bank estimates (Ajwad et al. 2014), the emigration rate was about 7% of the population in 2010 and the mobility rate was higher than the world average and those of other middle-income countries.

Parpiev (2015) reported that the new trends in labor migration are as follows: (1) migrants are becoming younger; (2) the education level is becoming lower; (3) the motivations to succeed in foreign countries are becoming stronger; (4) the duration of stay is increasing; (5) migration is becoming less seasonal; (6) migration is becoming feminized; and (6) family migration is becoming more popular.

These trends demonstrate the phenomenon of change in the behavior of citizens of Uzbekistan who, until recently, were less mobile; all of their hopes for the future were connected with their homeland. As a result of increasing migration, personal remittances have become significant contributors to Uzbekistan's GDP.

Personal remittances[23] received constitute about 5% of GDP, reaching as high as 11.59% in 2013 (Table 4.10).

Personal remittances are estimated by World Bank staff with consideration of IMF balance of payments data and World Bank–OECD (Organisation for Economic Co-operation and Development) GDP estimates.

According to UzDaily (2018), the Central Bank of Uzbekistan reported that in the first half of 2018, the volume of international money transfers to Uzbekistan amounted to

[23]Personal remittances comprise personal transfers and compensation of employees, consisting of all current transfers in cash or in kind made or received by resident households to or from nonresident households (https://data.worldbank.org/indicator/BX.TRF.PWKR.DT.GD.ZS?end=2016&locations=UZ&start=2006&view=chart).

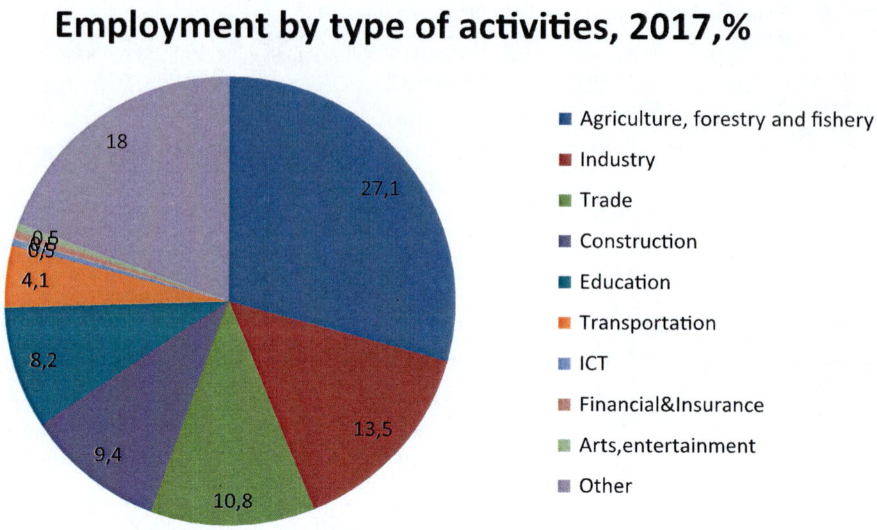

Fig. 4.8 Share of employment by types of economic activity in January–September 2017. (Compiled by the author, using data from the State Committee of the Republic of Uzbekistan on Statistics (public domain))

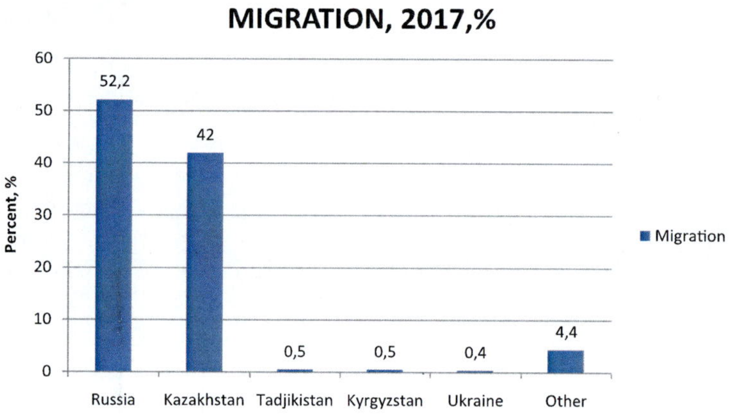

Fig. 4.9 Destinations of migration from Uzbekistan in 2017 (% of total). (Compiled by the author, using data from the State Committee of the Republic of Uzbekistan on Statistics (2017a, b) (public domain))

Table 4.10 Personal remittances received in 2006–2016

	Year										
Remittances	2016	2015	2014	2013	2012	2011	2010	2009	2008	2007	2006
% of GDP[a]	3.688	4.577	9.241	11.595	10.986	9.312	7.266	6.146	10.175	7.588	5.179
Remittances (US$ millions)[b]	–	3104	5828	6689	5693	4276	2858	2071	3007	1693	898

GDP gross domestic product, *IOM* International Organization for Migration
[a]Source: World Bank Group (2017a, b) © World Bank. http://hdl.handle.net/123123123123/123 License: Creative Commons Attribution license (CC-BY-4.0) via World Bank Group (public domain)
[b]Source: International Organization for Migration (http://gmdac.iom.int/remittance-inflows)

US$2.3 billion (Fig. 4.10), including US$600 million of imported foreign currency (Table 4.11).

4.5 The Economy

Historically, Uzbekistan's economy was of the agrarian type. At the end of the nineteenth century, cotton production became a high priority in the economic policy of the Russian Empire. This policy was strengthened during the Soviet era. Despite ongoing efforts to diversify crops, Uzbekistani agriculture remains largely centered on cotton. Uzbekistan is the world's seventh leading cotton-exporting country, the second in cotton fiber exports, and the sixth largest cotton producer in the world. A large proportion of Uzbekistan export revenues derive from cotton production. This export potential has resulted in establishment of several textile companies and a garment industry. According to O'zto'qimachilik, one of the policy priorities of Uzbekistan is further development of its textile industry.

Fig. 4.10 Structure of international money transfers received by the population in January–June 2018. (Compiled by the author, using data from Gazeta.uz (2018) and Central Bank of Uzbekistan (2018))

Table 4.11 International money transfers to and from Uzbekistan in January–June 2018

	Country (US$ millions)				
Transfers	Russia	Kazakhstan	Türkiye	South Korea	Israel
Received	1700	136.2	47.6	4.7	2.0
Sent	201.0	32.4	102.6	47.1	25.5

Source: Central Bank of Uzbekistan (2018)

Table 4.12 Growth of the economy in 2007–2016

	Sector (%)			
Growth	Agriculture	Industry	Construction	Services
Growth in 2007–2016	190	170	390	250
Average annual growth	6.5	5.6	15.0	9.7

Source: State Committee of the Republic of Uzbekistan on Statistics (2017a, b) via Statistika Qo'mitasi (public domain)

The threat of "Dutch disease"[24] exists for Uzbekistan's economy, which relies mainly on natural resources (cotton, gas, gold, and uranium). Today, this country needs structural transformation and diversification of its economy to overcome the disadvantage of its double-landlocked geographic location as well.

In the 1930s, industry was developed under a program of industrialization. During World War II, many industrial plants were transferred to Uzbekistan from the west of the Soviet Union. Today, this industry is a base for further industrial development. However, the state-owned industries were oriented toward the internal market of the Soviet Union. After the collapse of the Soviet Union, they could not compete with more advanced industries in other countries (Western European countries, China, and the USA). Several industries (e.g., the agricultural machinery industry and the aircraft industry) were closed, since they lost their market and faced bankruptcy. New ones (such as the automobile industry) were born. Since Uzbekistan gained its independence in September 1991, the government has maintained its Soviet-style command economy with subsidies and tight controls on production and prices. Uzbekistan's growth has been driven primarily by state-led investments. Exports of natural gas, gold, and cotton provide significant shares of foreign exchange earnings.

The value added to the GDP by agriculture changed over the year with the move from a planned economy to a market economy in 1989–1995. In 2010–2013, the share of the value added by agriculture to the annual growth in GDP remained the same as that of industry. The value added by agriculture to the GDP changed from 27.6% in 1987 to 18.3% in 2015 following a global trend toward shrinking of agriculture.

In 2007–2016, the growth in construction (3.9-fold, with average annual growth of 15.0%) and the growth in services (2.5-fold, with average annual growth of 9.7%) were greater than the growth in agriculture (1.9-fold, with average annual growth of 6.5%) and the growth in industry (1.7-fold, with average annual growth of 5.6%) (Table 4.12).

[24] "Dutch disease" is the negative impact on an economy of anything that gives rise to a sharp inflow of foreign currency, such as discovery of large oil reserves. The currency inflows lead to currency appreciation, making the country's other products less price competitive on the export market (https://markets.ft.com/glossary/term.asp?t=Dutch-disease).

Agriculture was one of the driving forces for economic development, and the annual growth in the value it added to GDP (6%) in 2010–2015 confirms this fact. However, in 2017, the annual growth in the value added by agriculture, forestry, and fishing dropped to 2% (Table 4.13; Fig. 4.11). The growth in this added value was about the same as the growth in added value of industrial production.[25] For comparison, industrial production growth in Uzbekistan was 6.6% in 2016 (Fig. 4.12). For compiling these charts (Fig. 4.11 and Fig. 4.12), the production, or value added, approach is used. The production, or value added, approach consists of calculating a sector's output and subtracting its intermediate consumption (the goods and services used to produce the output) to derive its value added.

In 2018, the GDP was USD 52.63 billion (in current rate) and the GDP per capita was USD 1,951.7.[26] Services contributed more to GDP than industry and agriculture (Table 4.14; Fig. 4.13).

According to official data,[27] the contribution to GDP of the mining industry rose from 17.1% in January–June 2017 to 25.1% in January–June 2018, while that of manufacturing fell from 72.5% to 66.8% over the same period. The greater growth in the contribution to GDP of services in January–June 2018 is a new trend in the country's economic development (Fig. 4.14).

Since Uzbekistan gained its independence in September 1991, the government has maintained its Soviet-style command economy with subsidies and tight controls on production and prices. Uzbekistan's economic growth has been driven primarily by state-led investments. Exports of natural gas, gold, and cotton provide significant shares of foreign exchange earnings.

Table 4.13 Value added in agriculture, forestry, and fishing

Year	Annual growth (%)
1988	9.82
1989	−3.456
1990	5.818
1991	2.217
1992	−6.469
1993	1.5
1994	−3.4
1995	2
1996	−5.7
1997	5.8
1998	4.1
1999	5.5
2000	3.2
2001	4.1
2002	6.0
2003	6.8
2004	10.1
2005	6.2
2006	5.305
2007	6.1
2008	4.5
2009	5.7
2010	6.8
2011	6.6
2012	7.0
2013	6.8
2014	6.9
2015	6.8
2016	6.6
2017	2.0

Sources: World Bank national accounts data and Organisation for Economic Co-operation and Development (OECD) national accounts data. License: CC-BY-4.0 via World Bank Data

4.6 Agriculture

Since Uzbekistan became independent, several agricultural reforms have been implemented to diversify its agricultural output from cotton production into production of wheat and other food products. The area under cotton was reduced and the structure of the sown area under agricultural crops was also changed, and these changes affected the increase in their production. Between 2000 and 2016, according to data from the State Committee of the Republic of Uzbekistan on Statistics (2017b), the volume of the gross agricultural output[28] was multiplied 2.7 times, the volume of crop production by 2.8 times, and livestock production by 2.6 times; the average annual growth was 6–7%. The structure of the sown area under agricultural crops was changed, and this affected the increase in their production (Table 4.15). At the same time, the production of industrial crops decreased from 40.0% to 36.0% and fodder crops from 11.4% to 9.0%. In 2016, cotton production was 98.6% of the raw cotton production in 2000. The area under food crops also expanded; in 2016, 21.0 Mt of fruit and vegetables were produced to meet the demands of the population.

The production of the main agricultural products per capita was multiplied by about 2–3 times (Table 4.16).[29]

In January–September 2017, the total volume of the products (services) of agriculture, forestry, and fishing was UZS 46.1943 trillion (US$1 was equal to UZS 8100 after devaluation on September 5, 2017) (Table 4.17).

[25] In Uzbekistan, industrial production measures the output of businesses integrated into the industrial sector of the economy, such as manufacturing, mining, and utilities.

[26] Source: State Committee of the Republic of Uzbekistan on Statistics (2018).

[27] Source: State Committee of the Republic of Uzbekistan on Statistics (2018).

[28] Gross agricultural output is the value of the volumes of crops and livestock production.

[29] Source: State Committee of the Republic of Uzbekistan on Statistics (2017a, b).

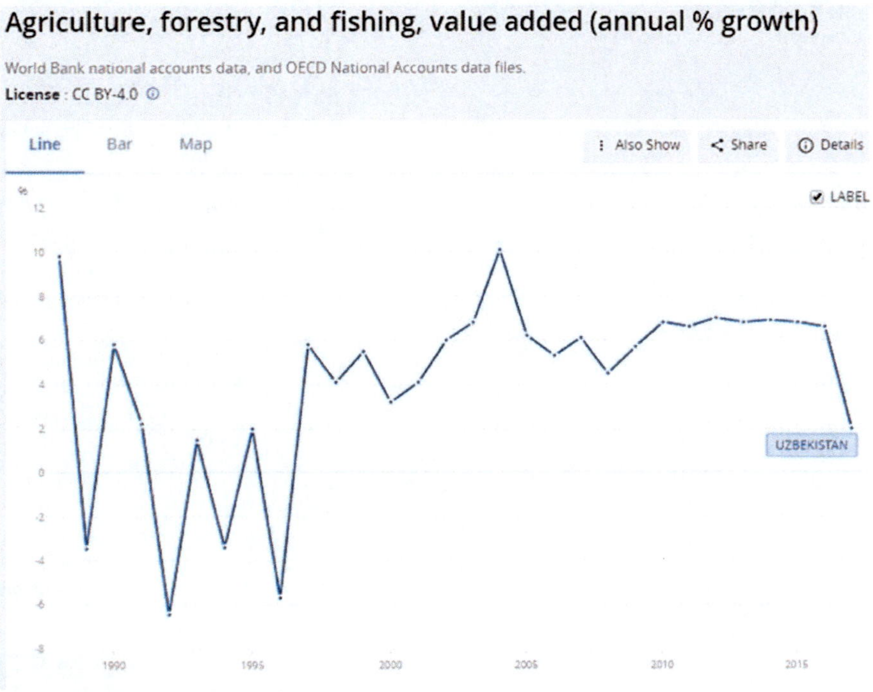

Fig. 4.11 Annual percentage growth in value added to gross domestic product (GDP) by agriculture, forestry, and fishing. (Source: World Bank Data (CC-BY-4.0) via World Bank)

Fig. 4.12 Value added by industry (including construction) in current US dollars. (Source: World Bank Data (CC-BY-4.0) via World Bank)

In January–September 2017, the total volume of products (services) of agriculture, forestry, and fishing rose in comparison with the same period of 2016.

The location of Uzbekistan in arid and semiarid climate zones leads to dependence of the main economy on the water and energy resources of Uzbekistan. Historically, irrigation is of high priority over other water uses. One of the challenges is seasonal fluctuations, leading to scarcity of water in many regions of the country in summer. Researchers (Hansjürgens 2016; Huttl et al. 2016) note that poor water management, in addition to outmoded infrastructure and inefficient use of water, creates problems for farmers.

Experts (Hansjürgens 2016) argue that partial systematic changes in a societal system may lead to failure of transformation from centralized planning to a market economy. Dukhovny and Stulina (2011) point to the additional challenges in water management in agriculture, and thus to less

Table 4.14 Macroeconomic indicators in January–June 2018

	GDP (UZS billions)	Growth since January–June 2017 (%)
GDP	152,533.4	4.9
Sector contributions to GDP		
Industry	39,865.2	6.1
Agriculture	21,863.9	2.7
Construction	9649.3	9.7
Services	12,499.3	1.7
Transportation and communication services	13,839.8	5.6
Other services	32,727.5	5.0

Source: State Committee of the Republic of Uzbekistan on Statistics (2018) via Statistika Qo'mitasi (public domain)
US$1 = UZS 7800 (source: Central Bank of the Republic of Uzbekistan (July 2018)
GDP gross domestic product

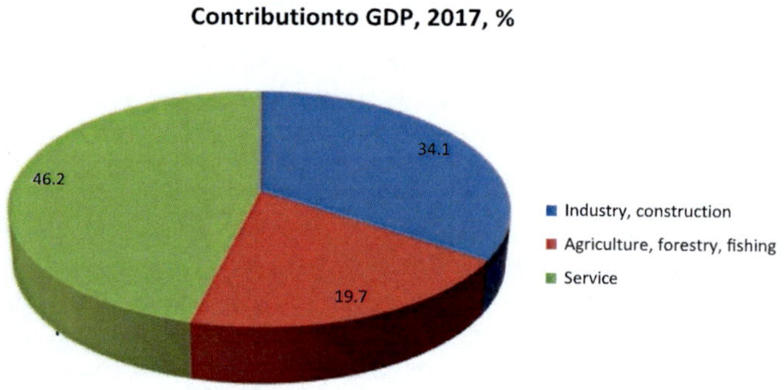

Fig. 4.13 Percentage contribution to gross domestic product (GDP) by type of activity in January–December 2017. (Compiled by the author, using 2018 data from the State Committee of the Republic of Uzbekistan on Statistics (public domain))

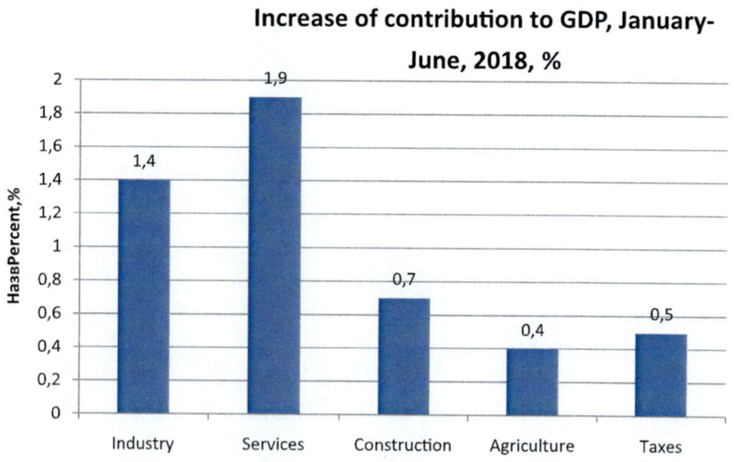

Fig. 4.14 Percentage contribution to gross domestic product (GDP) by kind of economic activity in January–June 2018. (Compiled by the author, using 2017 data from the State Committee of the Republic of Uzbekistan on Statistics via Statistika Qo'mitasi (public domain))

Table 4.15 Changes in production of agricultural crops between 2000 and 2016

Year	Crop production (%)			
	Grains	Potatoes	Vegetables	Gourds
2000	42.7	1.4	3.4	1.0
2016	45.6	2.3	5.5	1.6

Source: State Committee of the Republic of Uzbekistan on Statistics (2017a, b) (public domain)

Table 4.16 Changes in production of agricultural crops per capita between 2000 and 2016

Year	Crop production (kg)				
	Grains	Vegetables	Fruit	Meat	Milk
2000	159.0	107.0	32.0	34.0	147.0
2016	259.0	354.0	96.0	68.0	305.0

Source: State Committee of the Republic of Uzbekistan on Statistics (2017a, b) (public domain)

Table 4.17 Volume and growth of production (services) in agriculture between January–September 2016 and January–September 2017

	Crop, livestock production, hunting, and related services	Forestry	Fishing
Value (UZS billions)	45,982.3	63.4	148.6
Growth in 2016–2017 (%)	3.5	1.4	11.9

Source: State Committee of the Republic of Uzbekistan on Statistics (2017a, b) (public domain)

efficient use of water and land resources, and as a result of inadquate water pricing.

A strong linkage exists between irrigation and socioeconomic stability. The irrigated agriculture in this country is the major branch of food production. The main human settlements are located along big rivers and their tributaries. The existence of these settlements has been dependent on availability of water. It is water that has motivated progress in culture and sciences—an important factor in regional cooperation. Experts (Dukhovny and Stulina 2011; Khudayberganov 2011) point to the importance of cooperation on water governance at the regional level, since the annual per capita freshwater resource availability is decreasing. They agree that the management and cooperation mechanism in the region is not yet developed to a satisfactory level, leading to poor solutions to water-related problems in the Amu Darya river basin.

The trend toward a decrease in the water supply under the current practice of irrigation is a troubling sign from an economic point of view. An estimate by Yokubzod (2011) shows that the cost of putting new land under irrigation is increasing. In the 1980s, it was US$3000–8000/ha; by 2011, it was already up to US$12,000–15,000/ha, and by 2020, it was predicted to reach US$20,000/ha.

Table 4.18 Growth in food production between January–September 2016 and January–September 2017

Period	Growth in food production (%)				
	Potatoes	Vegetables	Fruit and berries	Grapes	Meat and milk
2016–2017	103.9	105.8	105.3	104.5	106.2

Source: State Committee of the Republic of Uzbekistan on Statistics (2017a, b) (public domain)

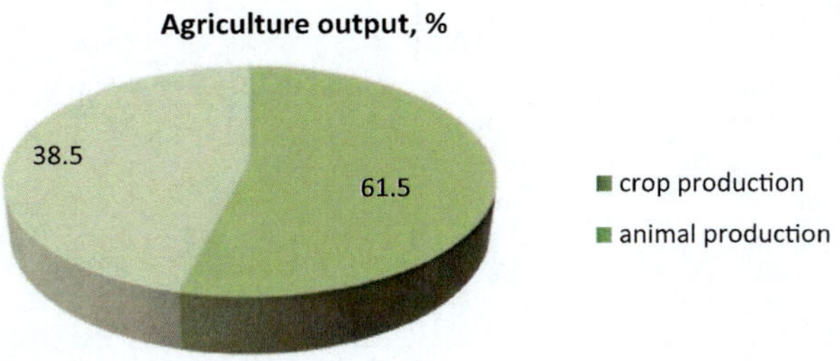

Fig. 4.15 Agricultural output in January–September 2017. (Compiled by the author, using 2017 data from the State Committee of the Republic of Uzbekistan on Statistics (public domain))

Food Security

Land use in Uzbekistan is composed of arable land (62.6% of all agricultural land), forest (7.7%), and other land (29.7%).

Uzbekistan occupies one of the leading positions in production of fruit, grapes, silk cocoons, karakul, and wool in Central Asia. In the 1990s, state-owned agricultural enterprises were reorganized into different private ones during structural and institutional transformations. This reform resulted in an increase of 1500 km^2 in the total area occupied by private farms, which now occupy more than 6500 km^2. Reforms occurred in land reallocation and crop production as well. Cotton planting was reduced and stabilized at 15,000 km^2. Grain self-sufficiency was achieved, plus stable production of over 5 Mt of fruit and vegetables.

Food production rose as well (Table 4.18). In 2017, crop production accounted for 61.5% of the total agricultural output and animal production accounted for 38.5% (Fig. 4.15).

In January–September 2017, the highest rates of growth in agricultural production were observed in the Samarkand (Samarqand in Uzbek), Surxondaryo, Bukhara (Buxoro in Uzbek), and Navoiy regions (Table 4.19).

In January–September 2017, the total value of crop production was UZS 27.8697 trillion (US$1 = UZS 7800)[30] (Table 4.20).

In January–September 2017, the value of animal production (Table 4.21) was UZS 17.4524 trillion.

In accordance with the program of the government of Uzbekistan for reforms in agriculture, lands under cotton are

[30] Source: Central Bank of the Republic of Uzbekistan (July 2018).

Table 4.19 Volume and growth in products (services) of agriculture, forestry, and fishing between January–September 2016 and January–September 2017

Region	Value of products (UZS billions)	Growth (%)
Republic of Karakalpakstan	1459.8	104.8
Andijon	4401.1	100.8
Bukhara	3337.2	105.8
Jizzakh	2444.7	103.2
Kashkadarya	3735.5	102.2
Navoiy	2180.2	105.6
Namangan	2493.1	104.6
Samarkand	6484.3	106.1
Surxondaryo	4297.0	106.0
Sirdaryo	1699.4	102.2
Tashkent	7231.9	101.9
Fergana	3752.1	101.5
Xorazm	2678.0	102.0

Source: State Committee of the Republic of Uzbekistan on Statistics (2017a, b) (public domain)

Table 4.20 Production of the main crops in January–September 2017

Region	Crop production (kt)			
	Grains	Potatoes	Vegetables	Fruit and berries
Republic of Karakalpakstan	182.2	44.3	192.5	33.1
Andijon	612.4	216.4	1132.9	381.7
Bukhara	580.2	181.9	591.7	204.1
Jizzakh	535.1	45.6	401.2	68.2
Kashkadarya	1002.1	149.1	448.3	118.3
Navoiy	274.7	60.9	186.8	70.8
Namangan	458.0	191.2	506.6	167.5
Samarkand	829.7	472.7	1472.3	300.1
Surxondaryo	635.7	261.7	822.0	113.2
Sirdaryo	555.1	49.1	279.0	36.0
Tashkent	651.7	330.9	1509.6	159.5
Fergana	761.6	253.4	760.1	350.1
Xorazm	302.0	100.2	392.2	143.4
Total	7380.5	2.357.4	8695.2	2146.0

Source: State Committee of the Republic of Uzbekistan on Statistics (2017a, b) (public domain)

significantly decreasing. Uzbekistan now has enough irrigated land to produce sufficient amounts of cereals for its population. Winter wheat is cultivated on irrigated lands. This cereal is grown for making bread. However, farmers are not encouraged to produce wheat of higher protein and quality grain. Today, wheat and flour of better quality are imported from Kazakhstan. Experts from the Food and Agriculture Organization of the United Nations (FAO) (Kirsten et al. 2011) have suggested implementing special measures, including offering adequate reimbursement for improving the quality of the grain and increasing its market and export value.

Table 4.21 Production of the main livestock products in January–September 2017

Region	Products			
	Meat (kt)	Milk (kt)	Wool (t)	Eggs (millions)
Republic of Karakalpakstan	71.6	207.2	935	188.6
Andijon	103.8	564.5	1196	333.4
Bukhara	159.5	633.7	2825	262.0
Jizzakh	143.6	440.6	3406	212.8
Kashkadarya	207.0	814.7	6969	258.1
Navoiy	110.4	317.0	3478	244.8
Namangan	105.5	501.9	1516	325.0
Samarkand	200.5	949.2	3167	1175.4
Surxondaryo	117.7	519.6	1845	217.2
Sirdaryo	51.3	252.5	537	121.9
Tashkent	179.3	678.4	2106	919.7
Fergana	107.5	–	980	325.0
Xorazm	116.5	800.4	1152	333.9
Total	1674.2	7327.9	30.1	4917.8

Source: State Committee of the Republic of Uzbekistan on Statistics (2017a, b) (public domain)
US$1 = UZS 7800 (source: Central Bank of the Republic of Uzbekistan (July 2018))

Another economic goal to be achieved is an increase in the productivity of each irrigated hectare. Previous projects demonstrated that capacity building and introduction of innovative modern production management systems led to an increase in fish production in the Surxondaryo region and milk production in the Andijan region.

In Uzbekistan, the highlands constitute 36,598 km^2 out of 425,400 km^2 (9% of the total land area), and they have considerable potential for agricultural development. According to the Annual Accumulated Cold Units[31] (AACU) classification, these lands are categorized mainly as cool and partly as cold (Roozitalab et al. 2013). In hilly and mountainous highlands, the main risk to agriculture is erosion, which may be exacerbated by the lithological properties of the local geology and various precipitation patterns.

Opportunities exist for diversified land use patterns through exploitation of microclimatic conditions, particularly related to differences in thermal regimes arising from different slopes and slope orientations, as well as possibilities for water harvesting. This approach could offer an additional land resource. The current land use/land cover and farming systems show the low level of using potential of these lands. Pastoral farming[32] is the main farming system.

[31] Annual accumulated cold units (AACU) refers to the total number of degrees Celsius below zero, based on the daily minimum temperature, summed over the entire year.

[32] Pastoral farming is based on mobility of flocks and herds, moving between more humid and drier areas, depending on the availability of grazing and water.

One of the advantages is that at least 70% of the highlands have growing periods of 7 months. In this country, more than 20% of the highlands are used for rain-fed crops. It is worth noting that a study on improvement of agricultural systems in the highlands is in progress. Researchers (Roozitalab et al. 2013) point to this potential for increasing food production with integration of technical and institutional options for increasing farm incomes and improving natural resource management.

4.7 Industry

Several sectors of the economy of Uzbekistan are well developed. A policy has been undertaken for diversification of the economy and growth of new industries. One of the goals of the government of Uzbekistan is to attract investment for modernization and technological re-equipment of enterprises that specialize in industrial processing.

According to the State Committee of the Republic of Uzbekistan on Statistics (2017a, b), industrial production growth in Uzbekistan increased by 6.6% in 2016 after average growth of 19.91% between 2004 and 2016.

One of the major contributors to the economy of Uzbekistan is industry, ranging from textiles to food processing, machine building, metallurgy, mining, hydrocarbon extraction, and chemicals. In terms of industry's percentage contribution to GDP, Uzbekistan ranked 46th out of 176 countries in 2016, with a percentage of 33.90%[33] (Countries of the World 2017).

In 2003, the government accepted Article VIII obligations under the IMF, providing full currency convertibility. However, since then, the policy of strict currency controls and tightening of borders have resulted in negative effects on economic activity.

Experts (Countries of the World 2017) have noted that because of the currency shortage and the worsening of Uzbekistan's trade and investment as a result of lower global commodity prices and an economic slowdown in neighboring Russia and China, the government has begun to revise its economic strategy. In 2017, the government started programs to improve the investment climate.

According to official data from the State Committee of the Republic of Uzbekistan on Statistics (2018), in January–September 2017, industrial enterprises produced goods to the value of UZS 99.0605 trillion. In the structure of industrial production, the largest share pertains to manufacturing, mining, and quarrying (Fig. 4.16). The share of small business in the total industrial output is 41.2%, while in manufacturing, it is 51%.

Industrial production classified by kinds of economic activity is represented in Table 4.22.

This program has been initiated by the government to encourage local manufacturers to produce finished products, components, and materials. This program is known as the localization program, and manufactured goods are known as localized products. Exports of such goods were valued at US$350.9 million in 2017. The increase in the production of finished products contributed to an increase in consumer

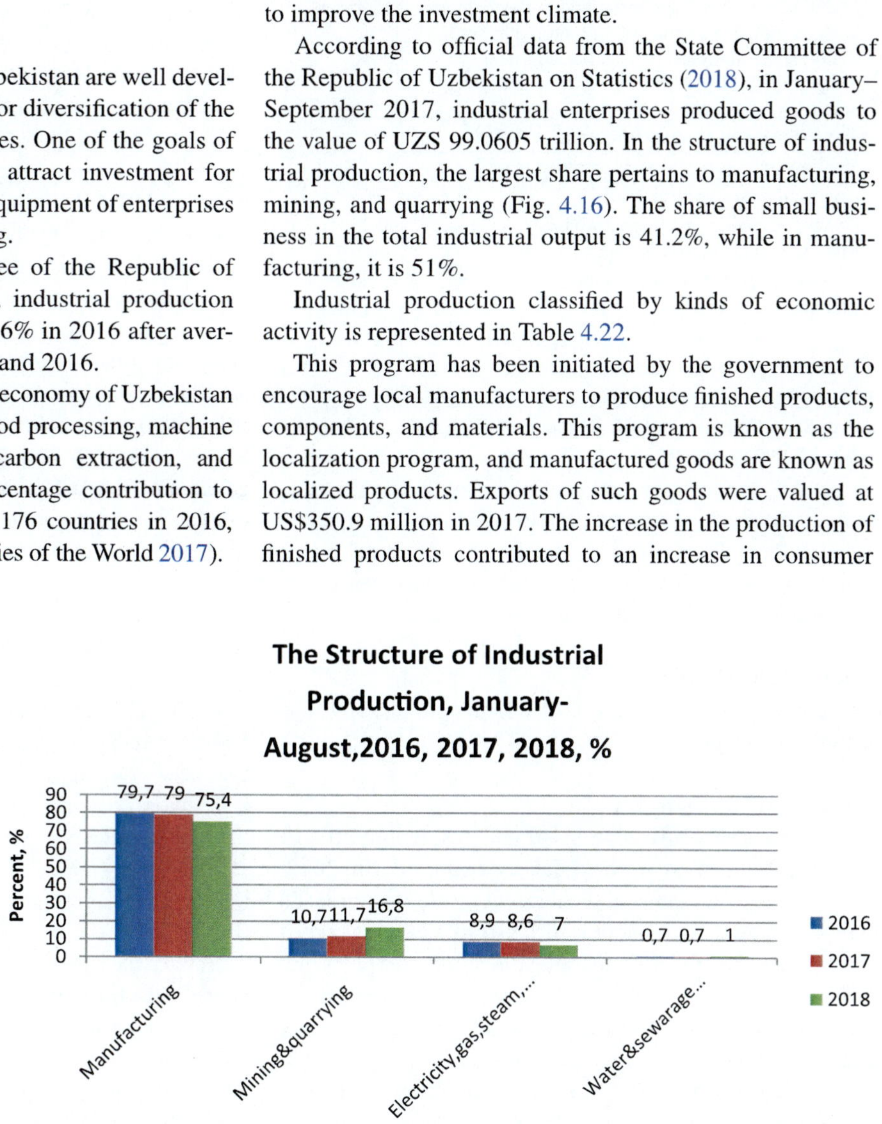

Fig. 4.16 Structure of industrial production classified by kind of economic activity (as a percentage of total industrial output). (Compiled by the author, using 2017 and 2018 data from the State Committee of the Republic of Uzbekistan on Statistics (public domain))

[33]This shows where production takes place in an economy; the distribution gives the percentage contributions of agriculture, industry, and services to the total GDP. Agriculture includes farming, fishing, and forestry. Industry includes mining, manufacturing, energy production, and construction. Services cover government activities, communications, transportation, finance, and all other private economic activities that do not produce material goods.

Table 4.22 Industrial production by different kinds of economic activity in January–September 2017

Activity	Value (UZS billions)
Total	99,060.5
Mining and quarrying	11,550.2
Coal, lignite, oil, natural gas	6812.2
Metallic ores	3674.0
Other mining activities	668.7
Manufacturing	78,259.9
Food, beverages, tobacco	21,238.5
Textiles, apparel, leather articles	14,650.5
Manufacture, repair, and installation of machinery and equipment; manufacture of motor vehicles, trailers, semi-trailers, other fabricated metal products	12,355.1
Goods made from wood, cork, straw, wicker; paper and paper-based products; furniture	2207.4
Printing and reproduction of recorded materials	904.2
Coke and petrochemicals	2663.4
Chemical, rubber, and plastic products	8724.6
Main pharmaceutical products	1160.3
Other nonmetallic mineral products	5140.7
Metallurgy	7847.7
Electricity, gas, steam, air conditioning supply	8529.7
Water supply, sewerage, waste management, remediation	720.7

Source: State Committee of the Republic of Uzbekistan on Statistics (2017a, b) (public domain)

goods from 39.8% of total in 2016 to 41.5% of total in 2017. The structure of industrial production of consumer goods changed as well. The share of nonfood consumer goods increased from 50.7% in 2016 to 53.6% in 2017.

Sectors of the Economy

The economy of Uzbekistan is diversified, and many new industries have grown.

The Oil and Gas Industry

Uzbekistan is self-sufficient in oil and gas. About 20% of the gas produced by Uzbekistan is exported. In 2018, exports of hydrocarbons and chemical products increased by 66.2%: gas exports increased 1.9-fold, chemical product exports by 7.3%, and plastic goods exports by 16.7%.[34]

According to data from the national holding company (NHC) Uzbekneftegaz (a state-owned company), "210 fields with free gas deposits and 125 oil fields contribute 15% to national GDP." The enterprises in this industry are united under NHC Uzbekneftegaz, comprising five Departments:

1. The Mubarek oil and gas production Department
2. Shurtan Gas Fields Department
3. Ustyurt Gas Production Department
4. Gazli Oil and Gas Production Department
5. Vodiy Oil and Gas Production Department

Uzbekneftegaz reports that "about 93% of the power system of the country is provided by this industry with the primary fuel resources. It is estimated that the annual production of gas is around 65 billion m^3 (eighth position in the world) and the annual production of liquid hydrocarbons is 5.9 million tonnes." In 2017, according to official data from the State Committee of the Republic of Uzbekistan on Statistics (2018), natural gas production was 56.4×10^9 m^3. Nearly 5% of the produced raw resources undergo refinery, with production of 197 kt of liquefied gas and 125 kt of polyethylene. The potential export value all produced gas is estimated at US$1.4 billion. The processing and production of finished products (polyethylene, polypropylene, and other petrochemical products) will allow the ready-made production value to be increased 10-fold.

Such renowned overseas companies as Lukoil and Gazprom (Russia), CNPC (China), the Korean National Oil and Gas Company (KNOC; Korea), Petronas Charigali (Malaysia), and others are operating in this field. In 2015, Russia's Gazprom announced that it would reduce its natural gas imports from Uzbekistan. However, Tashkent continues to export natural gas to China, and Chinese investments in the country have substantially increased.

4.7.1 The Chemical Industry

The chemical industry is one of the key industries in Uzbekistan. The development of this industry was related to the need for agricultural fertilizers. According to JSC Uzkimyosanoat, the first chemical enterprises were established when the Shursu sulfur mine started operation in the 1930s (Table 4.23). Today, different enterprises produce various chemical products at sites near the relevant resources (Fig. 4.17).

Today, the chemical industry in Uzbekistan is the most powerful in the Central Asian region. It is a top producer of chemical compounds and mineral fertilizers. Uzkimyosanoat (2018) reports that gas, oil, coal, sulfur, natural paraffin, table salt, various types of ferrous metal waste, processed raw cotton, and kenaf[35] are used as raw materials.

[34] Source: State Committee of the Republic of Uzbekistan on Statistics (2018).

[35] *Hibiscus cannabinus* is a plant in the Malvaceae family, also called Deccan hemp and Java jute. It requires a minimum of fertilizers, pesticides, and water. The history of its cultivation for food and fiber dates back about 4000 years. It is easily grown and is high in yield.

Table 4.23 Chemical industries in Uzbekistan

Enterprise	Year of founding	Products	Countries receiving exports
Shursu sulfur mine	1932		
Chirchik electrochemical plant (OJSC Maxam-Chirchiq)	1940	Ammonium nitrate, urea, ammonia, ammonia water, nitric acid, catalysts, ammonium carbonate, carbon dioxide, urea–ammonia mix, low-density nitrate of ammonium, catalysts for ammonia manufacture, magnesium sulfate, other	Russia, Kazakhstan, Tajikistan, Afghanistan, Turkmenistan, Kyrgyzstan, China, Iran, Türkiye
Fergana nitrogen fertilizer plant (JSC Farg'onazot)	1962	Ammonium nitrate, carbamide, ammonium sulfate	–
Navoiy chemical plant (JSC Navoiyazot)	1964	Ammonium nitrate, carbamide, ammonium sulfate	–
JV MAXAM-Uzbekistan	2001	Ammonium nitrate, diesel fuel, fuel oil, aluminum powder	–
OJSC Samarkankimyo	–	Phosphorus fertilizers, ammophos, superphos, simple ammoniates, superphosphate, ammonium superphosphate, ammonium sulphosphate, nitrocalcium phosphate	Tajikistan, Afghanistan, Kyrgyzstan, Iran
OJSC Kokand superphosphate plant	1935	Ammoniated superphosphate, acrylic water-dispersive emulsion, phosphorus fertilizers, ammonium superphosphate, ammonium sulphosphate, nitrocalcium phosphate	Tajikistan, Kyrgyzstan
OJSC Jizzakh Plastic	1997	Polyethylene film, polyethylene pipes (sizes from 315 to 630 mm in diameter), fittings trinitrophenol (TNP), connection details for polyethylene pipes	Kazakhstan, Tajikistan
OJSC Ammofos-Maxam	1969	Ammofos, ammonium sulfate, phosphate, superphosphate, ammonium sulfate, fiberboard, electrolytes, sulfuric acid, fodder phosphate of ammonium (for use only in cattle breeding, poultry farming, and fish culture), trisodium phosphate	Tajikistan, Afghanistan, China, Iran
UE Dekhanabad potassium fertilizer plant	–	Potassium chloride (GOST 4568-95), salt tech	Sri Lanka, United Arab Emirates, Iran, Kazakhstan, Georgia, Vietnam, Malaysia, China
JV-PC Elektrohimzavod	–	Plant protection chemicals, superphosphate, caustic soda	Tajikistan, Kazakhstan
UE Kungrad soda plant	2005	Soda ash, salt tech	Kazakhstan, Türkiye, Russia
Navoiy Hunan Pulp	2010	Hygiene products, children's diapers	Local
JV-PC Navoiy Beauty Cosmetics	2010	Cosmetics	Local
NeoPlant (Navoiy and Angren industrial zones)	2012	Industrial silicon	–

Source: JSC Uzkimyosanoat (2018) (public domain)

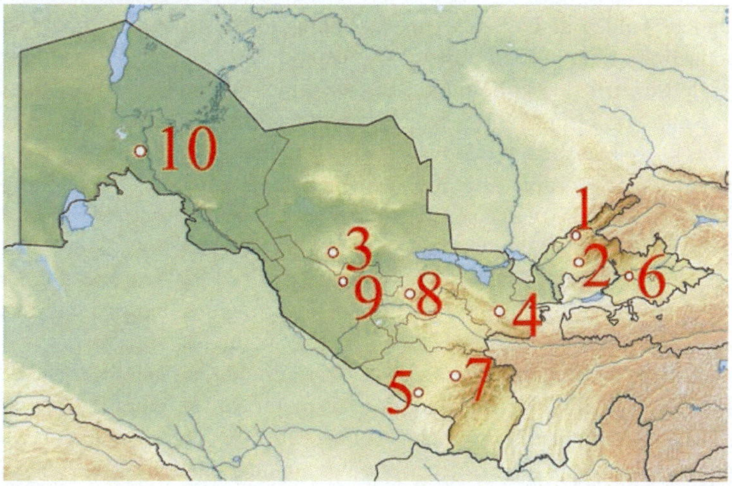

Fig. 4.17 The chemical industry. On the map, *1* indicates OJSC MAXAM (Chirchiq); *2* indicates Ammofos-Maxam (Almalyk); *3* indicates Navoiy Hunan Pulp and Navoiy Beauty Cosmetics (Navoi); *4* indicates Jizzakh Plastic; *5* indicates Shurtan Gas Chemical Complex; *6* indicates OJSC Kokand Superphosphate Plant; *7* indicates UE Dekhanabad Plant of Potassium Fertilizers; *8* indicates OJSC Samarqandkimyo; *9* indicates JV-PC Elektrohimzavod (Navoi); and *10* indicates UE Kungrad Soda Plant. (Source: CC-BY-SA 3.0 File: Uzbekistan location map.svg: NordNordWest derivative work Виктор_В (https://commons.wikimedia.org/wiki/File: Relief_Map_of_Uzbekistan.png), "Relief Map of Uzbekistan"; addition of objects with their names by Lola Gulyamova, https://creativecommons.org/licenses/by-sa/3.0/legalcode (public domain))

Like many other industries, the chemical enterprises are united under the state JSC Uzkimyosanoat. According to Uzkimyosanoat (2018) data, the enterprises are divided into four main production complexes by types of products:

1. Production of mineral fertilizers, organic and inorganic substances, and chemical reagents for energy power, gold mining, and the chemical industry
2. Production of organic chemicals, artificial fibers, and polymers
3. Production of chemical plant protection products
4. Production of soda ash

This industry attracts foreign investment for its development, modernization, and re-equipment. Some enterprises have been established with shares held by foreign investors from Austria, Russia, Türkiye, India, and China.

In 2007, new development of the industry started for modernization and re-equipment. Uzkimyosanoat (2018) reported that the industry began to supply its first new competitive products made with use of the most advanced technologies. Large deposits of natural gas, condensate, oil, sulfur, phosphorus, and sodium carbonate allow for long-term strategic planning in the industry and increases in its exports.

According to Uzkimyosanoat (2018), during re-equipment and modernization, the large chemical manufacturers (Samarqandkimyo, Farg'onaazot, Navoiyazot) focused their efforts on "green technologies" to cut harmful emissions and pollutants. In addition, all companies in the industry use sophisticated ecoprotection systems, minimizing the environmental impacts of their production processes.

In January–August 2018, Uzkimyosanoat (2018) reported that this industry produced goods valued at UZS 2.5 trillion.[36] This sum included production of 780.6 kt of mineral fertilizers (567.8 kt of nitrogen fertilizer, 92 kt of phosphorus fertilizer, and 120.8 kt of potassium fertilizer).

The chemical production is also managed by SGCC, which is part of NHC Uzbekneftegaz. According to SGCC, its annual production capacity is 3.9 billion × 10^9 m^3 (Shurtan Gas Chemical Complex 2018). SGCC produces polyethylene (125 kt/year), granulated sulfur (1500 kt/year), and finished products in the form of pipes and fittings for pressure pipes (more than 2.4 kt/year). Today, approximately 20% of polyethylene is exported to Italy, the Netherlands, Poland, Hungary, Lithuania, Latvia, and Türkiye, Iran, Pakistan, and China, Russia, Ukraine, Kazakhstan, Azerbaijan, Kyrgyzstan, and Tajikistan.

[36] US$1 = UZS 7800 [source: Central Bank of the Republic of Uzbekistan (July 2018)].

4.7.2 Electric Power

Uzbekistan is a major power generator in the Central Asian united energy system, which includes the energy power supply systems of Turkmenistan, Tajikistan, the Kyrgyz Republic, and South Kazakhstan. About 50% of the power-generating supply is concentrated in this country. Electric power in Uzbekistan operates under the state JSC Uzbekenergo (Fig. 4.18). According to data from Uzbekenergo, it "unites 53 enterprises and organizations, including 39 open joint stock companies, 11 unitary enterprises, 2 limited liability companies, as well as its affiliate company Energosotish" (Table 4.24). The biggest hydroelectric power plant (HPP), the Charvak HPP, is on the Chirchiq River in Tashkent region. HPPs are located on other rivers too. On March 27, 2019, a new HPP was commissioned at the Tuyabuguz reservoir in the Urtachirchik district of the Tashkent region. According to UzDaily,[37] this HPP will provide 1600 households with electricity (Figs. 4.19, 4.20, and 4.21).

Thermal power stations (TPSs) with a total capacity of 10.6 GW form the basis of Uzbekistan's energy system, five major TPSs have power units with capacity of 150 through 800 MW. The transmission lines of the company exceed a total length of 235,000 km. These transmission lines, with voltages of 220–500 KW, covering a total distance of 7500 km, provide stable operation of the electric power supply system.

According to 2018 data, in January–June 2018, consumers received electricity of 30.9 TW and heat energy of 3.95 Gcal. The power that was produced for export was valued at UZS 331.6 million.

For further development of this industry, Uzbekenergo is planning to construct new supply sources of energy, replace generation equipment that has already consumed the projected energy resources, meet the demand of the growing energy consumption estimates, introduce modern technologies for generating electricity, and reduce negative environmental impacts of energy-generating facilities. In this regard, four investment projects for construction of solar and wind power plants, including three solar power stations with a capacity of 100 MW each in the Samarkand, Surkhandarya, and Navoi regions and one wind station with a capacity of 102 MW in the Navoi region.

Various international firms and companies are involved in implementation of a number of huge investment projects. One of them involves generation and distribution of electric power: modernization of the Tashkent TPS with loans from

[37] UzDaily, March 27, 2019.

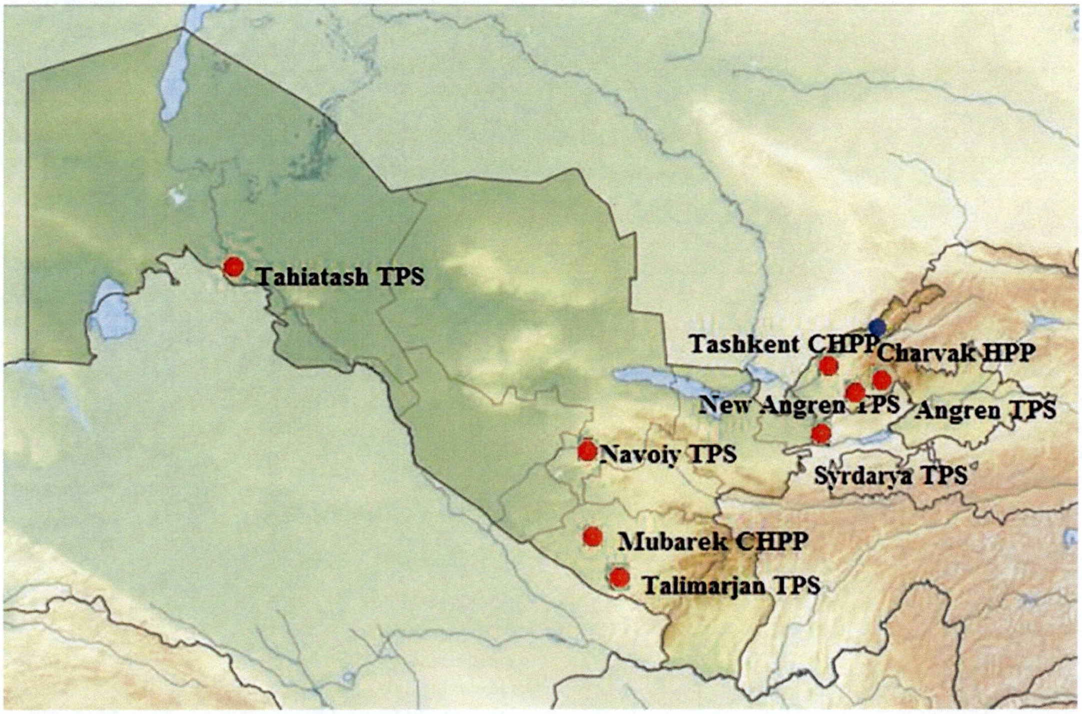

Fig. 4.18 The main power stations in Uzbekistan. *CHPP* indicates a combined heat and power plant, *HPP* indicates a hydroelectric power plant, *TPS* indicates a thermal power station. (Source: File: Uzbekistan location map.svg: NordNordWest derivative work Виктор_В via Commons Wikimedia ("Relief Map of Uzbekistan"); addition of objects with their names by Lola Gulyamova, https://creativecommons.org/licenses/by-sa/3.0/legalcode (public domain))

Table 4.24 The main power plants in Uzbekistan

Name	Installed power (MV)	Location
Hydropower plants		
Charvak HPP	620.5	Tashkent region, Chirchiq River
Xodjikent HPP	165	Tashkent region, Chirchiq River
Tuyamuyun HPP	150	Khorezm region, Amu Darya river
Andijon HPP	140	Andiaon region, Karadarya river
Farxad HPP	126	Syrdarya region, Syr Darya river
Gazalkent HPP	120	Tashkent region, Chirchiq River
Chirchiq HPP-7	84	Tashkent region, Chirchiq River
Tavaksay HPP-8	72	Tashkent region, Chirchiq River
Andijon HPP-2	50	Andijan region, Karadarya river
Gissar HPP	45	Kashkadarya region, Aksu River
Akkavak HPP-10	34.7	Tashkent region, Chirchiq River
Tupalang HPP	30	Surkhanndarya region, Tupalangdarya river
Hishrau HPP-2B	21.9	Samarkand region, Zerafshan River (Dargom Canal)
Ahangaran HPP	21	Tashkent region, Ahangaran River
Lower Bozsu HPP-4	17.6	Tashkent region, Chirchiq River, Lower Bozsu Canal
Aktepa HPP-9	15	Tashkent region, Chirchiq River, Aktepa Canal
Kadyrya HPP-3	13.2	Tashkent region, Chirchiq River, Bozsu Canal
Shahrixan HPP-5A	11.4	Andijan region, Shahrixansay Canal
Lower Bozsu HPP-3	11.2	Tashkent region, Chirchiq River, Bozsu Canal
Salar HPP-12	11.2	Tashkent region, Chirchiq River, Bozsu Canal
Kibray HPP-11	11.2	Tashkent region, Chirchiq River, Bozsu Canal
Lower Bozsu HPP-1	10.7	Tashkent region, Chirchiq River, Bozsu Canal
Akkavak HPP-2	9	Tashkent region, Chirchiq River, Bozsu Canal
Taligulyan HPP-3	8.8	Samarkand region, Zerafshan River (Dargom Canal)

(continued)

Table 4.24 (continued)

Name	Installed power (MV)	Location
Shahrixan HPP-7	7.1	Andijan region, Shahrixansay Canal
Lower Bozsu HPP-2	7	Tashkent region, Chirchiq River, Bozsu Canal
Burjar HPP	6.4	Tashkent region, Chirchiq River, Bozsu Canal
Iyertash PP	6.4	Samarkand region, Zerafshan River (Dargom Canal)
The Southern Fergana Canal HPP-3	6.3	Andijan region, the Southern Fergana Canal
Lower Bozsu HPP-6	4.4	Tashkent region, Chirchiq River, Bozsu Canal
Bozsu HPP-1	4	Tashkent region, Chirchiq River, Bozsu Canal
Shayhontohur HPP	3.6	Tashkent region, Chirchiq River, Anhor Canal
Urgut HPP	3	Samarkand region, Zerafshan River (Dargom Canal)
Taligulyan HPP-1	3	Samarkand region, Zerafshan River (Dargom Canal)
Kudash HPP	2.5	Fergana region, Sox River
Ertashsay HPP	2.2	Tashkent region, Ertashsay River
The Southern Fergana Canal HPP-1	2.1	Andijan region, Southern Fergana Canal
Pskem HPP	404	Tashkent region, Piskom River (under construction)
Kamchik HPP	18	Namangan region, Ahangaran River (under construction)
Tuyabuguz HPP	12	Tashkent region, Ahangaran River (under construction)
Thermal power stations and combined heat and power plants		
Syrdarya TPS (gas)	3050	Syrdarya region
Tashkent TPS	2230	Tashkent region
New Angren TPS (coal gas)	2100	Tashkent region
Talimarjan TPS (gas)	1700	Kashkadarya region
Navoi TPS (gas)	1643	Navoi region
Tahiatash TPS (gas)	730	Republic of Karakalpakstan
Angren TPS	634	Tashkent region
Fergana CHPP (gas)	305	Fergana region
Mubarek CHPP (gas)	60	Kashkadarya region
Tashkent CHPP (gas)	49.5	Tashkent region
Turakurgan TPS (gas)	900	Namangan region

Source: Uzbekenergo (public domain)
CHPP combined heat and power plant, *HPP* hydroelectric power plant, *TPS* thermal power station

Fig. 4.19 The dam at the Charvak hydroelectric power plant (HPP), Tashkent region

Fig. 4.20 The Charvak hydroelectric power plant (HPP) and the Charvak reservoir, Tashkent region

Fig. 4.21 The hydroelectric power plant (HPP) at Yallama, Syr Darya river, Tashkent region

Japan Bank for International Cooperation. Others involve construction of external energy power supply facilities for the city of Tashkent (developed with China Eximbank) and construction of high-voltage transmission lines (developed with Islamic Development Bank). In the near future, it is planned to attract investments worth around US$300 million for development of the industry.

4.7.3 Light Industry

This industry has a long history of development. Over time, availability of various natural resources has stimulated development of diversified production of cotton, silk, wool, and leather. The first industrial enterprises specializing in cotton processing were established in 1874. In 1920s, new enterprises were founded in different parts of the country (Table 4.25; Fig. 4.22).

Today, light industry is one of the priority directions for economic development. According to official data from the State Committee of the Republic of Uzbekistan on Statistics, in 2016, the contribution of light industry to GDP was 3.8%. It is a rapidly developing type of industry (Figs. 4.23, 4.24, 4.25, and 4.26), with annual growth of 18% and exports of 10%. However, experts (Strategic Textile industry 2018) note that its contribution to the economic growth of the country is still much lower than the available capacity.

The textile industry occupies a prominent place in the macroeconomic complex of Uzbekistan. According to data[38] from the O'zto'qimachiliksanoat Association (2017), one-third of all employees are involved in the textile industry. The contribution of this industry to GDP was 2.5% in 2016. The share of light industry in total production was about 15% in 2017. Cotton yarn, cotton fabrics, hosiery, knitwear, and ready-made garments are produced and exported. In Uzbekistan, 1297 textile producers that manufacture finished products and finished goods (textiles, garments, and knitwear) and silkworm farmers are united under the O'zto'qimachiliksanoat Association, which exports these products to 55 countries. The main trading partners are countries in the Commonwealth of Independent States (Table 4.26).

In 2017, the State Committee of the Republic of Uzbekistan on Statistics reported that textile industry enterprises exported products valued at US$1.1 billion. According to the O'zto'qimachiliksanoat Association in 2018, about 7000 enterprises are operating in Uzbekistan with capacity for production of 1.4 Mt of cotton fiber. Sixty percent of this production is used to meet the needs of the domestic textile industry. In January–April 2018, the State Committee of the Republic of Uzbekistan on Statistics (2018) reported that exports of cotton yarn from Uzbekistan were valued at US$424.9 million, representing a 15.7% increase in comparison with the same period in the previous year. This product remains the main one for sale abroad, accounting for 58.7% (worth US$249.4 million) of all textile exports from Uzbekistan. In addition, the top three exports of light industry goods also include cotton fabrics and knitted fabrics; in the first months of 2018, the State Committee of the Republic of Uzbekistan on Statistics (2018) reported that these were valued at US$20.4 million and US$20.6 million, respectively.

[38] All of the data are based on reports from the O'zto'qimachiliksanoat Association (2017, 2019).

Table 4.25 The main light industry enterprises

Industry	Location
Footwear manufacturers	Tashkent (founded in 1926)
	Samarkand
	Kokand
	Bukhara
	Yangiyul
	Chirchiq
	Andijan
	Fergana
Leather manufacturers	Tashkent (founded in 1928)
	Fergana
Garment manufacturer	Fergana (founded in 1926)
Silkworm factories	Fergana (founded in 1921)
	Samarkand (founded in 1927)
	Bukhara (founded in 1928)
	Margilan (founded in 1930)
	Namangan (founded in 1942)
	Urgench
	Shahrisabz
Textile factories	Fergana (founded in 1930)
	Tashkent (founded in 1936)
Cotton yarn factories	Nukus
	Andijan
Cotton spinning mill	Jizzakh
Knitting factories	Tashkent
	Fergana
	Kokand (founded in 1939)
	Samarkand
	Bukhara
	Jizzakh
Carpet factories	Khiva
	Almalyk
Nonwoven material factories	Namangan
	Pap (Namangan province)

Sources: O'zto'qimachiliksanoat 2019; Light Industry in Uzbekistan 2018 (public domain)

In 2019, the O'zto'qimachiliksanoat Association (2019) reported that Uzbekistan had been working on creation of cotton textile clusters. This model implies "the organization of a single production cycle, which includes the cultivation of raw cotton, primary processing and its further processing at cotton ginning (cleaning) enterprises with the production of final products." According to the O'zto'qimachiliksanoat Association (2019), 1409 km^2 of land was allocated for this production.

From the beginning of the 2000s, the volume of cotton exports was constantly decreasing; more cotton remained for further processing inside the country. According to the data of the Permanent Mission of the Republic of Uzbekistan to

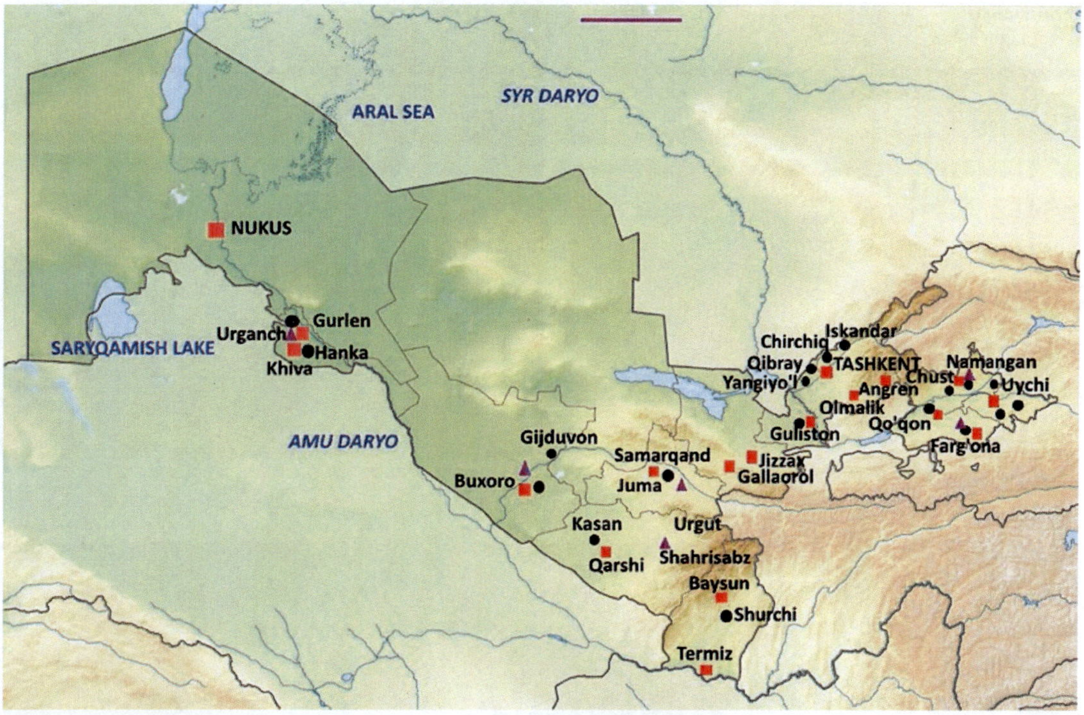

Fig. 4.22 The main light industry enterprises. On the map, indicates textile manufacturing, indicates footwear manufacturing, and indicates silk production. (Source: File: Uzbekistan location map.svg: NordNordWest derivative work Виктор_B via Commons Wikimedia "Relief Map of Uzbekistan"; addition of objects with their names by Lola Gulyamova, https://creativecommons.org/licenses/by-sa/3.0/legalcode (public domain))

Fig. 4.23 This product is manufactured by the silk industry in Fergana

Fig. 4.24 This product is manufactured in Namangan

Fig. 4.25 This traditional footwear is manufactured in Andijan

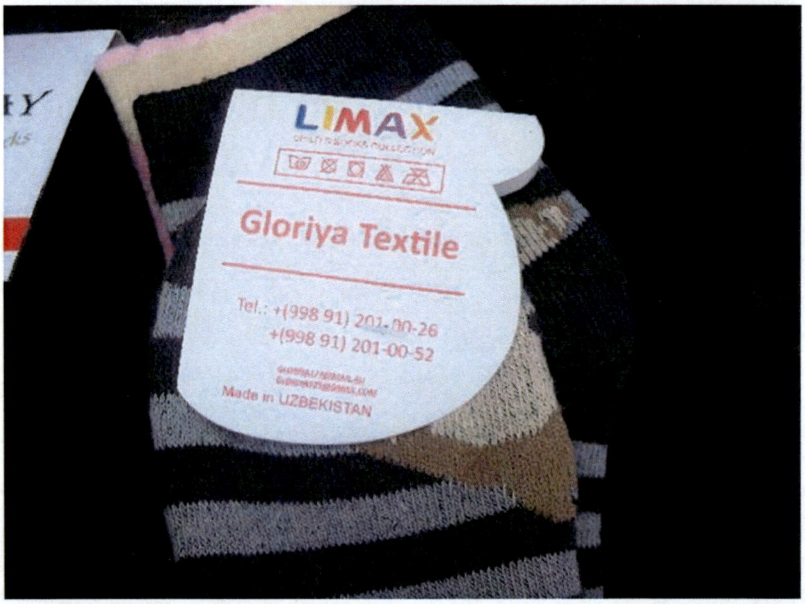

Fig. 4.26 This textile product is manufactured in Tashkent

Table 4.26 Shares of trading partners in textile industry exports in 2017

	Russia and CIS	South Asia	Europe	Middle East and Africa
Share of textile exports (%)	51	21	12	8

Sources: O'zto'qimachiliksanoat Association (2019)
CIS Commonwealth of Independent States

the United Nations in 2016, the proportion of cotton production that was processed by domestic enterprises rose from 7% in 1991 to 30% in 2016. It was planned to gradually reduce the volume of production of raw cotton by 2020, as well as government purchases, from 3.35 Mt to 3 Mt. The prospects for further development of this industry are enormous because this country produces 1 Mt of cotton fiber annually and only 30% is processed at the moment (Strategic Textile Industry 2018). The goal is to increase the volume of internal processing of cotton fiber to 50% and exports of textile products from US$800 million to US$1.5 billion. In Uzbekistan, the bulk of production capacity is concentrated in cotton spinning, so there is an excellent opportunity to invest in creation of spinning and sewing products. According to the O'zto'qimachiliksanoat Association (2019), countries such as Germany, Italy, South Korea, Türkiye, the USA,

India, the UK, and Austria are investors in these enterprises in Uzbekistan.

Uzbek legislation provides for a number of privileges and preferences. For example, enterprises that specialize in production of ready-made clothes (ready-made garments, knitwear, and leather products), hosiery, and shoes are exempted from all types of taxes and liabilities (except for value-added tax), and such enterprises are exempt from customs duties (except for customs formality fees) when importing technological and auxiliary equipment, production tools, and spare parts for their own use.

4.7.4 Production and Processing of Fruit and Vegetable Products

The favorable natural conditions, the strong dependence of the economy on agriculture over a long period of time, and the skills and traditions of farmers determine the significant potential of Uzbekistan's agricultural resources. According to the State Committee of the Republic of Uzbekistan on Statistics (2018), the annual agricultural production is "more than 1.7 million tonnes of gardening and wine growing products, around 5 million tonnes of vegetables and potatoes." Fertile soil, a favorable climate, and the skills and mastership of farmers contribute to the popularity of Uzbekistan in the external markets that sell its fruit and vegetables. In addition to grapes, apples, peaches, pears, watermelons, and melons, subtropical fruit such as figs, pomegranates, persimmons (*khurma*), and some types of citrus fruit are grown in this country.

According to O'zbekoziqovqatxolding, Germany, the Netherlands, Austria, the USA, Türkiye, Russia, Switzerland, Italy, and Spain are share investors in enterprises established for production and processing of fruit and vegetables. However, only 15% is currently processed, although the potential for further development is much greater.

4.7.5 Food

This industry is based on processing of local agricultural raw products. Enterprises in the fat-and-oil, confectionery, and reprocessing industries are united under the O'zbekoziqovqatxolding Fat-and-Oil and Food Industry Association. According to data from O'zbekoziqovqatxolding, it "unites 87 enterprises, including 20 joint-stock companies, 36 joint ventures, 28 private business and 3 unitary enterprises." As in other industries, foreign investors play a role in development of the food industry, while companies such as Nestlé (Switzerland) and Vimm-Bil-Dan-Produkty Pitaniya (Russia) are share investors. Spices, flavorings, infant food, dry breakfast confectionery, and sausage products are just a few of the many types of products created by the food industry.

4.7.6 The Electrical Engineering Industry

In Uzbekistan, this industry produces basic electric engineering products such as wire and cable products, electric lamps, electronic engineering products, power transformers, various electric devices, and engines. The relevant enterprises are united under the Uzeltekhsanoat Association. According to data from Uzbektekhsanoat (2019), "31 enterprises with various property forms, including 14 open joint-stock companies and 11 enterprises with foreign investments founded by the companies from Germany, Russia, UK, Türkiye, Switzerland and Italy" are members of this association. In Uzbekistan, production of copper, silver, gold, zinc, and secondary aluminum has been developed as well.

4.7.7 Construction Material Production

Uzbekistan is self-sufficient in production of construction materials. Significant reserves of construction materials are the basis for this industry. Construction materials enterprises are united under JSC Uzstroymaterialy. In 2019, JSC Uzstroymaterialy (2019) reported that "49 deposits of construction stone, 76 sand and gravel materials, 198 brick and tile raw materials, as well as milestone, construction sand, gypsum, haydite, chocks, talcum are discovered and function." They provide facilities with more than 100 types of basic construction products (cement; lime; gypsum; wall, roofing, and trimming materials (glass, marble, and granite plates); sanitary ceramics; and hydroisolation materials).

This industry is growing rapidly as a result of exemption from customs duties (except for customs formality fees) when importing equipment and accessories required for manufacturing purposes. Another reason is a boom in construction in Uzbekistan and neighboring countries. According to JSC Uzstroymaterialy (2019), foreign investors such as "Knauff (Germany), companies from Spain, USA, and Austria carry out their business in Uzbekistan." There is great potential for increasing production of cement, sanitary ceramics, ceramic plates, heat isolation materials, glues, lacquers, and paints.

4.7.8 The Automobile Industry

Before Uzbekistan became independent, it did not have an automobile industry. According to JSC Uzavtosanoat (2018), its history started in 1996 as the South Korea–based

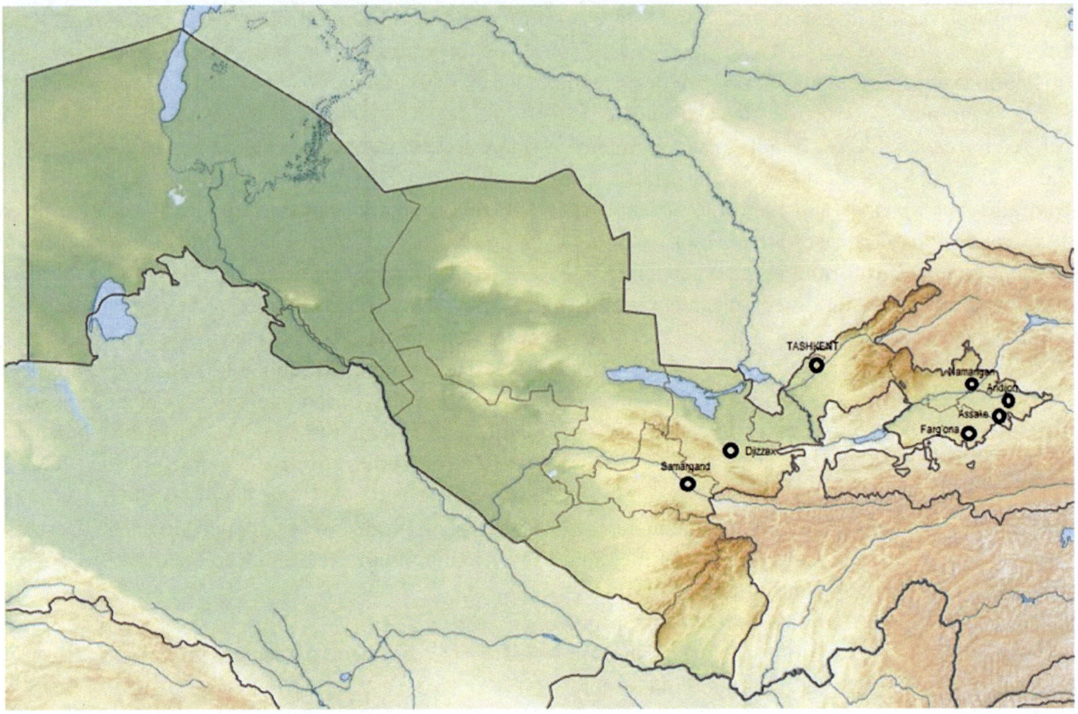

Fig. 4.27 The automotive industry in Uzbekistan. (Source: File: Uzbekistan location map.svg: NordNordWest derivative work Виктор_В (https://commons.wikimedia.org/wiki/File: Relief_Map_of_Uzbekistan.png), "Relief Map of Uzbekistan"; addition of objects with their names by Lola Gulyamova, https://creativecommons.org/licenses/by-sa/3.0/legalcode)

Daewoo GM (in 2008, it was renamed GM Uzbekistan). Today, JSC Uzavtosanoat is the biggest automobile manufacturer in Uzbekistan, uniting "20 enterprises." Joint stock company (JSC) Uzavtosanoat (2018) reports that the functioning in this sector joint ventures and foreign companies have been established with share investors from South Korea, Germany, Türkiye, Iran, the Czech Republic, and Belarus.

One such joint venture is located in the Fergana Valley, in the city of Assake. This industry is mainly supplied with advanced technology parts from South Korea and steel coil from Russia. According to JSC Uzavtosanoat (2018), in 2013, the Cooperative stock corporation (CSC) GM Uzbekistan produced 270,000 cars (Figs. 4.27, 4.28, 4.29, and 4.30), of which 150,000 were exported, and SamAuto produced 3300 cars, of which 2100 were exported. In 2017, according to the press service of CSC UzAutoSanoat, GM Uzbekistan exported automobiles worth US$1.3 million to Tajikistan. This industry unites several enterprises (Table 4.27).

GM Powertrain Uzbekistan is an engine plant in Tashkent, producing engines for use in GM's small passenger cars.

The MAN Auto-Uzbekistan Company is also a joint venture between the German MAN Nutzfrahrzeuge AG and OJSC UzAvtosanoat. It started operation in 2008.

As noted by JSC Uzavtosanoat (2018), SamKochAvto (or Samarkand Automobile Factory), with major investment

Fig. 4.28 A bus manufactured by JV MAN Uzbekistan

from the Turkish company Koç Holding in Samarkand, manufactures buses and cargo trucks. In November 2017, this plant celebrated the production of its 30,000th bus (Fig. 4.30). Technology from Isuzu Motors, Itochu Corporation, is used for production of gas-powered buses.

The target markets for the manufactured models are the Uzbek domestic market and markets in neighboring countries that were formerly part of the Soviet Union. UzAvtosanoat is the only truck manufacturer in Central Asia.

Fig. 4.29 A truck manufactured by JV MAN Uzbekistan

Fig. 4.30 A bus manufactured by SamAuto

Table 4.27 UzAutoSanoat enterprises in 2018

Enterprise	Location (city)	Year of establishment	Products
SamAuto	Samarkand	1990	City buses, water-carrier flushers, garbage trucks, cherry picker telescopic, board tilt
DAZ	Jizzakh	1998	Automobile batteries
UzChasys	Namangan	2008	Lighting systems
Avtooyna	Fergana	1998	Automobile glass
UzKoje	Xanabad	2003	Wiring harnesses
UzDongwon	Assake	1997	Exhaust systems
UzDong Yang	Andijan	1997	Interior parts
UzDongju	Andijan	1999	Paint
UzSaemyung	Andijan	1998	Fuel tanks, stamping parts
UzTongheung	Andijan	1997	Front and rear seats
UzKoram	Andijan	1997	Bumpers
Uzbekistan MB Buses	Tashkent	2011	City buses
JV MAN Auto Uzbekistan	Samarkand	2009	Trucks
GM Powertrain Uzbekistan	Tashkent	2008	Engines BDOHC 1.2 l, 1.5 l
CSC General Motors Uzbekistan	Assake	2008	Vehicles

Source: UzAutoSanoat (public domain)

According to Bloomberg (March 27, 2019), the Uzselkhozmash-Holding Company manufactures and markets agricultural machinery and equipment (tractors and trailers).

4.7.9 Metallurgy and Mechanical Engineering

One of the advanced industries of Uzbekistan is the ferrous and nonferrous metals industry, including production of copper, refractory and heat-resistant metals, and gold. The largest enterprises in this sector are the Navoiy and Almalyk Mining Metallurgy Combines. According to JSC Almalyk Mining Metallurgy Combine (Almalyk MMC) (2019), its production capacity is based on copper–molybdenum, lead–zinc, gold mining, and silver mining, situated in the Tashkent, Jizzakh, Namangan, and Surxondaryo regions. JSC Almalyk MMC (2019) reported that in July 2018, it was transferred to trust management by the SFI Management Group, remaining at state disposal. An Uzbek factory of refractory and heat-resistant metals (in the city of Chirchiq) works with tungsten ores and molybdenum discovered in the country. A metallurgy factory in the town of Bekobod (in Tashkent province) manufactures steel and rolled metal through processing of scrap metal.

4.7.10 Information and Communication Technology

A good prospect for future economic development is the rapid growth of information and communication technology (ICT). Access to affordable, reliable technologies is critical for achieving economic growth, industrial development, and manufacturing diversification, as well as for establishing export-oriented enterprises.

This industry is growing rapidly. In 2017, according to the State Committee of the Republic of Uzbekistan on Statistics, the total number of Internet users was 14,756,900, or about 50% of the population. In 2018, the share of Internet users grew by 1% (5.3 million) to 51% of the population of Uzbekistan, or 0.5% of world users. The State Committee of the Republic of Uzbekistan on Statistics reports that more than 20 million people in Uzbekistan use the Internet. One of the priorities is to introduce advanced ICT in remote areas.

According to the 2017 ICT development index (IDI) from the International Telecommunication Union[39] (ITU), Uzbekistan ranks 95th out of 176 countries, with an IDI value of 4.90 (it ranked 103rd in 2016, with a value of 4.48). The number of mobile cellular subscriptions per 100 inhabitants is 77.3, 75.4% of households have Internet access at home, 46.8% of individuals use the Internet, and 43.9% of households have a computer.

As in the developed countries of the world, the proportion of Internet users aged 15–24 years is relatively high (Figs. 4.31 and 4.32; Table 4.28).

Fig. 4.31 Information and communication technology (ICT) development in rural areas

Fig. 4.32 The younger generation are the biggest users of the Internet

4.8 Trade

For many years, the economy of Uzbekistan was oriented toward exports of mainly natural resources and low-value-added goods and imports of high-value manufactured goods. This policy led to a large trade deficit in the Soviet era. Today, Uzbekistan's economy is still suffering the same disease. The countries of the former Soviet Union (FSU) are Uzbekistan's main trading partners. Exports are concentrated and destined for those selected markets, making Uzbekistan vulnerable to price volatility and fluctuations in demand. The Central Bank of Uzbekistan has enjoyed the support of the Russian Federation, Türkiye, and China (its major trading partners). This has led to dependence on their relevant resources. Uzbekistan still remains a weak trade competitor in the world market. The economy needs structural transformations to become more competitive.

Data from the International Trade Center (2017) show a decrease in the trade-to-GDP ratio in 2011–2015. In 2011, exports of goods were 15.9% of GDP, but in 2015, they were only 8.5%. The World Bank reported that trade rose from 49.324% of GDP in 2015 to 68.519% of GDP in 2017 (Figs. 4.33 and 4.34 and Table 4.29).

[39]The International Telecommunication Union (ITU) is a specialized United Nations agency focusing on information and communication technologies.

Table 4.28 Information and communication technology (ICT) developments: changes in Internet usage between 2000 and 2016

	Internet users		Total population	Nonusers (Internetless)	Annual growth in Internet use		
Year	Number of users	Percentage of total population			Percentage	Number of users	Percentage of total population
2016	15,453,227	51	30,300,446	14,847,219	6.1	893,596	1.36
2015	14,559,632	48.7	29,893,488	15,333,856	13.4	1,725,485	1.44
2014	12,834,147	43.6	29,469,913	16,635,766	15.7	1,743,403	1.5
2013	11,090,744	38.2	29,033,361	17,942,617	6.2	648,409	1.54
2012	10,442,335	36.5	28,592,451	18,150,116	22.8	1,938,500	1.54
2011	8,503,835	30.2	28,158,395	19,654,560	53.3	2,955,882	1.51
2010	5,547,953	20	27,739,764	22,191,811	19	884,559	1.47
2009	4,663,394	17.1	27,338,109	22,674,715	90.5	2,216,056	1.43
2008	2,447,338	9.1	26,952,719	24,505,381	22.9	455,833	1.38
2007	1,991,505	7.5	26,586,701	24,595,196	18.8	315,021	1.31
2006	1,676,484	6.4	26,242,947	24,566,463	93.4	809,771	1.24
2005	866,713	3.3	25,922,239	25,055,526	30.4	202,018	1.15
2004	664,694	2.6	25,627,007	24,962,313	37.1	179,783	1.08
2003	484,911	1.9	25,353,547	24,868,636	78.6	213,466	1.06
2002	271,444	1.1	25,088,682	24,817,238	83.1	123,160	1.1
2001	148,284	0.6	24,814,629	24,666,345	24.9	29,531	1.21
2000	118,753	0.5	24,518,222	24,399,469	1499.4	111,328	1.34

Source: Internet Live Stats (2018) (public domain)

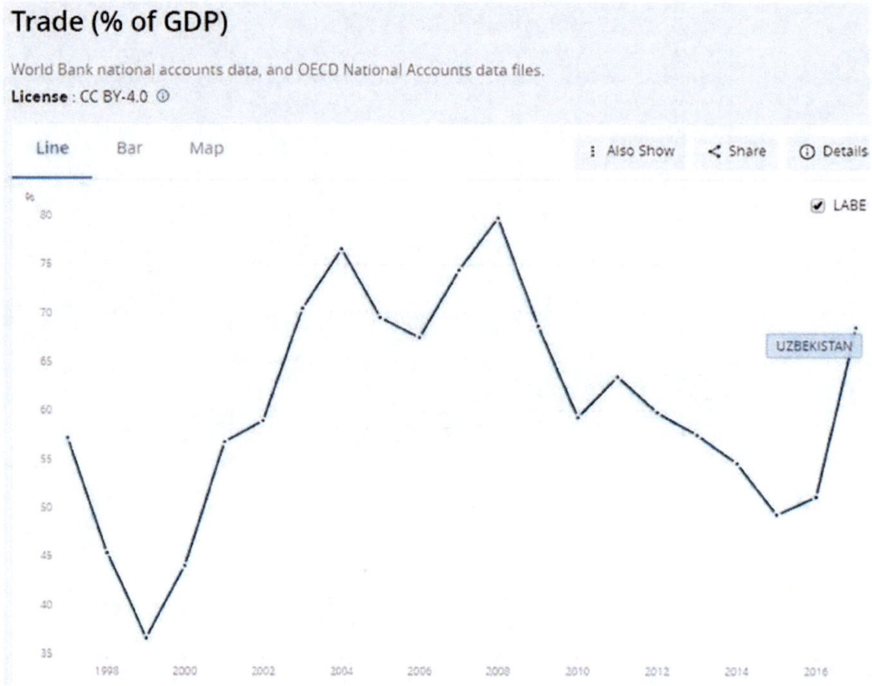

Fig. 4.33 "This indicator is used by WB to define the relative importance of international trade in the economy of Uzbekistan. The trade-to-GDP-ratio is the sum of exports and imports divided by GDP". International trade plays a crucial role for Uzbekistan. In 2020, the top exports of Uzbekistan are Gold ($6.52B), Non-Retail Pure Cotton Yarn ($936M), Petroleum Gas ($631M), Refined Copper ($587M), and Ethylene Polymers ($281M). They are exporting mostly to United Kingdom ($6.06B), Russia ($1.74B), China ($1.29B), Türkiye ($1.01B), and Kyrgyzstan ($757M). The top imports of Uzbekistan are Motor vehicles; parts and accessories (8701 to 8705) ($933M), Packaged Medicaments ($925M), Wheat ($598M), Refined Petroleum ($573M), and Cars ($475M), importing mostly from China ($4.88B), Russia ($4.29B), Kazakhstan ($2.14B), South Korea ($1.85B), and Türkiye ($1.18B). (Source: https://oec.world/en/profile/country/uzb) The country has been one of the most important suppliers of agricultural products, including cotton, fruits, and vegetables

Fig. 4.34 According to WB, Imports of goods and services consist of transactions in goods and services (purchases, barter, and gifts) from non-residents to residents. Imports of goods and services are goods which add to the stock of material resources of a country by entering its economic territory. In a case, when goods simply being transported through a country (goods in transit) or temporarily admitted (except for goods for inward processing) do not add to the stock of material resources of a country and are not included in the international merchandise trade statistics

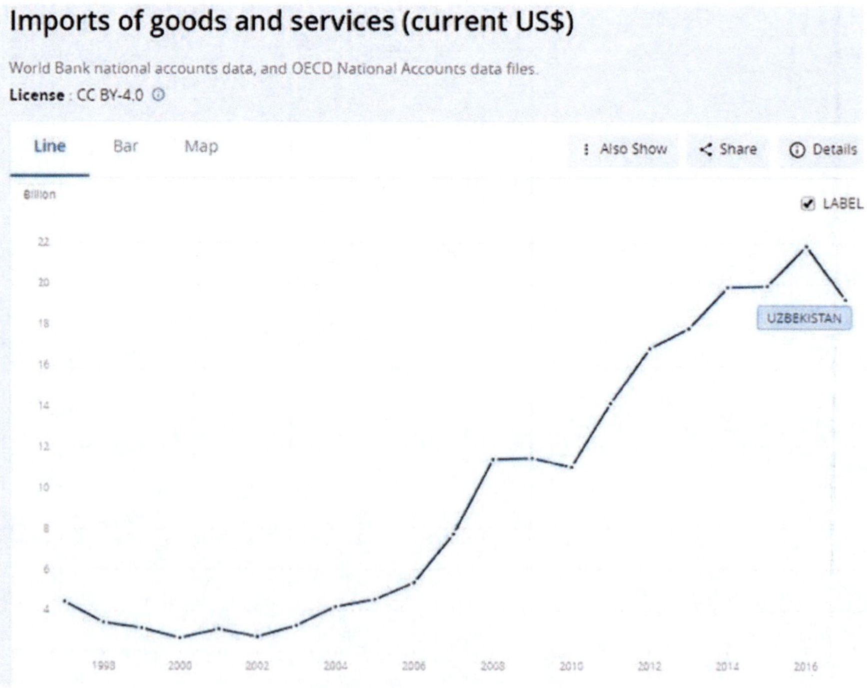

Table 4.29 Contributions of trade, exports, and imports of goods and services to gross domestic product (GDP)

Year	Trade (% of GDP)	Exports (% of GDP)	Import (% of GDP)
1997	57,199	27.109	30.089
1998	45,418	22.564	22.854
1999	36,619	18.179	18.44
2000	44,054	24.663	19.391
2001	56,776	29.675	27.101
2002	58,995	30.920	28.034
2003	70,459	38.238	32.222
2004	76,596	41.790	34.806
2005	69,525	37.809	31.716
2006	67,471	36.514	30.958
2007	74,414	39.671	34.743
2008	79,748	41.169	38.589
2009	68,673	34.663	34.011
2010	59,306	31.269	28.038
2011	63,489	32.651	30.838
2012	59,817	27.334	32.483
2013	57,480	26.596	30.883
2014	54,572	23.126	31.446
2015	49,324	19.596	29.729
2016	51,153	18.737	32.416
2017	68,519	29.796	38.722

Sources: World Bank national accounts data and Organisation for Economic Co-operation and Development (OECD) national accounts data. License: CC-BY-4.0 via World Bank Data

Table 4.30 Exports in January–June 2018

	Services	Oil and gas	Food products	Chemicals	Ferrous and nonferrous metals	Cotton	Machinery and automobiles	Other exports
Value (US$ millions)	1900	939.4	414.6	383.9	341.5	158.5	92.1	2200
Percentage of exports	29.8	14.5	6.4	5.9	5.3	2.5	1.4	34.2

Source: State Committee of the Republic of Uzbekistan on Statistics (2018) (public domain)

In January–May 2018, trade was worth US$14.7 billion, exports were worth US$6.5 billion, and imports were worth US$8.2 billion.[40]

In January–June 2018, according to official data,[41] the foreign trade turnover of Uzbekistan amounted to US$17.7 billion, including exports valued at US$7.8 billion and imports valued at US$9.9 billion. The foreign trade balance was US$2.1 billion.

Exports from Uzbekistan fell to US$12.871 billion in 2015 from US$14.084 billion in 2014 but increased to US$14.0 billion in 2017. Exports averaged US$3,609.85 million from 2006 to 2015, reaching a peak of US$14.084 billion in the fourth quarter of 2014 and a record low of US$1.3868 billion in the first quarter of 2006. Oil, natural gas, gold, cotton, and other natural resources dominate the country's exports. Other exports from Uzbekistan include machinery, equipment, and food products.

Uzbekistan ranks 84th out of 220 countries in the world on the basis of expected exports. Its exports continue to consist mainly of natural resources and goods with low added value (Table 4.30).

In January–May 2018, according to the State Committee of the Republic of Uzbekistan on Statistics (2018), exports of food products increased by 60.9% because of an increase in exports of fruit and vegetables by 73.8%. Exports of gas increased 1.8-fold, exports of chemicals increased by 2.6%, and exports of plastics increased by 18.5%. In 2017, exports of food products amounted to US$708.8 million, including fruit and berries (35.9%), vegetables (30.7%), grapes (22.5%), and products of processed fruit and vegetables (only 4.1%). Exports of textiles were valued at US$1.13 billion, an increase of 22.8% in comparison with 2016 and an increase in total exports by 8.1%.

In January–June 2018, according to the State Committee of the Republic of Uzbekistan on Statistics (2018), exports of energy carriers and petroleum products increased by 78.5%

[40]Source: State Committee of the Republic of Uzbekistan on Statistics (2018).

[41]Source: Statistics and Foreign Trade and Trade Statistics Department of the State Committee of Uzbekistan; https://www.gazeta.uz/ru/2018/07/17/trade/; accessed July 17, 2018.

in comparison with the same period in the previous year, including a 1.9-fold increase in exports of natural gas, a 5.3% increase in exports of chemical products, and an 18.7% increase in exports of plastics and articles made of plastic.

The volume of food products rose by 58.9%. Exports of fruit and vegetables amounted to US$463.2 million, a 73.3% increase in comparison with the same period in the previous year.

According to O'zto'qimachiliksanoat, exports of raw cotton decreased from 503.800 kt in 2015 to 115.600 kt in 2018. At the same time, textile product exports amounted to 8.4% of total exports and were valued at US$652 million, an increase of 16.7% in comparison with the same period in the previous year.

According to the State Committee of the Republic of Uzbekistan on Statistics (2018), in the structure of exports, services constituted 30.8% (US$2.4 billion), energy and oil products 16.4% (US$1.2 billion), food products 7.5% (US$582.5 million), chemical products 6% (US$466.8 million), ferrous and nonferrous metals 5.6% (US$439.2 million), cotton fiber 2.1% (US$166.1 million), machinery and equipment 1.3% (US$103.8 million), and other products 30.3% (US$2.3 billion) (Fig. 4.35).

Export Commodities

include energy products, cotton, gold, mineral fertilizers, ferrous and nonferrous metals, textiles, foodstuffs, machinery, and automobiles.

Exports: Partners

According to World Bank Trading Economics (2017a, b), the main export partners of Uzbekistan in 2017 were Switzerland (25.8%), China (17.6%), Kazakhstan (14.2%), Türkiye (9.9%), Russia (8.4%), and Bangladesh (6.9%) (Table 4.31). China was planning to increase the trade between the two countries from US$4.2 billion to US$10.0 billion in 2020. Gas, uranium, copper, cotton fiber, fertilizers, plastic goods, and food products were the major exported commodities.

In January–May 2018, a change occurred in the list of the main export partners: exports to Russia rose from 8.4% to 15.9% (Table 4.32).

In January–May 2018, exports of food products were valued at US$305.4 million. The shares of the main export partners are listed in Tables 4.33 and 4.34.

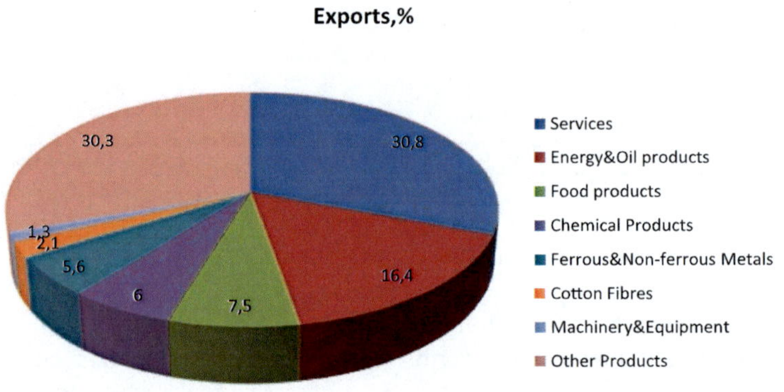

Fig. 4.35 As WB notes, Exports of goods and services represent the value of all goods and other market services provided to the rest of the world. They include the value of merchandise, freight, insurance, transport, travel, royalties, license fees, and other services, such as communication, construction, financial, information, business, personal, and government services. They exclude compensation of employees and investment income (formerly called factor services) and transfer payments" (Source: https://tradingeconomics.com/uzbekistan/exports-of-goods-and-services-percent-of-gdp-wb-data.html). The services play important role since Uzbekistan is a transit country for trade between West and East. Structure of exports. (Compiled by the author, using 2018 data from the State Committee of the Republic of Uzbekistan on Statistics (public domain))

Table 4.31 The main export partners in 2017

Country	Russia	Switzerland	China	Kazakhstan	Türkiye	Bangladesh
Percentage of exports	8.4	25.8	17.6	14.2	9.9	6.9

Source: Uzbekistan Exports (2017) (public domain)

Table 4.32 The main export partners in 2018

Country	Russia	Kazakhstan	Belarus	Ukraine	Kyrgyzstan	Tajikistan	China	Türkiye	South Korea	Afghanistan	Germany	Latvia	Iran
Percentage of exports	15.9	7.4	1.3	1.1	0.9	0.9	15.5	5.5	4.6	2	1.8	1.0	0.9

Source: State Committee of the Republic of Uzbekistan on Statistics (2018) (public domain)

Table 4.33 Food product export partners in January–May 2018

Country	Kazakhstan	Russia	Afghanistan	Kyrgyzstan	China	Vietnam	Türkiye	Pakistan	Iran
Percentage of exports	47.4	13.4	7.7	7.3	4.7	4.4	3.0	2.0	1.3

Source: State Committee of the Republic of Uzbekistan on Statistics (2018) (public domain)

Table 4.34 Fruit and vegetable export partners in January–June 2018

Country	Kazakhstan	Russia	Kyrgyzstan	Afghanistan	China	Vietnam	Türkiye	Pakistan	Iran
Percentage of exports	51.9	15.2	8.6	5.5	4.1	3.0	2.1	1.8	1.0

Source: State Committee of the Republic of Uzbekistan on Statistics (2018) (public domain)

Table 4.35 Investment in different kinds of economic activity in 2017

Activity	Oil and gas industry	ICT	Energy	Transport	Improvement of road infrastructure	Credit financing of small business	Light industry
Investment (%)	27.3	12.6	10.6	9.6	7.9	7.8	5.2

Source: Ministry of Foreign Economic Relations of the Republic of Uzbekistan (2017)
ICT information and communication technology

Table 4.36 Structure of exports in January–June 2018

	Services	Energy and oil products	Chemicals and products made from them	Ferrous and nonferrous metals	Cotton fiber	Machinery and equipment	Other products
Value	US$2.4 billion	US$582.5 million	US$466.8 million	US$439.2 million	US$166.1 million	US$103.8 million	US$2.3 billion
Percentage	30.8	7.5	6	5.6	2.1	1.3	30.3

Sources: State Committee of the Republic of Uzbekistan on Statistics (2018), Statistics and Foreign Trade and Trade Statistics Department of the State Committee of Uzbekistan (public domain); https://www.gazeta.uz/ru/2018/07/17/trade/; accessed July 17, 2018

Exports of fruit and vegetable products amounted to US$463.2 million, an increase of 73.3% in comparison with the same period in the previous year.

In January–June 2018, the main export destinations for fruit and vegetables from Uzbekistan were Kazakhstan (51.9% of the total), Russia (15.2%), Kyrgyzstan (8.6%), Afghanistan (5.5%), China (4.1%), Vietnam (3%), Türkiye (2.1%), Pakistan (1.8%), and Iran (1%).

In January–June 2018, exports of textile products constituted 8.4% of total exports and were valued at US$652 million, a 16.7% increase in comparison with the same period in the previous year. In January–May 2018, they rose by 15.3% in comparison with January–May 2017.

Foreign investment has been attracted to Uzbekistan's national economy from more than ten foreign countries, including China, Russia, Germany, South Korea, Japan, the Netherlands, the UK, and Malaysia, which contribute to the investment program. According to data from the Ministry of Foreign Economic Relations of the Republic of Uzbekistan, the foreign investments are for upgrading infrastructure (Table 4.35).

According to official data (Ministry of Foreign Economic Relations of the Republic of Uzbekistan 2017), in January–September 2016, 835 new entities were involved in export activities. Exports of 246 new types of goods occurred. At the same time, the geography of exports of Uzbekistan's domestic production was expanded into 52 new markets.

In January–June 2018, exports of energy carriers and petroleum products continued to increase. An increase of 78.5% was due to a 1.9-fold increase in natural gas exports in comparison with exports in January–June 2017. Exports of chemical products and products made from them increased by 5.3%, exports of plastics increased by 18.7%, and exports of food products (mainly fruit and vegetables) increased by 58.9% (Table 4.36).

Uzbekistan plans to become a member of the World Trade Organization (WTO) to gain more opportunities for adequate market access. For this, it is necessary to improve the country's regulatory and business environment, as well as its physical and quality infrastructure, and to raise its productivity (International Trade Center 2017). Data from the International Trade Center showed a decline in the trade-to-GDP ratio in 2011–2015.

Imports

In January–May 2018, imports were valued at US$8.2 billion,[42] compared with estimates of US$11.61 billion in 2016 and US$11.81 billion in 2015 (Tables 4.37, 4.38 and 4.39). The country ranked 92nd out of 222 countries in a worldwide comparison (Countries of the World 2017).

[42] Source: State Committee of the Republic of Uzbekistan on Statistics (2018).

Table 4.37 Structure of imports in January–June 2018

	Machinery and equipment	Services	Chemicals and products made from them	Food products	Ferrous and nonferrous metals	Energy carriers and petroleum products	Other products
Value	US$3.3 billion	US$2.4 billion	US$1.1 billion	US$845.1 million	US$782.8 million	US$487.5 million	US$889.5 million
Percentage	33.8	24.3	11.5	8.6	7.9	4.9	9

Sources: State Committee of the Republic of Uzbekistan on Statistics (2018), Statistics and Foreign Trade and Trade Statistics Department of the State Committee of Uzbekistan (public domain); https://www.gazeta.uz/ru/2018/07/17/trade/; accessed July 17, 2018

Table 4.38 The main import partners in 2018

Country	Russia	Kazakhstan	Belarus	Ukraine	Kyrgyzstan	Tajikistan	China	Türkiye	South Korea	Afghanistan	Germany	USA	Latvia
Percentage	16	8.1	1.2	1.2	1.0	0.9	15.9	5.3	1.5	1.9	1.9	1.5	1.3

Source: State Committee of the Republic of Uzbekistan on Statistics (2018) (public domain)

Table 4.39 Reserves of foreign exchange and gold in 2015–2018

Year	1 September 2018	1 November 2017	31 December 2016[a]	31 December 2015[a]
Reserves (US$ billion)	25.9	25.9	14.5	16

Compiled by the author, using data from the Central Bank of Uzbekistan (2019) (public domain)
[a]Estimate

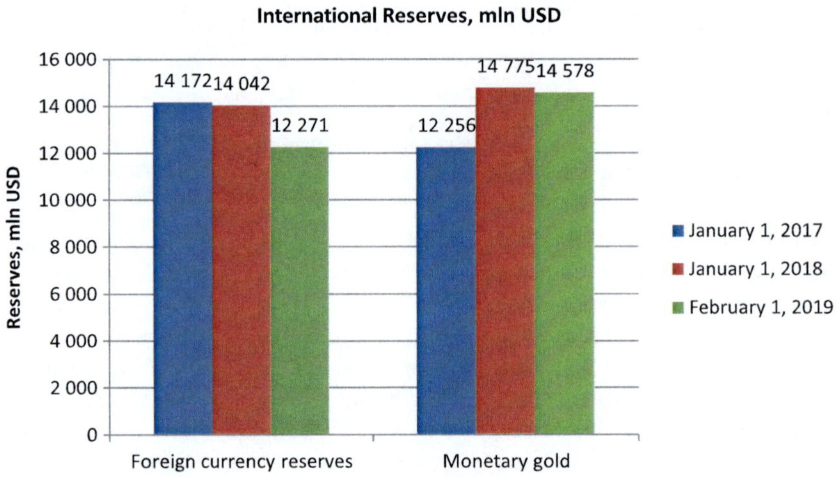

Fig. 4.36 International reserves of the Republic of Uzbekistan

Reserves of Foreign Exchange and Gold According to the Central Bank of the Republic of Uzbekistan, international reserves (Figs. 4.36 and 4.37) comprise (I) gross international reserves [including (I.1) foreign currency reserves and (I.2) monetary gold] and (II) net international reserves.

4.9 Transportation Systems and Networks

From ancient times, transportation, as a main driving force, has played a pivotal role in the economic development of this double-landlocked country. However, the transportation network is not yet well developed. As a result, many parts of the country do not have adequate access to national, regional, and global markets.

The first railroads were constructed by the Russian Empire at the end of the nineteenth century and at the beginning of the twentieth century. Today, the average density of railroads is 150 km per 10,000 km^2; these railroads form a structure for further densification of the transportation network and connect major cities in Uzbekistan. High-speed trains provide services between Tashkent, Samarkand, Buhkara, and Urgench (Fig. 4.37). The railroad system connects Uzbekistan with major transcontinental routes and with all of its neighboring countries.

The highway network complements the railroads and forms a transportation network that is a catalyst for economic growth. Abduvaliev (2008) noted that the network was being extended to the west by construction and reconstruction of the Kungrad–Beyneu highway in the Republic of Karakalpakstan (similar to the concept of the European route E-40) and to the east by the Andijon–Osh–Irkeshtam–Kashgar highway and the Uzbekistan–Kyrgyzstan–China highway. These projects are aimed at formation of new alternative transport corridors for carriage of goods via Euro-Asian links. Reichert (2018) points out that the Regional Roads Development Project is aiming to reduce road user costs on project roads and to have 65% of these roads in good/fair condition (Fig. 4.38).

Fig. 4.37 The Afrosiab high-speed train transports passengers between Tashkent, Samarkand, and Bukhara

The Great Uzbek Tashkent–Termiz Highway connects the major urban centers. It repeats the ancient caravan route from Tashkent to Samarkand, and it runs south almost to the border with Afghanistan. Another major route is the Zerafshan Highway, which connects such cities as Samarkand, Bukhara, and Navoiy. In general, many of the railroads and highways are dead ends. Only the Fergana Ring connects the cities of Andijon, Fergana, and Namangan, and it is the single finished stage of a planned intercity transportation circuit in Uzbekistan.

In December 2017, a new railroad from Bukhara to Miskin (Urgench) started up and replaced the existing Navoiy–Uchkuduk–Miskin connection. This improvement in the railroad network shortened the route by 75 km and provided an opportunity to link the cities of Tashkent, Samarkand, Bukhara, and Urgench with high-speed trains.

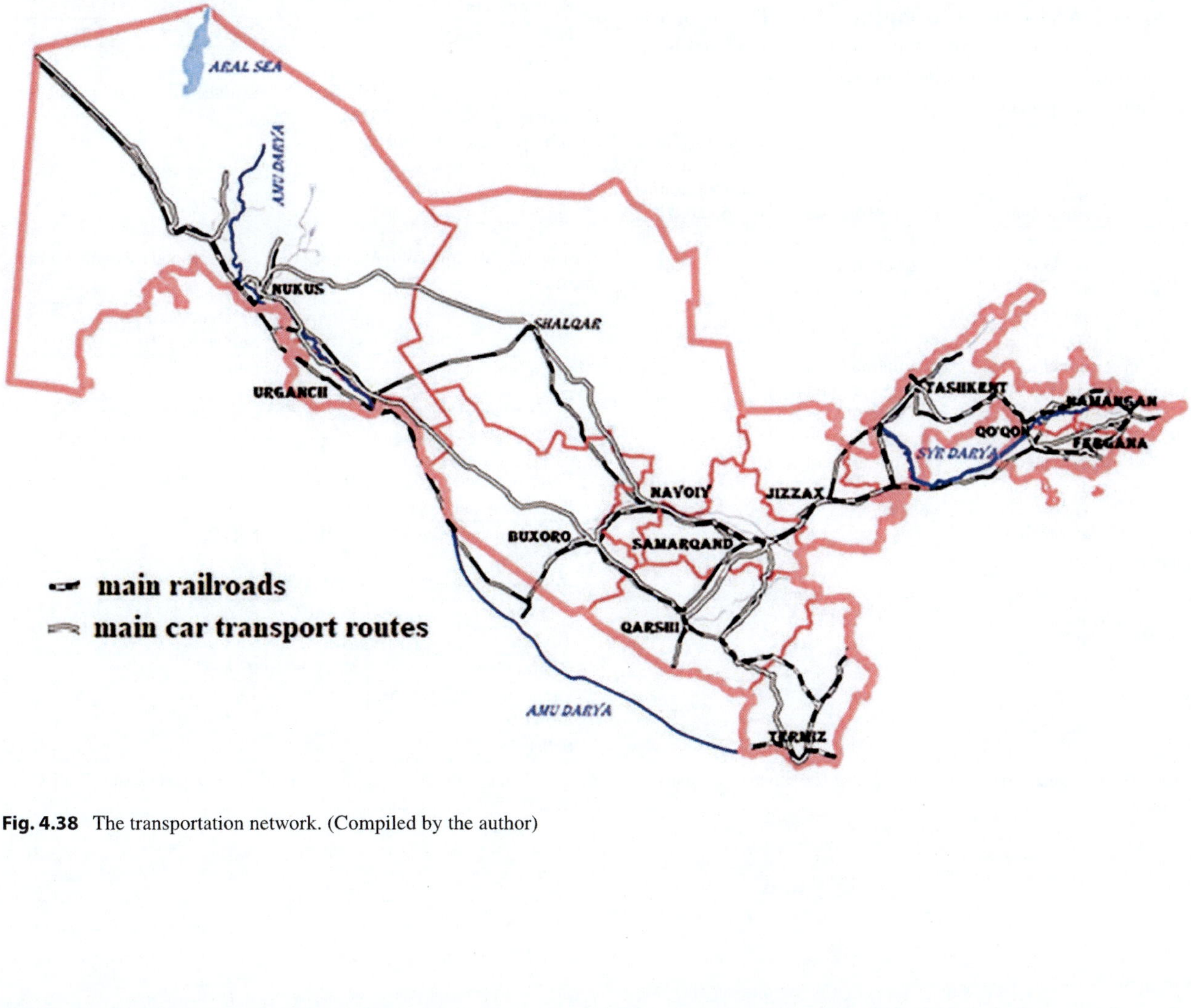

Fig. 4.38 The transportation network. (Compiled by the author)

Uzbekistan is still earning a significant share of its income from exports of raw materials (such as gas and cotton). Development of existing sectors such as cotton, food, and textile production is considerably dependent on development of the transportation infrastructure within this country and beyond. The Global Development Agenda considers development of international cooperation in the field of transport a key element. International partnership for creation of a sustainable transportation system is vital for implementation of the 2030 Agenda and Sustainable Development Goals (SDGs) in Uzbekistan.

The State Committee of the Republic of Uzbekistan on Statistics reports that in 2019 the length of communications of general purpose of railroad tracks (length in use) is 4600 km, that of roads with hard surfaces is 42,000 km, and that of the main gas pipelines is 13,800 km (Table 4.40).

According to the State Committee of the Republic of Uzbekistan on Statistics (2019) and UZAUTOTRNS (2019), road transport accounts for approximately 6% of international carriage of goods and 88% of domestic transportation of passengers and goods (Tables 4.41 and 4.42).

The volumes of the basic types of freight shipped by mainline railway transport are shown in Table 4.43 (January–September 2017).[43]

Gas Pipeline

The Central Asia–China Gas Pipeline, Line D China, was constructed by China, Kyrgyzstan, Tajikistan and Uzbekistan, at an investment cost of US$6.7 billion.

Table 4.40 Transportation in Uzbekistan

	Total length	Carriage of goods	Annual volume increment of transportation of goods
Railroads	4600 km (600 km of which are electrified)	75% of international carriage of goods	15–16%
Roads	180,000 km (including 4300 km of common use roads and 4600 km of international roads)	6% of international carriage of goods and 88% of internal transportation of passengers and goods	
Gas pipelines	13,800 km		

Source: State Committee of the Republic of Uzbekistan on Statistics (2019) (public domain)

Table 4.41 Freight transportation and freight turnover by transport type in 2016–2017

	2016	2017
Freight transportation	1132.5	1146.2
Railroad (Mt)	67.6	67.9
Air (kt)	26.5	26.4
Motor (Mt)	1.002.8	1013.13
Pipeline (Mt)	62.2	65.1
Freight turnover by transport type (t-km billions)	65.3	66.9
Railroad	22.9	22.9
Air	132.2	156.9
Motor	13.3	13.6
Pipeline	28.9	30.2

Sources: State Committee of the Republic of Uzbekistan on Statistics (2018)
t-km tonne-kilometers

Table 4.42 Passenger transportation and passenger turnover by transport type

	2016	2017
Passenger journeys (millions)		
Railroad	20.5	21.1
Road vehicles	5480.8	5591.3
Metro	53.5	61.6
Aircraft	2.1	2.2
Passenger turnover (pass-km billions)		
Railroad	4.0	4.3
Road vehicle	114.9	117.7
Metro	0.4	0.5
Aircraft	6.7	7.5

Source: (public domain) via State Committee of the Republic of Uzbekistan on Statistics (2019)
pass-km passenger-kilometers

Table 4.43 Volumes of basic types of freight shipped by mainline rail transport

	Volume (kt)	Percentage of January–September 2016
Total freight	50,525.9	100.5
Coal	3.317.4	132.8
Oil	7723.4	96.5
Ores	3754.8	102.3
Ferrous metals	617.1	89.5
Scrap iron and steel	594.1	110.1
Chemical and mineral fertilizers	3.010.4	85.3
Building materials	5396.3	99.7
Cement	3720.0	88.6
Timber	14.5	97.7
Cotton fiber	206.5	67.8
Grain and milling products	1227.1	134.8
Other	20,943.4	102.2

Source: State Committee of the Republic of Uzbekistan on Statistics (2017a, b) (public domain)

[43] Source: State Committee of the Republic of Uzbekistan on Statistics (2017a, b).

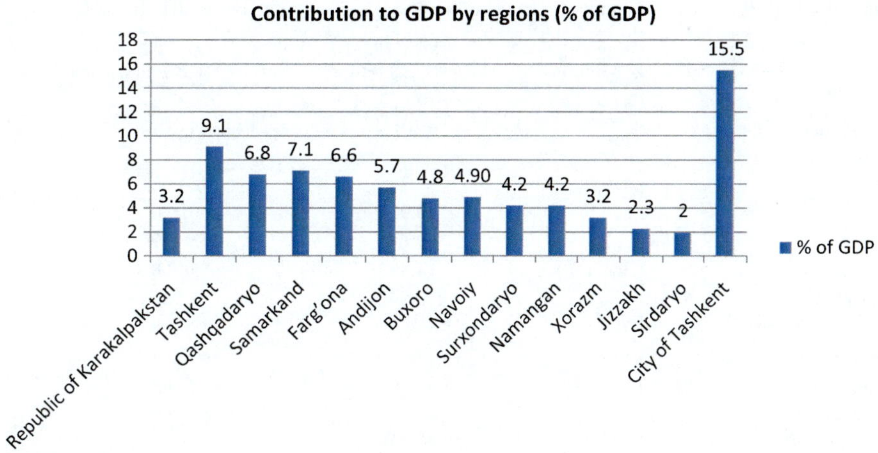

Fig. 4.39 Regional percentage contributions to gross domestic product (GDP) in 2019. (Compiled by the author using 2019 data from the State Committee of the Republic of Uzbekistan on Statistics (public domain))

4.10 Spatial Distribution of the Economy

The economy is also unevenly distributed across the country, as a result of the spatial mosaic of natural resources (land, water, and mineral reserves), urbanization, and transportation. The major portion of industrial production is concentrated in the more developed eastern part of the country. Historically, crafting, food production, trade, and business were linked to populous sites in the Fergana and Zerafshan Valleys, and the Xorazm and Tashkent regions. Today, many new industries (including gold mining, gas, and oil), are being developed in the Kyzylkum Desert, in the dried seabed of the Aral Sea, and in Kashkadarya province.

Structural transformation of the economy is now increasing the role of the different regions in Uzbekistan's economy. Today, growth in GDP in the regions is a result of the efficiency of targeted regional programs aimed at further diversification of the economy (Fig. 4.39). The economy of the city of Tashkent is developing with higher growth rates, attracting more government investment and producing more products. The growth in GDP and the gross regional product (GRP)[44] of the city of Tashkent are significantly greater than those in other regions. The GRP per capita is significantly higher in the city of Tashkent and the Navoiy region (Table 4.44).[45]

In 2016, the contributions made by industries in the city of Tashkent and in the Tashkent region to GDP were 15.8% and 10.3%, respectively. Industrial production in the regions of Kashkadarya (oil and gas), Samarkand (automobiles), and Fergana (chemicals, petrochemicals, textiles, food, machinery, and building materials) offers a wide range of opportunities for further development (Table 4.45). The contributions to GDP from industries in these regions exceed 50%.

The economies of the different regions are characterized by different driving forces (Fig. 4.30). The major contributor to the growth of regional products is services, replacing the shrinking agricultural sector.

Development of agriculture is related not only to natural conditions but also to historical traditions and attitudes to the environment. More vegetables are produced in the Tashkent, Samarkand, and Andijon regions. The Andijon, Fergana, and Samarkand regions are the leading producers of fruit and grains, and the Samarkand, Kashkadarya, and Tashkent regions are the leading producers of meat and milk. The Republic of Karakalpakstan and Sirdaryo province are at the bottom of the list, with less promising results in terms of agricultural production (Table 4.46).

4.11 Prospects for Future Development

Structural economic transformation is critical for Uzbekistan to reduce the negative impacts of its geographic position. Diversification of the economy and of exports is one of the priorities of national policy to increase the country's independence in extraction and simple processing of commodities, with increasing value addition in manufacturing and agricultural production. There are good prospects for Uzbekistan in service-based growth (for example, from tourism), and attraction of foreign development investment in high-value-added sectors.

Uzbekistan is a resource-rich country and is well placed for increasing domestic resources. According to an estimate by UN experts (United Nations Economic and Social Council 2017), in landlocked countries, resource-based rents may yield 37–47% of GDP amounts of development finance. Significant overseas development assistance (ODA) is needed for transit infrastructure, corridor, trade capacity, and

[44]The gross regional product (GRP) is a monetary measure of the market value of all final goods and services produced in a region during the year.

[45]Source: State Committee of the Republic of Uzbekistan on Statistics (2017a, b).

Table 4.44 Gross regional product[a] (GRP) in 2016 and 2017, and growth rates (2017 to 2000)

Region	GRP in 2017 (UZS billions)	GRP in 2016 (UZS billions)	GRP per capita in 2017 (UZS thousands)	GDP (GRP) growth in 2017–2000[b]
Republic of Uzbekistan	249,136.4	199,325.1	7692.4	–
Republic of Karakalpakstan	8285.2	6518	4527.7	356.1
Andijon	14,479.0	11.266	4847.2	350.0
Bukhara	12,283.9	10.924	6616.0	295.3
Jizzakh	5762.2	4907	4388.9	403.9
Kashkadarya	17,366.1	14,897	5568.8	285.9
Navoiy	12,317.4	10,542	12,961.6	202.1
Namangan	10,552.5	8867	3943.8	359.7
Samarkand	18,319.5	14,060	4970.4	384.1
Surxondaryo	10,633.0	9173	4274.3	339.7
Sirdaryo	5.066.4	4190	6259.5	282.9
Tashkent	23,525.3	20,529	8268.1	279.1
Fergana	16,773.1	13,923	4669.0	250.3
Xorazm	8232.9	7094	4597.6	312.3
City of Tashkent	38,642.9	31,483	15,805.5	460.9

Source: State Committee of the Republic of Uzbekistan on Statistics (2019)
[a]GRP is the main indicator of the system of national accounts, which characterizes the final result of the production activity of institutional units—residents located in the economic territory of the region for a certain period of time. GRP is calculated by the production method, as the sum of the gross value added by types of economic activity in basic prices and net taxes on products
[b]These 2017 values are percentages of the values in 2000

Table 4.45 Value added to gross domestic product (GDP) by different types of activity

	Value added to GDP (%)			
Region	Services	Agriculture	Industry	Construction
City of Tashkent	61.1		31.6	7.3
Republic of Karakalpakstan	44.2	15.3	32.9	7.6
Tashkent	33.5	23.4	39.3	3.8
Kashkadarya	34.4	22.7	35.3	7.6
Samarkand	42.8	32.5	18.8	5.9
Fergana	45.4	23.6	24.8	6.2
Andijon	40.3	33.3	20.8	5.6
Bukhara	39.8	31.0	19.7	9.7
Navoiy	23.1	18.4	53.7	4.8
Surxondaryo	43.8	40.3	8.2	7.7
Namangan	48.0	30.8	14.7	6.7
Xorazm	44.6	30.7	17.4	7.3
Jizzakh	42.4	35.2	14.3	8.1
Sirdaryo	32.7	37.4	25.0	4.9

Source: State Committee of the Republic of Uzbekistan on Statistics (2019) (public domain)

energy and ICT development in the case of oil and gas exports. Today, an inflow of foreign development investment (FDI) from the East and the West is essential for transportation and energy sector development. The UN experts (United Nations Economic and Social Council 2017, p. 18) have pointed out that the available financing by source decreased in 2010–2014. This is a challenge for policy making and economic strategy development.

The future of the economy of Uzbekistan is requires particular focus on transit policy issues, transit cooperation, infrastructure development and maintenance, trade facilitation, and structural transformation, including increased investment in human resources, technology adaptations, and local innovations.

Uzbekistan is still earning a significant share of its income from exports of raw materials (such as gas and cotton). Development of existing sectors such as cotton, food, and textile production is considerably dependent on development of the transportation infrastructure within this country and beyond.

Travel and Tourism

The rich nature and history of Uzbekistan may provide a good opportunity for developing different types of tourism

Table 4.46 Production of the main agricultural products in January–September 2017

	Production (kt)						
Region	Grains	Potatoes	Vegetables	Fruit and berries	Meat	Milk	Growth (%)
Republic of Karakalpakstan	182.2	44.3	192.5	33.1	71.6	207.2	104.8
Tashkent	651.7	330.9	1509.6	159.5	179.3	678.4	101.9
Kashkadarya	1002.1	149.1	448.3	118.3	207.0	814.7	102.2
Samarkand	829.7	472.7	1472.3	300.1	200.5	949.2	106.1
Fergana	761.6	253.4	760.1	350.1	107.5	648.2	101.5
Andijon	612.4	216.4	1132.9	381.7	103.8	564.5	100.8
Bukhara	580.2	181.9	591.7	204.1	159.5	633.7	105.8
Navoiy	274.7	60.9	186.8	70.8	110.4	317.0	105.6
Surxondaryo	635.7	261.7	822.0	113.2	117.7	519.6	106.0
Namangan	458.0	191.2	506.6	167.5	105.5	501.9	104.6
Xorazm	302.0	100.2	392.2	143.4	116.5	800.4	102.0
Jizzakh	535.1	45.6	401.2	68.2	143.6	440.6	103.8
Sirdaryo	555.1	49.1	279.0	36.0	51.3	252.2	102.0

Source: State Committee of the Republic of Uzbekistan on Statistics (2017a, b) (public domain)

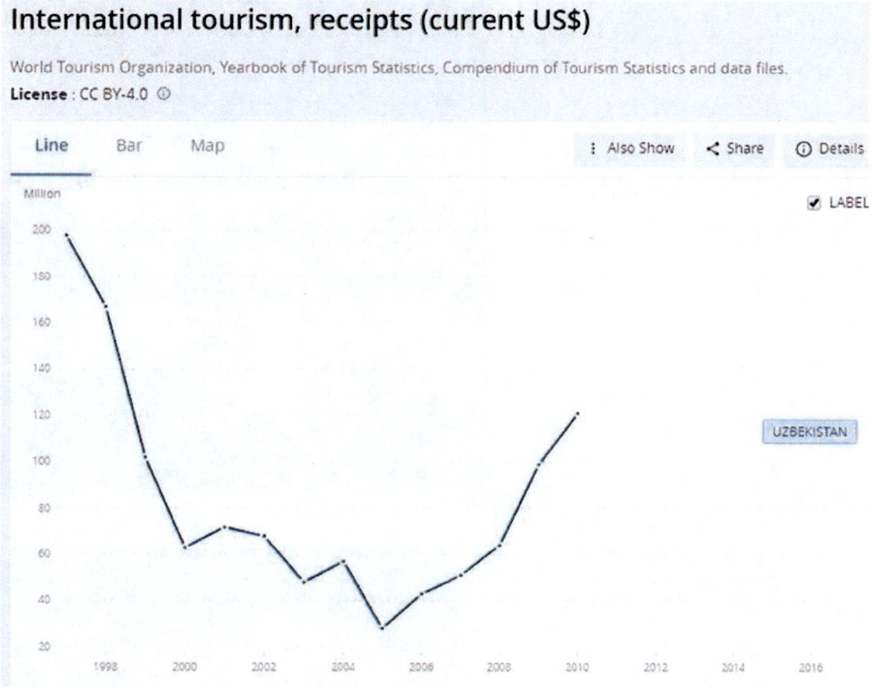

Fig. 4.40 International tourism receipts are defined by World Trade organization as expenditures by international inbound visitors, including payments to national carriers for international transport. These receipts include any other prepayment made for goods or services received in Uzbekistan. Receipts from inbound tourism are recorded on the credit side by Uzbekistan. This indicator is used to define the contribution of tourism to the growth of GDP. The chart shows a trend of the average annual growth rate of foreign visitors to Uzbekistan

with the purpose of creating new jobs, ensuring economic diversification and development of regions, and increasing the incomes and quality of life of the population. This industry is developing rapidly in historical cities such as Bukhara, Khiva, and Samarkand. Small enterprises contribute to the development of this area of economy.

According to data from the World Travel & Tourism Council (2018), the direct contribution of this sector to GDP was US$275.1 million (0.9% of GDP) in 2017. Hasanovich notes that despite its potential, the contribution of tourism to GDP was only 1.8% in 2015. The social impact is in creating new jobs. In 2017, this sector directly supported 98,500 jobs (0.8% of total employment), and its total contribution to employment, including jobs indirectly supported by the industry, was 305,000 jobs (2.4% of total employment). According to World Bank estimates of receipts, international tourism contributes to the economy "through payments to national carriers for international transport, any other prepayment made for goods or services received in the destination country, receipts from same-day visitors" (Fig. 4.40).

The popularity of different kinds of tourism is a sign of changes in lifestyle and living standards. The State Committee of the Republic of Uzbekistan on Statistics

(2018) reported that in 2016, 44% of tours within the country were sold to citizens of Uzbekistan, 9.9% to CIS citizens, and 31.3% to citizens of other foreign countries. Today, more people can afford to travel for a vacation within Uzbekistan as well as abroad. New tourism businesses are a rapidly developing type of economic activity. For example, in Xorazm province, the number of such entities increased from 13 in 2001 to 59 in 2015. They provided services to 63,400 persons, valued at US$1,977,500 in 2015.

Religious tourism, in the form of pilgrimages, is a widespread form of tourism among the local people. A lot of places all over the country are now available as destinations for pilgrimages. The attraction of such tourism is not only acquisition of knowledge about nature and the environment but also discovering interesting theories about the origins of particular sites. Beautiful legends are full of fabulous fantasies and interesting interpretations of the value of natural or man-made objects. And, like all oriental tales, these legends tell of fascinating facts and miracles. Objects of tourism include holy places, mausoleums (Figs. 4.41, 4.42, 4.43, 4.44, 4.45, 4.46, and 4.47), historical monuments (Figs. 4.48, 4.49, and 4.50), unusual natural objects such as caves

Fig. 4.41 The mausoleum of Imam al-Bukhari, Samarkand region

Fig. 4.42 The mausoleum of Said Mir Kulol, Bukhara region

Fig. 4.43 The mausoleum of the mother of Naqshbandi, Bukhara region

Fig. 4.44 The mausoleum of Naqshbandi, Bukhara region

Fig. 4.45 The mausoleum of Sheikh Mohammad Sodiq, Kamashi, Kashkadarya province. (Image courtesy of Magfirat Muminova)

Fig. 4.46 The Mausoleum of Mahmud Paxlavon, Khiva

(Fig. 4.51), huge stones and stone gates (Figs. 4.52 and 4.53), trees (Fig. 4.54), and lakes. The mountains attract people for ecological and sporting tourism.

One program has started in the Western Tien Shan in the Tashkent region. It is planned to create all possible opportunities for extreme and mountain tourism, including skiing, snowboarding, paragliding and hang-gliding, cycling, motorcycling, and extreme sports, as well as environmental and rural tourism. The Western Tien Shan—with numerous tourist routes and trails, waterfalls, caves, gorges, walnut groves, and orchards—was added to the United Nations Educational, Scientific and Cultural Organization (UNESCO) World Heritage List in 2016 as a natural site of universal value. This site is a place of many endemic animals and plant species, and four natural lakes (Urengach, Bodak, Shovur, and Ihno) within the Chatkal–Ugam National Park.

The tourism and travel business offers a lot of opportunities for further development. However, its poor infrastructure and lack of an economically and socially justified concept for its development pose further challenges.

Fig. 4.47 The Palace of Amir Temur (Tamerlane) Aksaroy in Shakhrisabz, Kashkadarya region. (Image courtesy of Magfirat Muminova)

Fig. 4.48 The reservoir (*sardoba*) in Ithcan Kala, Khiva. The tradition of saving water in a *sardoba* is widespread in Uzbekistan

Fig. 4.49 The Kalta Minor Minaret (Short Minaret) in Itchan Kala is the symbol of the city of Khiva

Fig. 4.50 Pilgrimage to a cave named after David (Dovud). (Image courtesy of Magfirat Muminova)

Fig. 4.51 The local people respect huge natural objects such as stones and trees, which serve as objects of pilgrimage, on the Ugam–Chatkal Ridge, Tashkent region

Fig. 4.52 The Tamerlane Gates rock portal is between the Nurota and Turkestan hills in Jizzakh region. This was a strategic point on the Silk Route on the way to the Zerafshan Valley

Fig. 4.53 This tree in Bukhara is more than 300 years old

References

Abduvaliev A (2008) Head of Automobile and River Transport Agency, Republic of Uzbekistan Transport strategy of the Republic UZBEKISTAN and prospects of development of Trans-Asian and Euro-Asian transit carriages. Meeting of Ministers of Transport of the Euro-Asian region https://www.unece.org/trans/ministeri-alitc70/min_agenda.html. Retrieved on September 17, 2018 from https://www.unece.org/fileadmin/DAM/trans/MinisterialITC70/documents/Presentation_UZBEKISTAN_E.pdf

Ajwad MI, Abdulloev I, Audy R, Hut S, de Laat J, Kheyfets I, Larrison J, Nikoloski Z, Torracchi F (2014) The skills road: skills for employability in Uzbekistan. World Bank, Washington, DC

Asian Development Bank (2019) Data library Uzbekistan: by numbers. Retrieved January 12, 2019 from https://data.adb.org/dashboard/uzbekistan-numbers

Asian Development Outlook 2017 (2017a) Transcending the middle-income challenges. Retrieved from https://www.adb.org/countries/uzbekistan/economy. Available from April 16, 2017

Asian Development Outlook 2017 (2017b) Retrieved on May 15, 2018 from https://doi.org/10.22617/FLS178632-3, https://www.adb.org/publications/asian-development-outlook-2017-middle-income-challenge

Contessi NP (2016) Central Asia in Asia: charting growing trans-regional linkages. J Eurasian Stud 7:3–13. Elsevier https://doi.org/10.1016/j.euras.2015.11.001. Available online from September 2, 2015 http://elsevier.com/locate/euras

Countries of the World. Uzbekistan. Economy (2017) Source: 2017 CIA World Fact Book and other sources. Retrieved on July 15, 2018 from http://www.theodora.com/wfbcurrent/uzbekistan/uzbekistan_economy.html

Dukhovny VA, Stulina G (2011) Water and food security in Central Asia. In: Madramootoo C, Dukhovny VA (eds) Water and food security in Central Asia, NATO science for peace and security series C: environmental security. Springer, 1–24 pp. https://doi.org/10.1007/978-90-481-9974-7_3

Hansjürgens B (2016) Theory, the market and the state: agricultural reforms in post socialist Uzbekistan between economic incentives and institutional obstacles. In: Hüttl RF et al (eds) Society – water – technology, water resources development and management. https://doi.org/10.1007/978-3-319-18971-0_12

Hüttl RF, Bens O, Bismuth C, Hoechstetter S (eds) (2016) Society – water – technology, water resources development and management, a critical appraisal of major water engineering projects. Springer International Publishing. https://doi.org/10.1007/978-3-319-18971-0_12

International Monetary Fund (2019) Datasets. World Economic Outlook (April 2019). Retrieved on November 18, 2019 from https://www.imf.org/external/datamapper/NGDPDPC@WEO/OEMDC/ADVEC/WEOWORLD

International Organization for Migration (2018) Migration facts and trends. South-eastern Europe, Eastern Europe and Central Asia. Retrieved On November 17, 2019 from http://publications.iom.int/system/files/pdf/migration_facts_and_trends_seeeca.pdf, https://www.iom.int/countries/uzbekistan

International Trade Center (2017) Retrieved on JUly 17, 2018 from http://www.intracen.org/country/uzbekistan

Internet Live Stats (www.InternetLiveStats.com). Elaboration of data by International Telecommunication Union (ITU), World Bank, and United Nations Population Division. Retrieved on September 22, 2018 from http://www.internetlivestats.com/internet-users/uzbekistan/

Khudayberganov Y (2011) Security and water resources management problems and experience in the Amu Darya River basin. https://doi.org/10.1007/978-90-481

Kristen K, Ibragimov N, Lamers JPA (2011) Food security: quantity and quality matters for eating good bread in Uzbekistan. In: Madramootoo C, Dukhovny V (eds) Water and food security in Central Asia, NATO science for peace and security series C: environmental security. Springer, Dordrecht. https://doi.org/10.1007/978-90-481-9974-7_13

Light industry in Uzbekistan (2018) Uzbekistan business directory 2018. Light industry – list of companies and organizations operating in Tashkent and Uzbekistan in the sphere of activity Light industry. Retrieved August 5, 2018 from https://www.goldenpages.uz/en/rubrics/?Id=37

"O'zto'qimachiliksanoat"association (2017) Retrieved March 12, 2018 from http://uztextile.uz/uz/news

"O'zto'qimachiliksanoat"association (2019) Discussed issues of accelerating the introduction of clusters in the cotton-textile industry. Retrieved March 08, 2019 from http://uzts.uz/post/obsujden-wopros-uskoreniya-wnedreniya-klasterow-w-khlopkowo-tekstilnoi-otrasli

Parpiev Z (2015) Who is behind remittances? A profile of Uzbek Migrants UNDP Uzbekistan. Retrieved on September 16, 2018 from http://www.uz.undp.org/content/uzbekistan/en/home/ourperspective/ourperspectivearticles/2015/03/05/who-is-behind-remittances%2D%2Da-profile-of-uzbek-migrants.html

Reichert JA (2018) Disclosable version of the ISR – regional roads development project – P146334 – sequence no: 07 (English). World Bank Group, Washington, DC. http://documents.worldbank.org/curated/en/907291544018173519/Disclosable-Version-of-the-ISR-Regional-Roads-Development-Project-P146334-Sequence-No-07

Roozitalab MN, Serghini H, Keshavarz A, Eser V, de Pauw E (eds) (2013) Elements of Research Strategy and Priorities for Sustainable Agricultural Development of Highlands in Central, West Asia and North Africa. Review report. International Center for Agricultural Research in the Dry Areas. Retrieved onSeptember 17, 2018 from http://www.icarda.org/publication/working-papers

Shurtan Gas Chemical Complex (2018) Website. Retrieved on July 15, 2019 from https://sgcc.uz/en/page/complex/about

Strategic Textile (2018) In J Economic Review № 1 2018 (I Russian Стратегический текстиль//Экономическое обозрение № 1б 2018). Retrieved March 15, 2018 from http://uzts.uz/post/strategicheskii-tekstil

The Ministry of Foreign Economic Relations of the Republic of Uzbekistan (2017) Statistics. Retrieved from http://www.mfer.uz/en/export/statistics/

The State Committee of the Republic of Uzbekistan on statistics (2017a) Analysis of macroeconomic indicators of the Republic of Uzbekistan over the years of independence (1991–2016). Retrieved on July 17, 2018 from https://stat.uz/en/about-committee/435-analiticheskie-materialy-en1/2064-analysis-of-macroeconomic-indicators

The State Committee of the Republic of Uzbekistan on statistics (2017b) Retrieved from https://stat.uz/en/219-ofytsyalnaia-statystykaen/demografiya-i-trud-en/4333-human-resources-and-employment-in-2017

The State Committee of the Republic of Uzbekistan on statistics (2018) Retrieved on July 15, 2019 from https://stat.uz/en/press-center/news-of-committee/news-of-committee-20198/3559-4465-foreign-trade-4

The State Committee of the Republic of Uzbekistan on statistics (2019) Retrieved on December 15, 2019 https://stat.uz/en/official-statistics/services

Travel and Tourism. Economic Impact 2018. Uzbekistan (2018) World Travel & Tourism Council. Retrieved on October 10, 2018 from https://www.wttc.org/-/media/files/reports/economic-impact-research/countries-2018/uzbekistan2018.pdf.

UN. Economic and Social Council. Economic and Social Commission for Asia and the Pacific. Note by Secretariat (2017) Regional Implementation of the Vienna Programme of Action for Landlocked Developing Countries for the Decade 2014–2024. Retrieved August 16, 2018 from https://www.unescap.org/commission/73/document/E73_3E.pdf

UNDP Human Development Report (2016) Retrieved on August 18, 2019 from http://hdr.undp.org/en/countries/profiles/UZB

Uzavtosanoat (2018) Accessed on August 19, 2019 from http://uzavtosanoat.uz/en

Uzbekistan Exports (2017). Retrieved on August 17, 2019 from http://www.tradingeconomics.com/uzbekistan/exports

Uzkimyosanoat (2018) Retrieved on August 16, 2019 from https://uzkimyosanoat.uz/ru/company/history

World Bank (2017a) Trading economics. Uzbekistan Industrial Production. Retrieved September 20, 2018 from http://www.tradingeconomics.com/uzbekistan/industrial-production

World Bank (2017b) Data. Retrieved September 19, 2018 from https://data.worldbank.org/indicator/BX.TRF.PWKR.DT.GD.ZS?locations=UZ)

World Bank (2018a) Trading Economics. Retrieved September 20, 2019 from https://tradingeconomics.com/uzbekistan/gdp-per-capita-ppp/forecast%20)?poll=2018-12-31

World Bank (2018b) Trading economics. Retrieved September 19, 2019 from http://www.tradingeconomics.com/uzbekistan/employed-persons/forecast

World Bank (2018c) Global economic prospects. Darkening Skies. Regional outlooks. Europe and Central Asia. Retrieved September 16, 2019 from http://www.worldbank.org/en/publication/global-economic-prospects

World Bank Data (n.d.) https://data.worldbank.org/indicator/NY.GDP.MKTP.CD?locations=UZ

World Bank Data (2018) Macroeconomic indicators. Retrieved January 15, 2019 from https://data.worldbank.org/country/uzbekistan?view=chart

World Bank Group (2016a) Doing business economy rankings. World Bank, Washington, DC. Retrieved August 17, 2018 from http://www.doingbusiness.org/rankings

World Bank Group (2016b) Doing business economy profile 2017: Uzbekistan. World Bank, Washington, DC. Retrieved August 17, 2018 from https://openknowledge.worldbank.org/handle/10986/25644 License: CC BY 3.0 IGO. http://hdl.handle.net/10986/25644

Yokubzod S (2011) Food security and the improvement of water use efficiency. In: Water and food security in Central Asia, NATO science for peace and security series C: environmental security. Springer, 53-pp. https://doi.org/10.1007/978-90-481-9974-7_3

5 Human Geography

Abstract

This chapter discusses spatial perspectives on demographic, social, cultural, economic, and political changes in Uzbekistan. It includes information about patterns of regional and national populations. Analysis of the data provides understanding of the basic demographic perspectives, and this is helpful for explaining processes contributing to population pressures, growth, and migration, as well as the distribution, composition, determinants, and consequences of population trends.

Keywords

Population · Demography · Ethnicity · Religion

5.1 Geography of the Population

The natural resources, the favorable geographic locations of many cities at the crossroads of trade routes on the Great Silk Route, their convenience and well-developed infrastructure for trade, and their prosperity have made them the objects of many invasions and conquests over several thousand years.

Numerous tribes and nationalities have contributed to the history and culture of Uzbekistan. Their movement to Uzbekistan has resulted in the mosaic of its ethnicity and the spatial distribution of its population.

Several states, including Uzbekistan, were established in 1924–1925 during the national delimitations in Central Asia. Arbitrary borders were created that did not correspond to the ethnic and linguistic demarcations in this region. Initially, the demarcation between states was based on an idea of Pan-Turkism with the purpose of uniting the Turkic-speaking population into the Turkestan Socialist Federative Republic (1918–1927). Later, this was replaced by the idea of breaking down that republic into several smaller ethnic and regional units. Today, this unfair division echoes during redemarcation in the Fergana Valley.

The new realities of the globalized economy are new challenges for the region and especially for double-landlocked Uzbekistan. These issues need to be discussed if this region is to overcome the tyranny of its geography and avoid being marginalized in a fast-globalizing world (Hakimogli 2014).

5.1.1 Demography

According to official data,[1] the population of Uzbekistan was 32,653,900 (0.42% of the total world population) on January 1, 2018—16,533,900 (50.6%) urban and 16,120,000 (49.4%) rural (Table 5.1). Uzbekistan ranks 44th in the list of countries by population and is the most populous country in the Central Asian region. The State Committee on Statistics of the Republic of Uzbekistan[2] reported that in January–June 2018, the population grew by 243,400; the growth rates were highest in Surkhandarya region (2.1%); Kashkadarya and Samarkand (Samarqand in Uzbek) regions (both 1.9%); and Jizzakh, Namangan regions, and the city of Tashkent (all 1.8%). The lowest growth rate was in Tashkent region (1.1%). In 2017, the urban and rural populations grew at the same average annual rate of 1.7%. Higher rates of urban population growth occurred in (2.0%), Kashkadarya (2.4%), and Khorezm regions (4.9%; Table 5.1).

According to official estimates, the birth rate was lower in January–September 2017 (523,600) than in January–September 2016 (532,400). In 2017, the annual birth rate was 22.1 per 1,000 people (less than in 2016, when it was

[1]Source: State Committee on Statistics of the Republic of Uzbekistan, open data (2018).

[2]Source: State Committee on Statistics of the Republic of Uzbekistan, open data (2018).

L. Gulyamova, *The Geography of Uzbekistan*, World Regional Geography Book Series,
https://doi.org/10.1007/978-3-031-07873-6_5

Table 5.1 Population of Uzbekistan

Region	Percentage on January 1, 2018	Population (thousands), January–October 2017			Population growth (%), January–October 2017		
			Urban or rural			Urban or rural	
		Total	Urban	Rural	Total	Urban	Rural
Republic of Karakalpakstan	5.6	1834.5	902.2	932.3	+1.4	+1.2	+1.5
Andijan	9.2	2998.5	1570.4	1428.1	+1.7	+2.0	+1.3
Bukhara	5.7	1862.8	696.4	1166.4	+1.5	+0.9	+1.8
Jizzakh	4.1	1319.3	619.6	699.7	+1.9	+1.7	+2.0
Kashkadarya	9.6	3131.2	1351.4	1779.8	+1.9	+2.4	+1.5
Navoiy	2.9	954.1	465.7	488.4	+1.6	+1.4	+1.8
Namangan	8.3	2687.0	1735.6	951.4	+1.8	+1.5	+2.2
Samarkand	11.4	3701.5	1386.2	2315.3	+1.9	+1.3	+2.2
Surxondaryo	7.7	2500.5	889.1	1661.4	+2.1	+1.8	+2.2
Sirdaryo	2.5	812.5	348.9	463.6	+1.6	+1.2	+1.8
Tashkent	8.8	2853.3	1390.6	1462.7	+1.2	+1.0	+1.3
Fergana	11.1	3605.5	2042.2	1563.3	+1.6	+1.3	+1.9
Xorazm	5.5	1795.9	598.3	1197.6	+1.6	+4.9	+0.0
City of Tashkent	7.6	2455.3	21,455.3		+1.6		
Republic of Uzbekistan	**100.0**	**32,511.9**	**16,451.9**	**16,060.0**	**+1.7**	**+1.7**	**+1.7**

Source: [CC-BY-4.0] via State Committee on Statistics of the Republic of Uzbekistan, open data (2018)

Table 5.2 Birth rates and population growth during different periods in 2016–2018

Region	Birth rate (per 1000 people)			Population growth (%), January–July 2018
	January–September 2016	January–September 2017	January–December 2017	
Republic of Uzbekistan	22.8	21.7	22.1	
Republic of Karakalpakstan	20.7	19.5	20.7	+1.4
Xorazm	20.7	19.5	21.0	
Bukhara	20.6	19.5	20.0	
Fergana	21.9	20.8	21.0	
Tashkent	20.2	19.2	19.2	+1.2
Kashkadarya	24.7	23.8	22.4	+2.0
Namangan	23.5	22.6		+1.8
Andijan	22.9	22.1		
Navoiy	22.3	21.7	19.8	
Samarkand		23.8	22.9	+2.0
Surxondaryo		25.6	25.9	+2.1
Sirdaryo		21.9		
Jizzakh		23.8	23.6	

The total population of Uzbekistan in 2017 was 32.5 million
Source: [CC-BY-4.0] via State Committee on Statistics of the Republic of Uzbekistan, open data (2018)

22.8). The significant declines in the birth rates in many regions were evidence of social and economic development, and of changes in women's roles and status. However, this indicator was not the same in different regions (Table 5.2).

In 2016, the population of Uzbekistan rose by 450,375 (1.49%) in comparison with the population a year before. In 2015–2016, the natural growth was positive; the number of births exceeded the number of deaths by 491,045. However, as a result of emigration, the population declined by 40,669. The total sex ratio was 0.988 (988 males per 1000 females), which was lower than the global ratio of 1.016 (United Nations Department of Economic and Social Affairs 2017).

In 2017, the population of 32,511,900 was more than eight times what it had been in 1926 (4,621,000; Table 5.3).

Table 5.3 Population growth in 1926–2017

Year	Population (thousands)
1926	4621.0
1939	6346.9
1959	8118.0
1970	11,799.0
1979	15,390.8
1989	19,780.0
1991	20,607.7
2017	32,511.9

Source: [CC-BY-4.0] via State Committee on Statistics of the Republic of Uzbekistan, open data (2018)

Table 5.4 Annual population growth in 1974–2017

Year	Population growth (%)	Year	Population growth (%)	Year	Population growth (%)	Year	Population growth (%)
1974	+2.633	1985	+2.507	1996	+1.913	2007	+1.423
1975	+2.681	1986	+2.493	1997	+1.885	2008	+1.605
1976	+2.744	1987	+2.471	1998	+1.609	2009	+1.687
1977	+2.779	1988	+2.447	1999	+1.078	2010	+2.823
1978	+2.785	1989	+2.421	2000	+1.384	2011	+2.684
1979	+2.75	1990	+2.622	2001	+1.266	2012	+1.472
1980	+2.689	1991	+2.132	2002	+1.224	2013	+1.562
1981	+2.624	1992	+2.344	2003	+1.164	2014	+1.687
1982	+2.573	1993	+2.272	2004	+1.154	2015	+1.744
1983	+2.535	1994	+1.963	2005	+1.163	2016	+1.739
1984	+2.516	1995	+1.807	2006	+1.22	2017	+1.679

Source: (Public domain) via the World Bank under License CC-BY-4.0

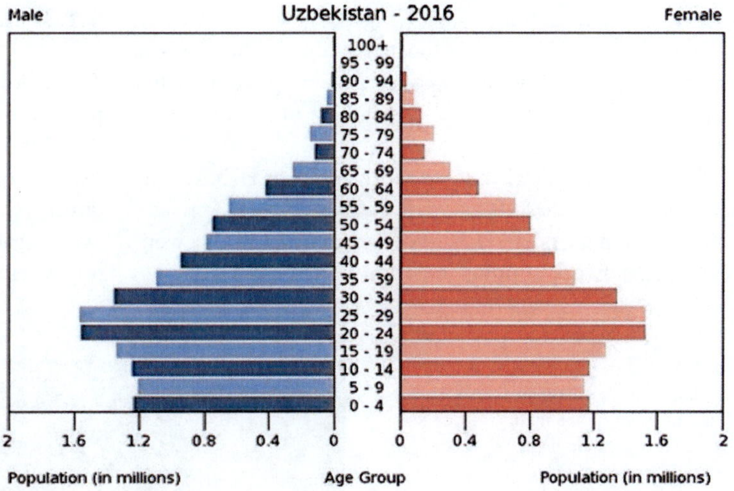

Fig. 5.1 Age composition of the population, 2016. (Source: (Public domain) Creative Commons CC0 License; via Wikimedia)

Between 1974 and 2017, the annual population growth declined from 2.633% to 1.679% (Table 5.4).

Experts from the United Nations Development Programme (UNDP 2018, p. 57) have forecast that Uzbekistan's population will grow to 36.7 million by 2030 if yearly growth of 1.4% (the medium fertility variant) occurs.

Age Structure

Human resources are the main treasure of Uzbekistan. This nation is young. In 2017, the median age of the population was 26.3 years, and 26.5% of the population was under 15 years of age. The population of working age (15–64 years) was 21.6 million, while only 1.456 million people (808,800 women and 647,800 men) were older than 65 years (State Committee on Statistics of the Republic of Uzbekistan 2018, p. 71).

The population pyramid is of an expanding type, which is common for developing countries with relatively short life expectancy (Fig. 5.1).

According to the State Committee on Statistics of the Republic of Uzbekistan,[3] the population under working age declined between 1991 and 2017. The committee reported that this proportion of the total population was 43.1% in 1991, 31.3% in 2011, and 30.1% in 2017. At the same time, the working-aged proportion of the total population rose from 49.2% to 60.5%. According to the committee, the age dependency ratio is 45.3%, or less than half of the working part. This means that the pressure on the productive population is relatively low; they can easily provide goods for themselves and cover expenditures on children and aged persons. From this perspective, the situation is comfortable. The child dependency ratio is 38.5%, and the aged dependency ratio is 6.8%.

Life expectancy (for both sexes) at birth is 72.5 years (69.5 years for males and 75.7 years for females), which is

[3]Source: State Committee on Statistics of the Republic of Uzbekistan, open data (2017).

above the average indicator for the global population (71 years).[4]

The literacy rate of the population[5] is one of the highest in the world: 99.9% of the adult population (aged 25 years and above) are able to read and write (UNDP 2018, p. 64).

5.1.2 Spatial Distribution

The population of Uzbekistan is unevenly distributed in the country. Historically, settlements were founded along the main rivers. The population density is higher at oases and in the foothills in the eastern part of the country. Proximity to water resources influences the spatial distribution of settlements. This results in the highest density of population within oases at sites close to water resources and in places with extensive irrigation systems (Fig. 5.2).

Settlements are frequently located at alluvial fans, where water draining from mountainous terrain emerges onto flatter plains. The best example is the distribution of settlements in the Fergana Valley (Fig. 5.3).

The Fergana Valley is known as "the pearl of Uzbekistan" because of its fertile soils and rich water resources for irrigation. It is the most populous part of Central Asia, with a well-developed network of urban and rural settlements.

The population density is far lower in western Uzbekistan than in the Fergana Valley. Some parts of this area, particularly the Republic of Karakalpakstan (Qaraqalpoghiston), are under threat of depopulation caused by the environmental crisis of the Aral Sea.

The average population density is 71.5 persons/km^2, it reaches 900 persons/km^2 in some rural parts of the most populous Fergana Valley and declines to zero in the uninhabited Kyzylkum Desert. The highest population density in Central Asia is in the Fergana Valley, in the Andijan, Namangan, and Fergana provinces (Table 5.5).

In 2000–2018, the average population density increased by 32.3%. The highest rates of growth occurred in Kashkadarya and Surxondaryo provinces (Table 5.5).

5.1.3 Ethnicity

The colorful picture of the modern ethnic mosaic of the population is the result of different waves of migration in the nineteenth and twentieth centuries. They include resettlement of people from the Russian Empire to Turkestan during colonization, a Bolshevik invasion in the 1920s, evacuations from the western part of the Soviet Union during World War II, industrialization in the 1930s and in the

Fig. 5.2 The density of population and settlements is higher at oases on the main rivers. The oases are depicted in dark gray-green. (This image is in the public domain because it is a screenshot from NASA's globe software World Wind using a public domain layer, such as Blue Marble, MODIS, Landsat, SRTM, USGS or GLOBE. https://commons.wikimedia.org/wiki/Atlas_of_Uzbekistan#/media/File:Uzbekistan_satellite_photo.jpg. Poulpy, based on work by NASA – World Wind)

[4]Source: State Committee on Statistics of the Republic of Uzbekistan, open data (2017).

[5]The estimation data on the literacy of Uzbekistan's population is based on data published by the United Nations Educational, Scientific and Cultural Organization (UNESCO) Institute for Statistics (accessed on March 13, 2016).

Fig. 5.3 The Fergana Valley. (Source: User MGA73bot2 NASA (public domain) via Wikimedia)

Table 5.5 Change in population density, 2000–2018

Region	Population density (persons/km²)		
	2000	2018	Growth, 2000–2018 (%)
Republic of Karakalpakstan	9	11.1	+23.3
Andijan	520.5	700.4	+34.5
Bukhara	35.2	46.4	+31.8
Jizzakh	46	62.5	+35.8
Kashkadarya	75.8	110.2	+45.4
Navoiy	7.1	8.6	+21.1
Namangan	260	362.8	+39.5
Samarkand	158.9	221.8	+39.6
Surkhandarya	86.4	125.1	+44.8
Syrdarya	149.3	190.6	+27.7
Tashkent	153.6	187.6	+22.1
Fergana	397.7	535.5	+34.6
Khorezm	217	298.3	+37.5
City of Tashkent	6472.2	7380	+14.0
Republic of Uzbekistan	**54.6**	**72.7**	**+32.2**

Source: [CC-BY-4.0] via State Committee on Statistics of the Republic of Uzbekistan, open data (2019)

1960s–1970s, and reconstruction after the Tashkent earthquake in 1966.

The Central Asian region is an example of human evolutionary history because of its central geographic location in Eurasia and the rich history of human population movements between Asia, the Middle East, India, and Europe. Irwin et al. (2010, pp. 195–204) note that "the mitochondrial DNA (mtDNA) haplogroup[6] composition for some regions of Uzbekistan shows that the two most common haplotypes in the entire data set were shared by four of the five subpopulations and thus, the basic diversity measures were very consistent among populations." According to research (Irwin et al. 2010, pp. 195–204), "AMOVA[7] results mirrored the summary statistics, indicating that 99.55% of the mtDNA variation is accounted for by differences within the population, while only 0.45% of the variation is due to differences between them." The haplogroup compositions vary in the different parts of the country (Fig. 5.4).

According to the last census conducted in the Soviet era (1989), more than 100 nationalities lived in Uzbekistan at that time. After dissolution of the Soviet Union, the ethnic structure changed and many people of other nationalities/ethnicities left this country: Crimean Tatars moved to Crimea, Germans to Germany, Jews to Israel, Russians to Russia, and so on.

Today, Uzbeks make up the majority (83.8%) of the population; about 11% are representatives of other Central Asian nations (Karakalpaks, Kazakhs, Kyrgyz, Tajiks, and Turkmen), and fewer than 7% are from other nations/ethnicities, including Russians, Ukrainians, Koreans, Jewish, and others[8] (Table 5.6; Fig. 5.5).

In urban areas, the ethnic mix is more diverse.

[6]The mitochondrial DNA haplogroup is a group of genes in an organism that are inherited together solely from the mother.

[7]AMOVA (analysis of molecular variance) is a statistical model of molecular variation in a single species, typically biological.

[8]Source: State Committee on Statistics of the Republic of Uzbekistan.

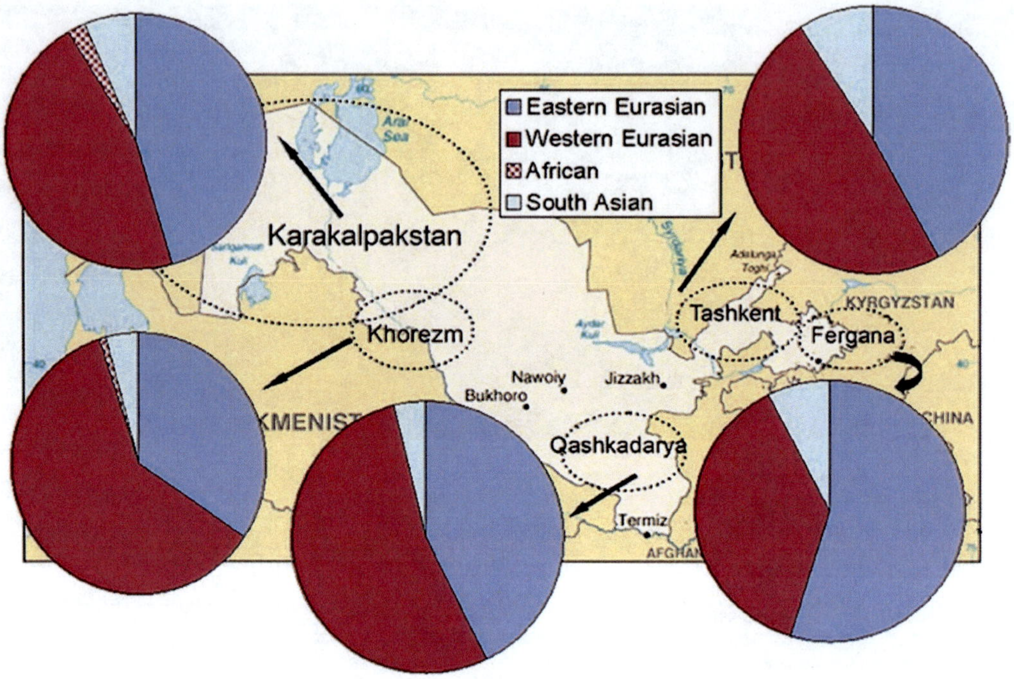

Fig. 5.4 Mitochondrial DNA haplogroup distributions among autochthonous Uzbek subpopulations. (Source: Irwin et al. (2010). Copyright Clearance Center RightsLink https://s100.copyright.com/AppDispatchServlet. License Number 4430714035733 Sep 16, 2018)

Table 5.6 Ethnic structure (on January 1, 1991, and on January 1, 2017)

	1991		2017	
Ethnic group	Thousands	Percentage	Thousands	Percentage
Uzbeks	14,995.3	72.8	26,917.7	83.8
Karakalpaks	431.9	2.1	708.8	2.2
Tajiks	980.7	4.8	1544.7	4.8
Kazakhs	845.3	4.1	803.4	2.5
Russians	1593.8	7.7	750.0	2.3
Kyrgyz	182.6	0.9	274.4	0.9
Tatars	414.6	2.0	195.0	0.6
Turkmen	126.6	0.6	192.0	0.6
Koreans	183.7	0.9	176.9	0.6
Ukrainians	146.8	0.7	70.7	0.2
Others	706.4	3.4	486.9	1.5

Source: [CC-BY-4.0] via State Committee on Statistics of the Republic of Uzbekistan, open data (2017)

Ethnic Minorities

It is supposed that Bukharian Jews began settling in Central Asia in the tenth century BCE. Several theories about the origin of Bukharian Jews are based on the deep history of the Middle East. According to Kotler (1994), one of these theories refers to the times of the Assyrian captivity of Israel in the seventh century BCE, when descendants of the Napthali and Issachar tribes were exiled. According to another theory, Bukharian Jews were exiled from Persia in the fifth century BCE. However, they are considered one of the oldest ethnoreligious groups in Central Asia. The Silk Route served as a bridge for migration of different ethnic and religious groups, including Jews from other Eastern and Western countries.

In Bukhara (Buxoro in Uzbek), in times when non-Muslim constructions were strictly forbidden, Jews and Muslims shared space in the same building, named Magoki Attoron, which translates as the "Mosque in a Pit" (Fig. 5.6). It has Zoroastrian inscriptions on its facade. This building is an example of the religious and ethnic tolerance of the nations at the crossroads of the Silk Route. This tolerance has always been the major principle of social life in Uzbekistan.

In the eighteenth century, there was considerable discrimination against Bukharian Jews, and some were even forcibly

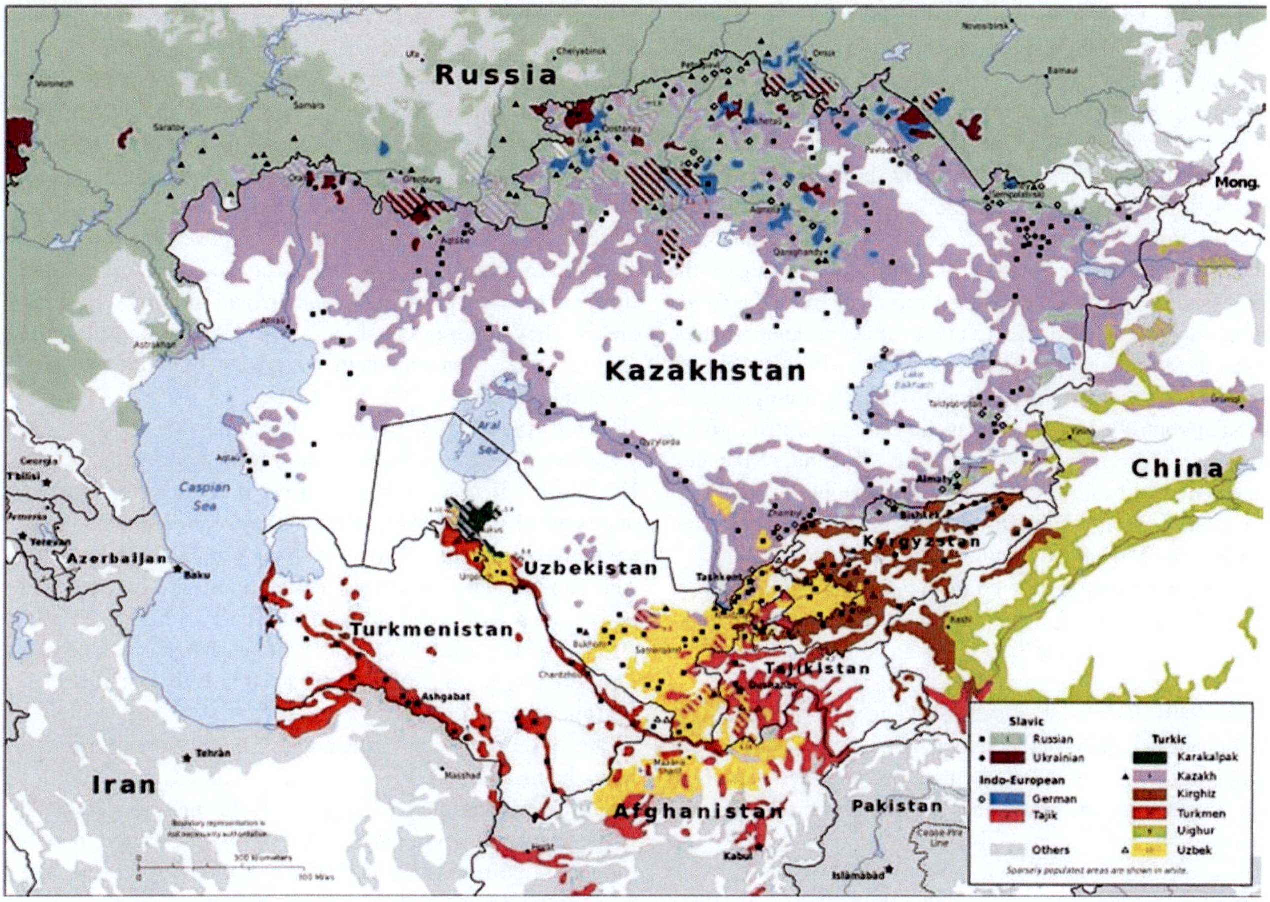

Fig. 5.5 Major ethnic groups in Central Asia. (Source: User Pmx (CC-BY-SA 3.0, https://creativecommons.org/licenses/by-sa/3.0) via Wikimedia)

Fig. 5.6 Magoki Attoron (the "Mosque in a Pit") in Bukhara

converted to Islam in the Emirate of Bukhara (Kotler 1994). They were called the *Chala* (an Uzbek term meaning "neither this nor that") because they outwardly practiced Islam but secretly retained their Jewish traditions. The impossibility of their return to the true Jewish faith, due to the fatal consequences associated with leaving the Islamic faith, led to isolation of the *Chala*. The inclusion of Turkestan in the Russian Empire led to more freedom for Bukharian Jews, who became free to practice Judaism. As a result of the oppressive religious policy in the Emirate of Bukhara, *Chala* communities emerged in the cities of Samarkand, Khiva, Kokand, Margilan, and Shakhrisabz. Even after the Russian conquest of Central Asia, they had to illegally immigrate into Russian-controlled areas such as those incorporated into the Turkestan governorate Khiva and Kokand Khanates. Conversion to Islam also caused *Chala* people to lose their self-identification as Jewish, resulting in their full assimilation into the Muslim population.

The community attempted to preserve its own traditions and culture (Figs. 5.7 and 5.8). Bukharian Jews spoke Bukhori, a dialect of the Tajik–Persian language with certain linguistic traces of Hebrew.

In 1940–1945, a lot of Ashkenazi Jewish refugees from the European regions of the Soviet Union and Eastern Europe settled in Uzbekistan. Many of them became successful doctors, actors, dancers, musicians, singers, film producers, and artists.

Koreans were forcibly relocated from the Soviet Far East to the Central Asian region in 1937–1938 (Levinson 1994). This was another example of ethnic repression. Chey (1994) notes that Koryo-saram, or Koryoin, is the name that ethnic Koreans use to refer to themselves in the former Soviet Union, including Uzbekistan. Adaptation to the arid climate was difficult for ethnic Koreans. They were rice farmers and fishermen, but opportunities for these activities were limited. The largest number of them settled in rural areas in the Tashkent and Samarkand regions, and in the Fergana Valley and Karakalpakstan. Estimates (Pohl 1999) suggest that 40,000 deported Koreans died in 1937–1938. However, the hardworking Koreans were able to recover their original standard of living within 3 years. The Koryo-saram became involved in rice production and irrigation works in rural areas. They are respected as good rice farmers all over Central Asia. Many of them have been honored for their work in agriculture.

From the 1970s to the 1980s, the the growth in Uzbekistan's Korean population slowed down significatly. Koreans began to actively move from rural areas to large cities, especially to Tashkent, where many of them settled in the southern areas and suburbs (such as Kuilyuk and Sergeli). The Koreans switched from agricultural activity to urban activities (science, education, health care, culture, and art). After dissolution of the Soviet Union, Uzbek Koreans had nowhere else to go, and so they stayed in Uzbekistan, which becomes their homeland.

According to official data, about 176,900 Koreans resided in Uzbekistan in 2017. The largest Korean communities today are concentrated in the city of Tashkent (about 60,000), in the Tashkent region (about 70,000), in the Sirdaryo region (about 11,000), in Karakalpakstan (about 8000), in the Samarkand region (about 6000), and in Khorezm (about 5000).

In 1939, the Soviet government decreed that Korean was to be taught as a second language, but from 1945 onward, it

Fig. 5.7 A *Chala* family in Bukhara. (Image courtesy of Shavkat Boltaev, Bukhara)

Fig. 5.8 A folk music ensemble of Bukharian Jews. (Image courtesy of Shavkat Boltaev, Bukhara)

completely stopped the teaching of this language. Only the older generations knew the Korean language; subsequent generations lost the use of it. There is a difference in the linguistic practices of the different generations. Koreans who migrated to Central Asia speak the northeastern Hamghian dialect, which differs markedly from the Khangugin dialect of South Korea. For 70 years of life in Uzbekistan, the language of the local Koreans absorbed some borrowings from Russian and Uzbek. In rural areas, many representatives of the older generation still speak their native language. Almost all Koreans in Uzbekistan are fluent in Russian, and most (especially those in rural areas) know the Uzbek language well.

The Korean language was taught at the Tashkent State Pedagogical Institute from 1956 to 1964; the teaching was restarted in 1988. Teaching of the Korean language has been offered at the Tashkent State Institute for Oriental Studies since 1993, at the Tashkent State University of World Languages since 1991, and at Samarkand State University since 2001. Korean is taught as a second foreign language at the University of World Economy and Diplomacy, Urgench State University, Karakalpak State University, Karshi Pedagogical Institute, Bukhara State University, Namangan State Pedagogical Institute, and Andijan State University. The students at the Center for Korean Education are of different ethnicities, including Koreans, Uzbeks, and Russians (Hwang and Lee 2008). In 2018, a new faculty was opened in association with the Hanguky University of Foreign Languages at the Tashkent State Institute for Oriental Studies, teaching the following subjects: Korean philology, economy and politics of Korea, and history and culture of Korea. It was intended that the students would be able to participate in joint educational programs with Korean partners, and that their diplomas would be recognized in both countries.

The majority of Korean seniors are communists, and atheism is widespread among Koreans. However, they practice Christianity and Buddhism. Many customs and rituals have been modified, simplified, and preserved, with some changes. They were influenced by Russian culture and the culture of the peoples the Koreans lived among. Many Koreans were baptized, which also led to a cultural change. A major ritual of Koreans is visiting cemeteries three times each year. The rituals for the first birthday (*tol*) and the sixtieth birthday (*hwangab*) and observance of the ancient lunar calendar during traditional holidays (*Hansik* in spring and *Chusok* at the autumn equinox) have been preserved in their traditional forms.

The Lyuli are one of the Romani (Gypsy) communities who live in Central Asia. As noted by Crowe et al. (1994), they are also known as the Multoni, Mughat, Churbat, or Jugi.

Little information exists about this ethnic group, since their community is extremely closed toward non-Lyuli. They speak a Tajik dialect and practice Islam, as well as Christianity. They refer to themselves as Mughat, which is derived from the Old Persian word *magi* (fire worshiper). Multoni ethnic groups live in several villages in the Surkhanndarya region. Crowe et al. (1994) suppose that their origin is related to the city of Multan in Pakistan, from where they emigrated in around 1380. They were wandering musicians and entertainers.

Roma are divided into two separate, ethnically different groups: the Lyuli (Uzbek)—also known as the Jugi (Tajik)—and the Mazangs. Members of these groups isolate themselves from the native population of Central Asia. They are only superficially Muslims (Sunni of the Hanafi school) and preserve many pre-Islamic rites and beliefs. Gypsies are bilingual (speaking Uzbek and Tajik), but the majority consider Tajik to be their native language. They also use a secret language (mainly Iranian), *lavz-i mugat*, with a special vocabulary.

Today, the Gypsies are one of the world's marginal groups, surviving by begging and by recycling scrap metal or plastic bottles.

It is supposed that about 12,000 Lyuli live in Uzbekistan. Their living standards are low, and their children do not receive a good education.

In a study (Marushiakova and Popov 2016) of Gypsies in Central Asia and the Caucasus, it was noted that the aggregation of Central Asian Gypsies includes different communities. These communities are divided into local groups on the basis of their origins, language, identity, lifestyle, and main occupations, but they themselves make distinctions between their groups. These communities are undergoing constant ethnic development. It is not surprising that in Russia, where Lyuli are migrating to, they are mistakenly identified as Tajik, since they have become assimilated into Tajik culture and their dress code and lifestyle have changed over time.

Iranians

Ethnic Iranians (Ironis) live in Samarkand and Bukhara. They practice Shi'a Islam, representing about 1% of all Muslims in Uzbekistan.

For the majority of Shi'a Muslims, Ashura marks the climax of the Remembrance of Muharram and commemorates the death of Husayn Ibn Ali (the grandson of Prophet Muhammad) at the battle of Karbala on the tenth day of the Muharram (the first month in the lunar Islamic calendar). The Remembrance of Muharram (or Mourning of Muharram) is a set of rituals (associated with Sunni and Shi'a Islam), which take place in *hussainia* (congregation halls) (Figs. 5.9 and 5.10). This Day of Ashura serves to define Shi'a communal identity. The difference between the ways in which Sunni and Shi'a Muslims commemorate this event is the absence of solemn mourning among the Sunni communities. Sunnis observe this event as a time of remembrance, rather than mourning.

The Persians of Uzbekistan number approximately 40,000. They are part of the cluster of Persian people within the Persian–Median affinity bloc. Globally, this group totals 41,699,200 people in 28 countries. Their primary language is Iranian Persian.

According to one study (Bennigsen and Wimbush 1985), in 1979, 30.7% of Ironis spoke Persian, 9.8% spoke Russian, and 60.5% spoke various Turkic languages. They preserve their own traditions, rituals, culture, and language.

Arabs

Arabs settled in Uzbekistan during the Muslim conquests in the seventh and eighth centuries. Later, Timur brought craftsmen from Damascus. They came as soldiers, artisans, intellectuals, and nomads. The Arabs lived in isolated communities and did not favor intermarriage with the local population. These customs helped them to remain a separate people and preserve their language in a multilingual milieu. This ethnic group came to Uzbekistan early in the tenth century, and they became isolated from the rest of the Arabic-speaking world and from the classical Arabic language.

Today, most descendants of the Arabs speak the local language and Uzbeki Arabic, which has structural elements from the Tajik and Uzbek languages. In Uzbekistan, the dialect of Bukharian Arabic is influenced by Tajik, while Kashkadarya Arabic is influenced by the Turkic language. They live in rural areas of Bukhara, Samarkand, Kashkadarya, and Surxondaryo provinces. The names of villages testify to

Fig. 5.9 Women pray in a congregation hall (*hussainia*) in Bukhara during the Ashura holiday. (Image courtesy of Shavkat Boltaev, Bukhara)

Fig. 5.10 A woman lights candles in Bukhara on the night of Ashura. (Image courtesy of Shavkat Boltaev, Bukhara)

their presence—for example, Arabxona, Arablar, Arabcha, Katta Arab, and Mir Arab (Fig. 5.11). The Arab population is about 1100.

According to the official data, 1100 Arabs live in these villages. Arabs wear the same clothes as Uzbeks, eat the same food, have the same customs and traditions, and practice Sunni Islam. Today, Arabs are assimilated. Their community is an example of a multilingual mosaic, which is very common for many places in Uzbekistan.

The Uzbek language belongs to the Karluk (or southeastern) dialects, in the Chagatai (Turki) branch of the Turkic language group. Anatolian Turkish is a variant of the Oghuz (or southwestern) dialect, which is spoken as an official language in Türkiye. Arabic and Persian words were removed from Anatolian Turkish, whereas Uzbek kept them. Loanwords are of Russian origin in Uzbek, while they are of French in Turkish.

Under the influence of the Persian language, the vowel "a" is rounded to "o"; this is the most noticeable distinction between the Uzbek language and other Turkic languages. The Persian influence is seen also in the Uzbek vocabulary, phraseology, and pronunciation. Several dialects are spread over the country, i.e., the Tashkent, Fergana, Khorezm, Chimkent–Turkestan, and Surkhanndarya dialects.

The modern Uzbek language is understandable by other Turkic-speaking nations. At the same time, other languages are also widespread in places where foreign-language speakers have settled in the Central Asian nations, in schools, and in the media.

After dissolution of the Soviet Union, when Uzbekistan became independent, many people from other nations left the country, leading to changes in the ethnic balance and employment patterns.

5.2 Urbanization

The history of Uzbekistan's urbanization is similar to those of many countries in the world. However, it has its own factors that have made this process different. This country is one of the oldest centers of urbanization in the world. The main cities of Uzbekistan (Bukhara, Qarshi, Khiva, Samarkand, Shahrisabz, Tashkent, and Termez) are more than 2500 years old. The ruins of many ancient cities (such as Afrosiab, Ahsi, Pop, Nasaf, Kesh, and Poykend) are evidence of the history of Uzbekistan's urbanization.

Urban development in the region and in Uzbekistan has a long history, starting several millennia ago. The Silk Route just connected existing trade routes and cities in one network. Exchange of knowledge and ideas, art and science, and information and technologies was the contribution of Central Asian cities to humankind. Khayde calls the Great Silk Route the first "information superhighway" in times when it was essential and important to unite states and cities under one umbrella of common rules and agreements. Many centuries ago, people understood this and acted in accordance with it to gain more prosperity and sustainable development. The predecessors of people in modern Uzbekistan were skillful craftsmen, smart merchants, and good engineers and planners. Cities in the Fergana Valley did not have strong armies, since they did not have any intentions to invade other lands. They were always very peaceful and tolerant but patriotic, and their fight for independence and freedom cost them dearly, as their cities were disastrously destroyed many times by invaders and internecine strife.

Rapid growth of the urban population occurred in 1865–1917 as a result of migration from the Russian Empire to the

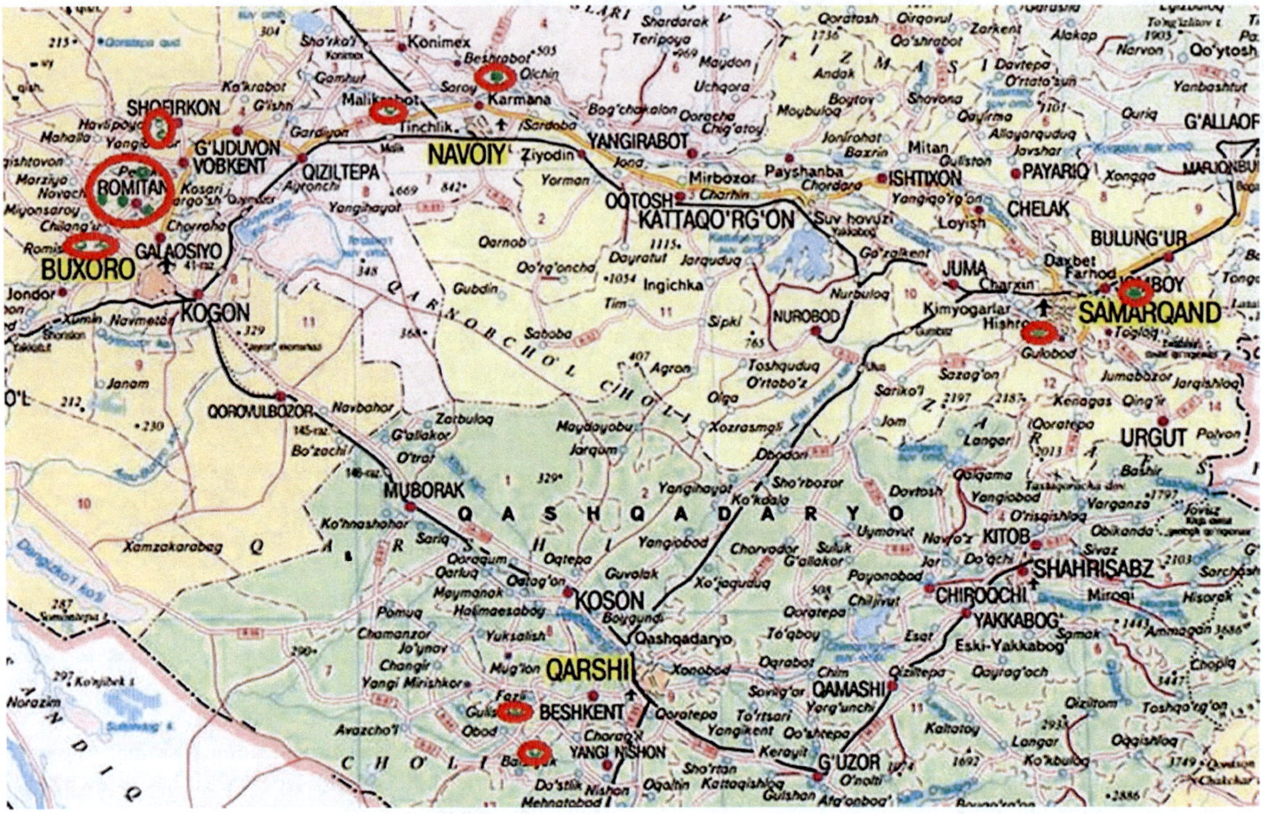

Fig. 5.11 On this map, settlements with a majority of Uzbeki Arabs are marked in *green* within the areas marked in *red*

cities of Uzbekistan. Second and third waves of migration occurred during Uzbekistan's industrialization in the 1930s and 1970s. The rapid growth of the urban population was also a result of incorporation of rural settlements and towns into cities. During World War II, the population of Uzbekistan decreased by at least 1,000,000. In 1972, new demographic standards changed the population requirement for city status from 10,000 to 7,000 people. This technically increased the number of cities, and each year, about ten more villages became classified as towns. In 2009, more new standards were established that changed the population requirement for town status from 3,000 to 2,000 people. Increases in the numbers of cities and the urban population led to pseudourbanization.

In 1950s, the rural population constituted 70% of the total population. In the 1950s–1990s, the rural population was still dominant. Today, the share of the urban population is somewhat higher (50.6%) than that of the rural population (49.4%). According to United Nations (UN) estimates (World Bank Data), this situation will continue for the next 10 years.

Uzbekistan has one city with more than 1 million people, 16 cities with between 100,000 and 1 million people, and 92 cities with between 10,000 and 100,000 people. On April 1, 2018, the population of city of Tashkent was approximately 2.475 million.[9] The 17 cities constitute a significant portion of the urban economy.

The distribution of cities in Uzbekistan largely corresponds to Zipf's law: there is one megacity (Tashkent) and three big cities: Namangan, Samarkand, and Andijan. (Table 5.7).

The populations of small cities (with fewer than 50,000 inhabitants) and medium-sized cities (with 50,000–100,000 inhabitants) account for more than one-third of the urban population. However, they generate less than one-third of the national industrial output. They lack urban functions and have become "big villages." This has led to strengthening of the differences between regions and to widening of the gaps between the big cities and small towns. Economic activity is concentrated in large cities, which are not prepared to provide migrants from underdeveloped towns and regions with jobs, services, and housing. Researchers (Center for Economic Research 2013) point to the threat of unplanned urban development as a result of population influx into cities.

[9]Source: State Committee on Statistics of the Republic of Uzbekistan, open data (2018).

Table 5.7 Populations of major cities in Uzbekistan on January 1, 2018

City	Population (thousands)
Tashkent	2464.9
Namangan	600.2
Samarkand	529.6
Andijan	427.5
Bukhara	276.4
Nukus	310.9
Karshi	267.7
Kokand	245.5
Chirchiq	149.4
Fergana	278.5
Jizzakh	171.7
Urgench	140.2
Termez	143.8
Margilan	228.3
Navoi	135.6
Angren	183.6
Almalyk	127.5

Source: [CC BY -4.0] public domain via State Committee on Statistics of the Republic of Uzbekistan (2018), pp. 94–106

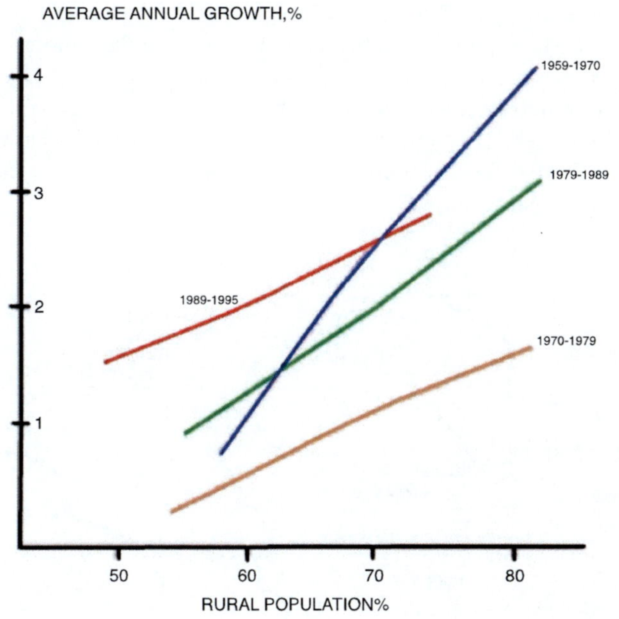

Fig. 5.12 Average annual growth rate of the population versus the share of the rural population

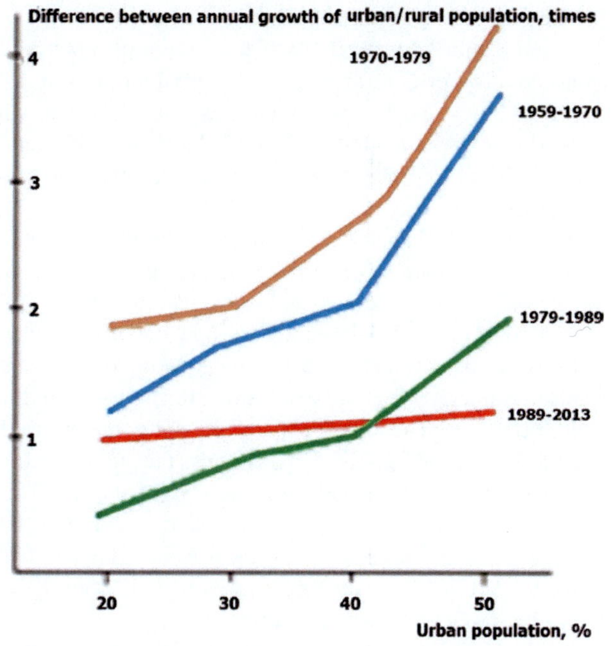

Fig. 5.13 Difference between the average annual growth of the urban and rural populations versus the share of the urban population

A decline in the average annual growth rate is a significant feature of Uzbekistan's demography. In 1959–1979, there was a significant difference in average annual growth rates between the urban and rural populations. This difference has lessened over time (Fig. 5.12).

This difference depends on the share of the urban population; in more urbanized areas, it has reached 400% as a result of huge migration to Uzbekistan's cities during the industrialization that occurred in 1970–1979. In 1989–2009, the average annual growth rate was the same in urban and rural areas (Fig. 5.13). This trend continued in the 2010s; in 2017, the urban and rural populations grew at the same average annual growth rate of 1.7%.

During the past 50 years, the urban population has grown from 3 million to 12 million.[10] About half of the urban population lives in secondary cities with fewer than 500,000 inhabitants (United Nations Department of Economic and Social Affairs 2014). Outdated infrastructure, limited work opportunities, and loss of urban functions have made these cities socially and economically vulnerable. They face even more challenges for sustainable development. The rapid growth in the urban population is not supported by adequate job creation, urban infrastructure development, or environmental protection. The imbalance in the opportunities of cities to meet the needs of their citizens is one of the lessons learned from "pseudourbanization"; experts (Center for Economic Research 2013) have pointed to the risk of "false urbanization" without a comprehensive urbanization policy.

The scarcity of land and water resources is the main challenge for further growth of the cities at the expense of agricultural land without a smart land use policy. The excessive labor in rural areas is another challenge for the economy of Uzbekistan.

[10]World Bank data (http://data.worldbank.org/indicator/SP.URB.TOTL?locations=UZ&page=3), accessed on April 27, 2017.

The city of Tashkent has had the strongest influence on the urbanization of Uzbekistan. This is the most advanced agglomeration in Central Asia. This agglomeration unites about 30 cities and towns, which make up a huge urban network. Other agglomerations (Andijan, Bukhara, Fergana–Margilan, and Samarkand) are developing rapidly as well.

The city of Tashkent is an example of city growth due to annexation of suburbs. This is a common feature of the spatial growth of many cities in Uzbekistan. Figures 5.14, 5.15, 5.16, 5.17, 5.18, 5.19, and 5.20 show the urban sprawl in 1986–2000. In 1860 (Fig. 5.14), the location of Tashkent was determined by water resources in a fertile valley of the Chirchiq River. The city's expansion along major trade routes and around markets illustrates the monocentric model of development and allocation of resources. This model was common for many cities in the region. After incorporation of Turkestan into the Russian Empire, two separate parts of the city were developed in accordance with the policy of two governing systems (Fig. 5.15; Tashkent Boundaries in 1913). However, the monocentric model came back (Fig. 5.16). An earthquake in 1966 caused serious damage in the central part of the city, which was close to the earthquake's epicenter (Fig. 5.17; Tashkent Boundaries in 1967). New development on the periphery resulted in even growth of the city in all directions, repeating its previous growth along ancient trade routes (Figs. 5.18, 5.19, and 5.20; Tashkent Boundaries in 1981–2000). Today, the city incorporates land that was previously agricultural before being urbanized (Fig. 5.21; Tashkent_Subdivision 2014).

According to research (Angel et al. 2016), the city of Tashkent is expanding its territory faster than its population; thus, its residential density is decreasing. Angel et al. (2016) report that the built-up area density changed from 88.31 persons/ha in 1990 to 55 persons/ha in 1999 and 54 persons/ha in 2013. They argue that this type of spatial development demonstrates unplanned, chaotic management of the urban space. Another perspective is that gentrification is causing a decrease in residential density in selected parts of the city where banks, hotels, and administrative buildings have replaced housing.

Andijan and Namangan regions in the Fergana Valley are the most populous in Central Asia. The urban expansion there has the same character as that in many other places in the world. The urban sprawl is also chaotic and unplanned. The changes in urbanized areas between 1934 and 1989 (Figs. 5.22, 5.23, 5.24, and 5.25) demonstrate the preferences of urban land use management in connection with access to the main natural resources for agriculture: water and fertile land. These major factors influence the spatial distribution and grouping of settlements. This multistage process corresponds well to cellular automaton–based modeling of land use and urban sprawl when the required resources are sufficient, as was the case in 1934–1989. But this modeling is unsuccessful in the case of limited fertile land resources, as has now happened in the Fergana Valley.

Fig. 5.14 Tashkent in 1860. (Source: Map by G Cosnahan (CC-BY-SA 4.0) via Wikimedia Commons G Cosnahan (https://commons.wikimedia.org/wiki/File:Tashkent_History_1860.jpg), change from .jpg to .tiff by Lola Gulyamova, https://creativecommons.org/licenses/by-sa/4.0/legalcode)

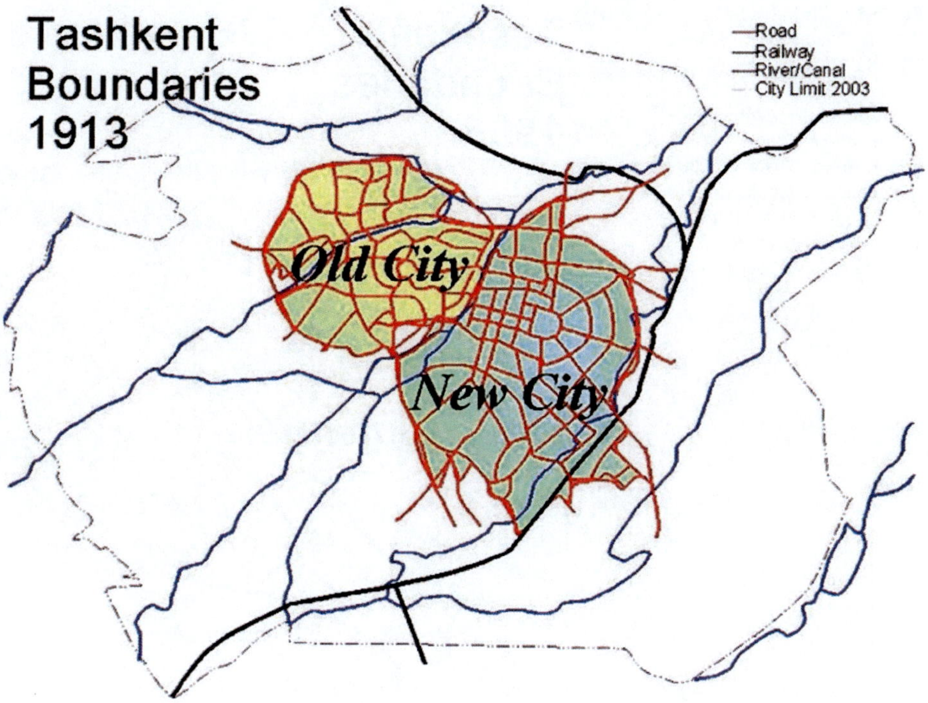

Fig. 5.15 Tashkent in 1913. (Source: Map by G Cosnahan (CC-BY-SA 4.0) via Wikimedia Commons G Cosnahan (https://commons.wikimedia.org/wiki/File:Tashkent_History_1860.jpg), change from .jpg to .tiff by Lola Gulyamova, https://creativecommons.org/licenses/by-sa/4.0/legalcode)

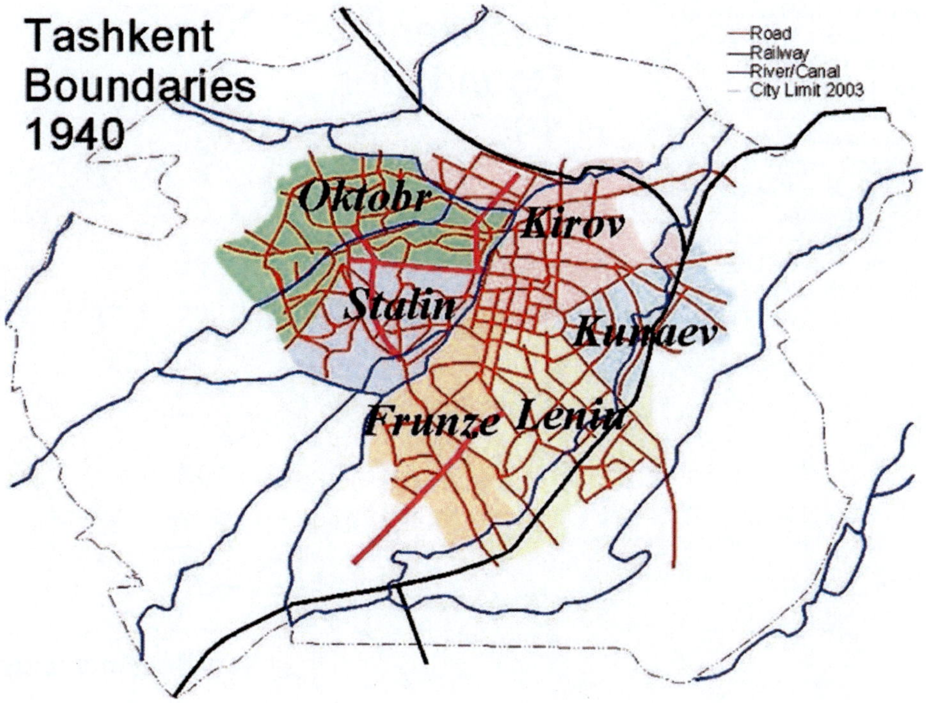

Fig. 5.16 Tashkent in 1940. (Source: Map by G Cosnahan (CC-BY-SA 4.0) via Wikimedia Commons G Cosnahan (https://commons.wikimedia.org/wiki/File:Tashkent_History_1860.jpg), change from .jpg to .tiff by Lola Gulyamova, https://creativecommons.org/licenses/by-sa/4.0/legalcode)

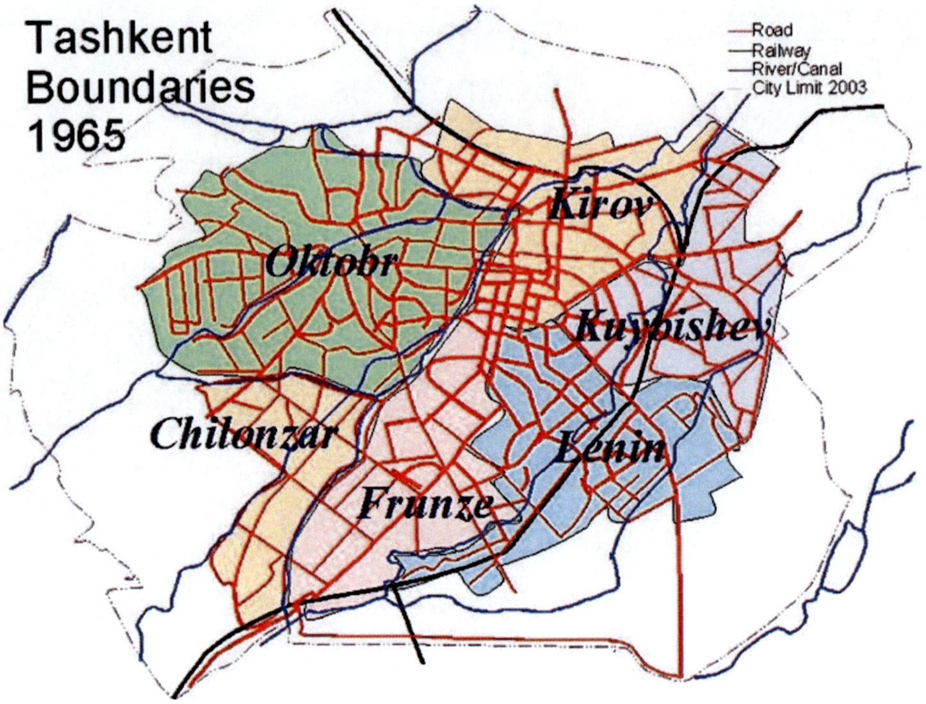

Fig. 5.17 Tashkent in 1965. (Source: Map by G Cosnahan (CC-BY-SA 4.0) via Wikimedia Commons G Cosnahan (https://commons.wikimedia.org/wiki/File:Tashkent_History_1860.jpg), change from .jpg to .tiff by Lola Gulyamova, https://creativecommons.org/licenses/by-sa/4.0/legalcode)

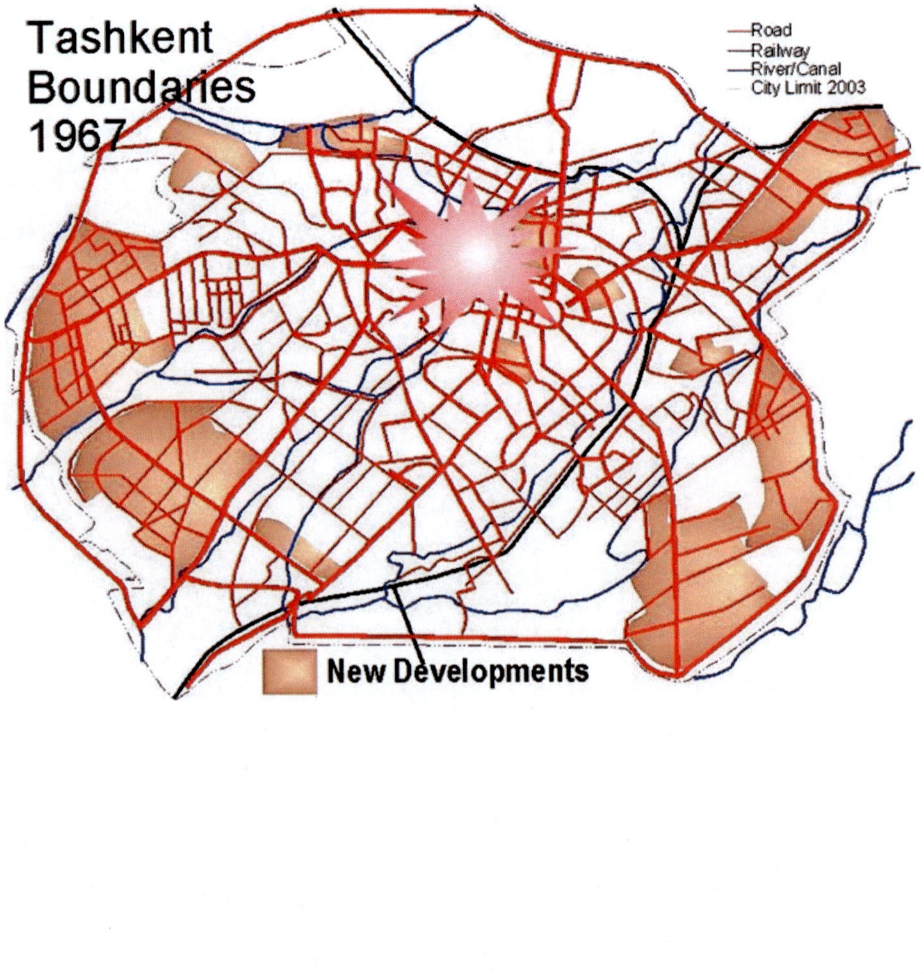

Fig. 5.18 Tashkent after an earthquake in 1967. (Source: Map by G Cosnahan (CC-BY-SA 4.0) via Wikimedia Commons G Cosnahan (https://commons.wikimedia.org/wiki/File:Tashkent_History_1860.jpg), change from .jpg to .tiff by Lola Gulyamova, https://creativecommons.org/licenses/by-sa/4.0/legalcode)

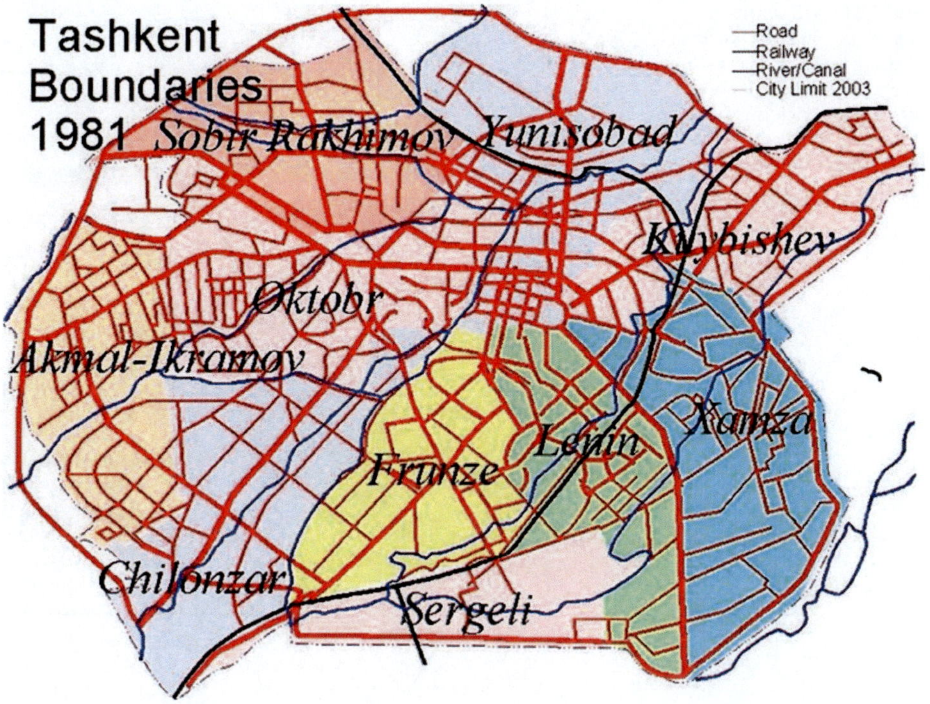

Fig. 5.19 Tashkent in 1981. (Source: Map by G Cosnahan (CC-BY-SA 4.0) via Wikimedia Commons G Cosnahan (https://commons.wikimedia.org/wiki/File:Tashkent_History_1860.jpg), change from .jpg to .tiff by Lola Gulyamova, https://creativecommons.org/licenses/by-sa/4.0/legalcode)

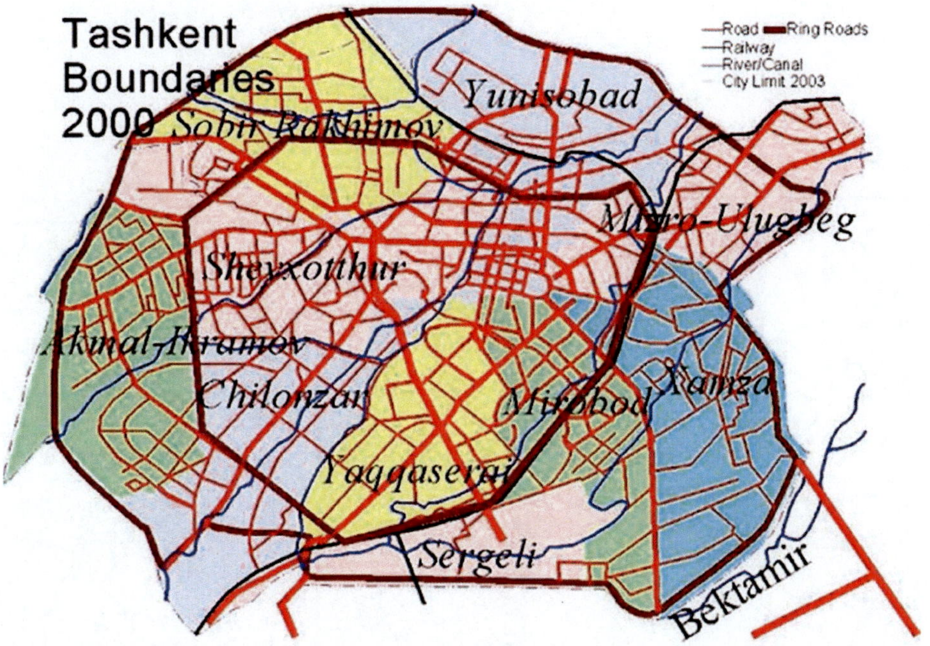

Fig. 5.20 The growth of Tashkent (1860–2003). (Source: Map by G Cosnahan (CC-BY-SA 4.0) via Wikimedia Commons G Cosnahan (https://commons.wikimedia.org/wiki/File:Tashkent_History_1860.jpg), change from.jpg to .tiff by Lola Gulyamova, https://creativecommons.org/licenses/by-sa/4.0/legalcode)

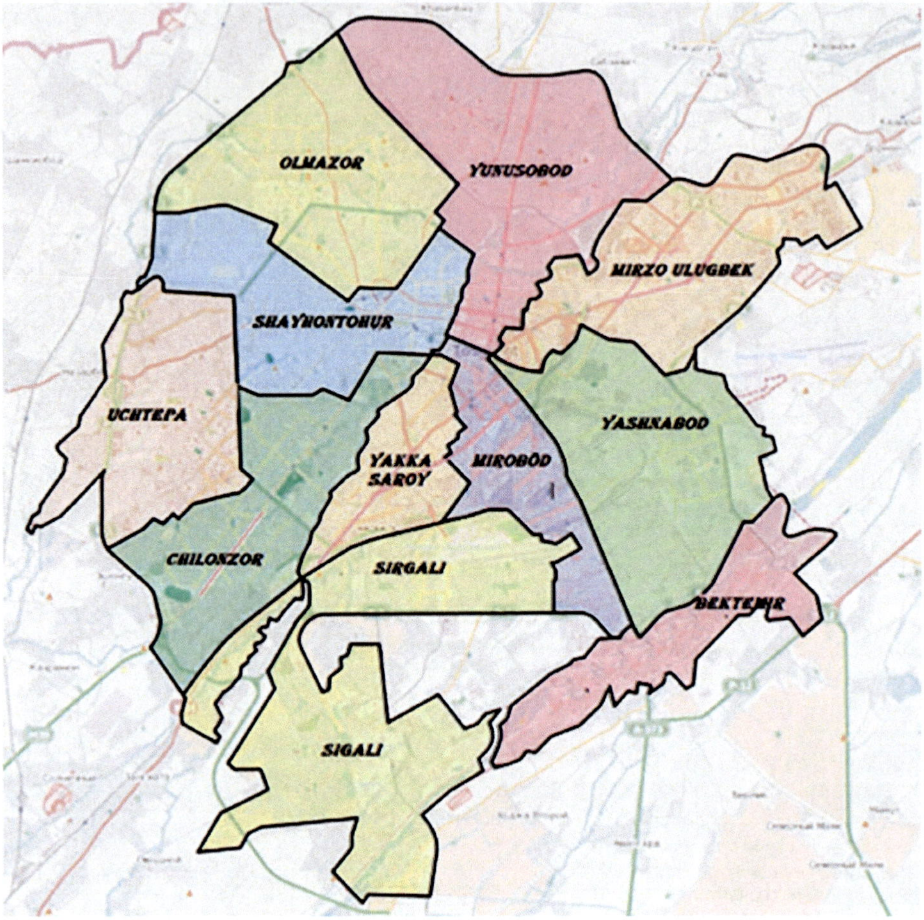

Fig. 5.21 A Tashkent subdivision in 2014. (Source: User: retired electrician. This image is based on an open street map which is CC-SA-2.0—own work, via Wikimedia Commons)

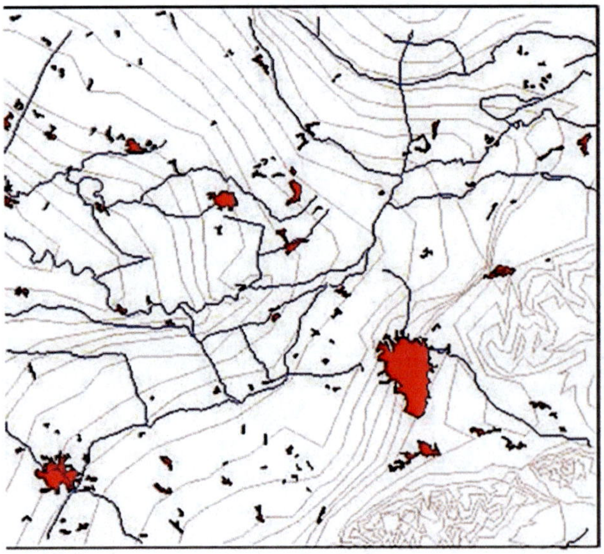

Fig. 5.22 Urban expansion in the Andijan and Namangan provinces in 1934

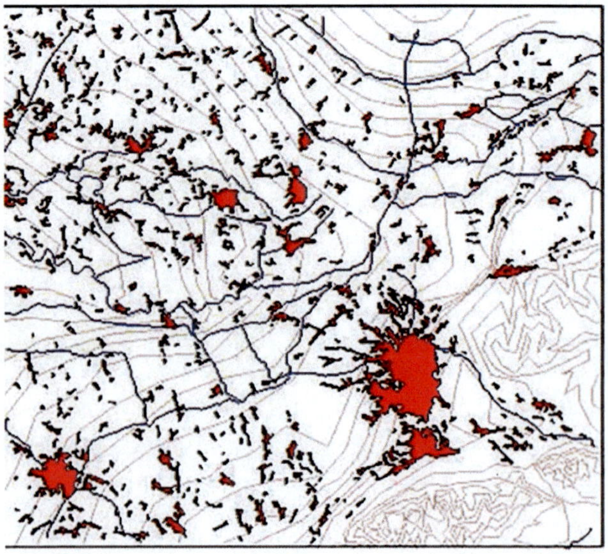

Fig. 5.23 Urban expansion in 1959

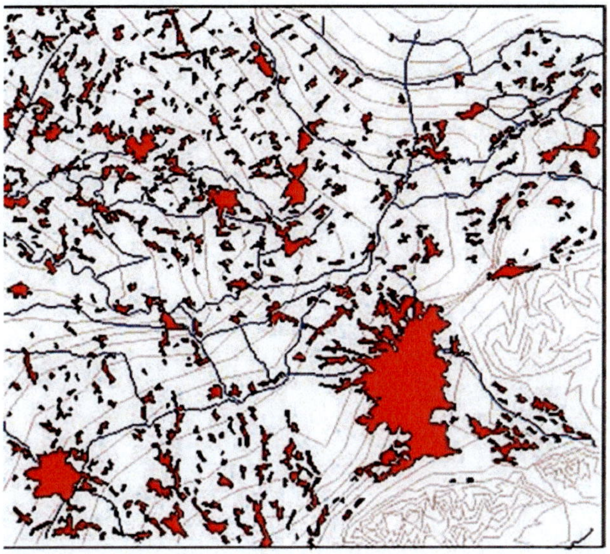

Fig. 5.24 Urban expansion in 1979

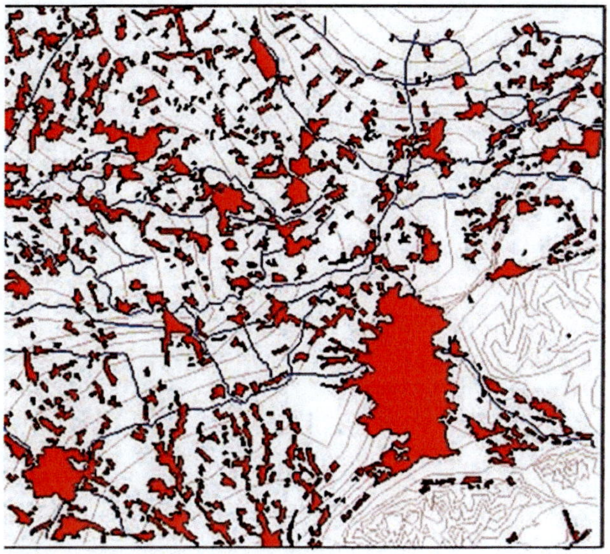

Fig. 5.25 Urban expansion in 1989

Fig. 5.26 Average annual growth of the rural population versus population density in the Fergana Valley

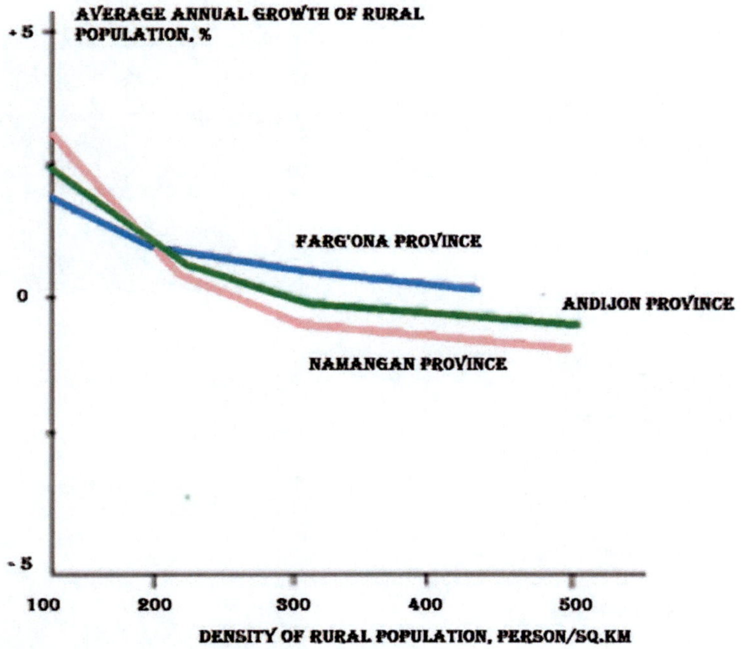

The average annual growth of the rural population is declining in the Fergana Valley in places with a population density of more than 500 persons/km^2 (Fig. 5.26). Adverse living conditions in these places force people to migrate to urban centers in Uzbekistan and abroad.

The government of Uzbekistan's policy for rural–urban transformation is implemented under the Housing Improvement Rural Development (HIRD) program. The objective of this program is to achieve more inclusive growth through improved living standards and livelihood opportunities.

HIRD was started in 2009 to (1) implement a wide range of measures for further development of industrial and social infrastructure in rural areas; (2) improve the efficiency of the agricultural sector; (3) accelerate establishment of modern enterprises for deep processing of agricultural products; (4) increase employment growth in rural areas, mainly among young people; (5) generate higher incomes; and (6) improve the quality of life of rural residents. This program envisages provision of a full range of utility services and loans for low-income borrowers, young families, teachers, health care workers, farmers, and rural entrepreneurs. In 2015, doctors and teachers were the main beneficiaries of this program (Fig. 5.27). New housing development is widely spread all over the country (Figs. 5.28 and 5.29).

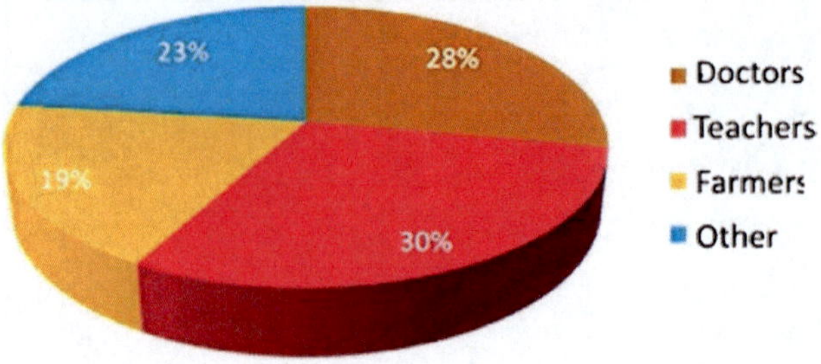

Fig. 5.27 Beneficiaries of the Housing Improvement Rural Development (HIRD) program (2015). (Source: Home buyer survey conducted by the author, February 4–15, 2015)

Fig. 5.28 This house in Samarkand region was built under the Housing Improvement Rural Development (HIRD) program (2015)

Fig. 5.29 New housing development in Syrdarya region (2015)

5.3 Urban Development

Uzbekistan's urban development is closely related to its location on the trade routes of the Great Silk Route (Fig. 5.30). The main cities in modern Uzbekistan have their roots in the ancient centers of trade and craft of Bactria, Sogdiana, Transoxiana (*ma wara-al-nahr* in Arabic), and the Great Khorezm (Xorazm) and Timurid Empires.

Transportation has played an important role in urban development since ancient times when the Silk Route stimulated the flourishing of cities. Many of them are modern centers of the spatial distribution of the population and of the cultural and economic life of Uzbekistan. However, the transportation network is not yet well developed. Not all parts of the country have adequate access to national, regional, and global markets.

The railroad system was constructed by the Russian Empire at the end of the nineteenth century and at the beginning of the twentieth century; it constitutes a skeleton for further densification of the transportation network. Many cities were founded along this railroad—for example, Kagan in Bukhara region, Fergana in the Fergana Valley, Termez (Termiz) on the border with Afghanistan, and Urgench in Khorezm region. Roads complement the railroads and form a transportation network as a catalyst for urban growth. The Great Uzbek Tashkent–Termez Highway connects the main urban centers. It starts in Tashkent and runs south almost to the border with Afghanistan. Another main route is the Zerafshan Highway connecting such cities as Samarkand, Bukhara, and Navoi (Navoiy). However, many roads and railroads are dead ends. The Fergana Ring is the single finished cycle of a planned intercity transportation circuit and connects the cities of Andijan, Fergana, and Namangan.

By the sixth to the fourth centuries BCE, settlements had already been founded near water resources. The first of them were located close to streams and small rivers; later, the settlements spread along the main rivers (Ribakov 1985; Derevyanko et al. 2013).

The further urban development of the Central Asian region, including Uzbekistan, is now influenced by the rising economies of China, India, and South Korea. These countries are interested in infrastructure development and reconstruction of the Great Silk Route. Their efforts are being manifested as investments in roads and in information and communication technology (ICT) development. Researchers (Patnaik 2016) have pointed out that the geographic space of Central Asia is the main attraction for many external powers.

5.3.1 Multifunctional Cities and Agglomerations

The locations of the main cities on the Great Silk Route determined their cyclical development, when prosperity alternated with destruction during invasions and conquests. Many cities disappeared; some of them are in ruins. The cities of modern Uzbekistan reflect the development of the

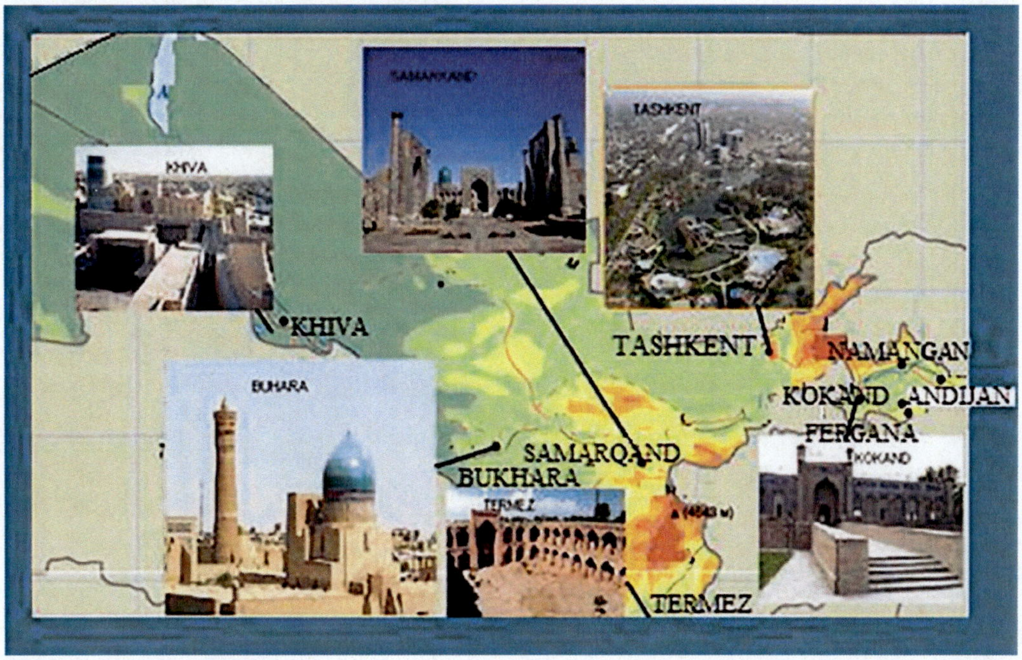

Fig. 5.30 Ancient cities on the Great Silk Route

nation as well. The skills of survival and rebirth have been created in the cities. The cities served as the cultural, political, and trade centers of the region. Such famous cities as Samarkand, Bukhara, Tashkent, Andijan, Khiva, and Kokand remind us of the long history that they have accrued over many centuries.

Main Cities on the Great Silk Route

Tashkent is the capital of this country and the biggest city in the region. According to the Open Data Portal of the Republic of Uzbekistan,[11] between 2002 and 2019, the population of the city increased by 13% from 2,136,600 to 2,510,800 (on January 1, 2019; Table 5.8).

The economic center of the region is rapidly developing in Tashkent. New plans indicate intensive development of its infrastructure, including roads, banks, and hotels (Figs. 5.31 and 5.32).

Different faiths coexist in Tashkent, which continues the tradition of tolerance on the Silk Route (Figs. 5.33 and 5.34).

The history of Tashkent is an example of development of a city on the border between areas with settled and nomadic populations. It has been destroyed several times and has been relocated from one place to another. The modern city is located on the banks of the Chirchiq River. Its surface repeats the overall surface slope of the river valley from northeast to southwest, and the difference between the highest and lowest points within the city is about 200 m.

Researchers (Alimova and Filanovich 2007) estimate that Tashkent is more than 2200 years old. It is mentioned under such names as Chach or Shash in many historical sources. Initially, this city was described in eastern sources in the second century BCE. The ancient settlement of Shashtepa Hill was founded in the third century BCE on an easily irrigated flat and wide plain in the lower reaches of the Chirchiq River. The walls of the first citadel and the ruins of the second one have been excavated on the other bank of the Jun Channel in the southern part of Tashkent. It was a big settlement with a high fortress. This ancient settlement was one of the settlements of the Burgulyuk culture.[12] This culture is named after a discovery on the bank of Burgulyuksay in the valley of the Akhangaran River in the Tashkent region (Ribakov 1985). Its people cultivated soft wheat and barley, as well as being engaged in cattle breeding. In the seventh and sixth centuries BCE, however, this settlement was abandoned. In the third century BCE, it was occupied by nomads. The ruins of Shashtepa in the southern part of modern Tashkent are evidence of the well-developed infrastructure and fortification of that ancient city (Fig. 5.35).

Tashkent has been destroyed several times by nomadic tribes. In 713 CE, it was largely destroyed by the warriors of Kuteyb Ibn Muslim. The greatest devastation was caused by the army of Genghis Khan in 1219.

Table 5.8 Population growth in the city of Tashkent in 2002–2019

Year	Population (thousands)	Growth (%)
2002	2,136.6	
2005	2,135.7	−0.04 since 2002
2010	2,234.3	+4.62 since 2005
2015	2,372.4	+6.18 since 2010
2017	2,409.4	+1.56 since 2015
2019	2,510.8	+4.21 since 2017

Source:[CC-BY-4.0] via Open Data Portal of the Republic of Uzbekistan

Fig. 5.31 A business center in modern Tashkent. (Image courtesy of Magfirat Muminova)

[11]The user has the right to use (including reuse) open data freely, indefinitely, free of charge and without limitation of the territory of use, including the right to copy, publish, distribute, modify open data and combine them with other information, use open data for commercial purposes, use to create software products and applications (Source: Open Data Portal of the Republic of Uzbekistan).

[12]The Burgulyuk people lived in the most ancient known settlements in the ninth to seventh centuries BCE.

Fig. 5.32 A new mosque. (Image courtesy of Magfirat Muminova)

Fig. 5.33 The Sacred Heart Catholic Cathedral was constructed in Tashkent in 1912

In the history of the city, several steps of its development are defined: (1) the early history, (2) the pre-Islamic period, (3) the Islamic period, (4) the Mongol conquest, (5) the Kokand Khanate; (6) the Russian Czarist period, (7) the Soviet period, and (8) independence.

All of these periods contributed to the state of the city, and today one may find traces of some of these periods. For example, the Kalkauz Canal was constructed under the governance of Kuteyb Ibn Muslim, and the Kukeldash Madrassa was founded during the Kokand Khanate (Fig. 5.36). Many cities in Uzbekistan have histories similar to this.

The modern city of Tashkent was founded to the northwest of Shashtepa Hill. Its rich natural resources and location favorable for trade contributed to its revival. It served as one of the main trading points on the Great Silk Route. Tashkent has always been a city of rich merchants and crafts-

Fig. 5.34 The building of the Synod of the Central Asian Metropolitan Area in Tashkent

Fig. 5.35 The ruins of the Shahstepa settlement. (Source: (Public domain) Carpodacus (CC-BY-SA 3.0) via Wikimedia; https://creativecommons.org/licenses/by-sa/3.0)

Fig. 5.36 The Kukeldash Madrassa in Tashkent

men; its main buildings were mostly caravanserais, mosques, and mausoleums. The earliest buildings that still survive are mausoleums built along caravan routes. Today, Tashkent is the center of the Tashkent metropolitan area, with different functions and good prospects for further development.

The whole picture of urban development in Uzbekistan cannot be fully described without mentioning the cities in the Zerafshan Valley, Fergana Valley, and Lower Amu Darya river valley. Some of them are included on the United Nations Educational, Scientific and Cultural Organization (UNESCO) World Heritage List.

The historic centers of Khiva, Bukhara, and Samarkand were added to the UNESCO World Heritage List in 1990, 1993, and 2001, respectively, as having outstanding universal value.

Bukhara

Excavations show that this site was inhabited about five millennia ago. This ancient city is more than 2000 years old. Bukhara was governed by different rulers under various cultures, traditions, religions, and beliefs. The history of this city is rich with many fascinating stories about the development of human life and philosophy over time. Today, many artifacts tell of this. The most ancient synagogue in Central and Middle Asia is in Bukhara. A lot of examples of tolerance and coexistence of various cultures, religions, and traditions demonstrate the rich spirit of the people of Bukhara. Actually, this city, like many others in Uzbekistan, has been part of different empires during its history. It was closely connected to other cities beyond its close neighborhood, with a shared cultural life and economic life. Thus, Bukhara contributed to the development of metropolitan areas many years ago. It was one of the major trade and cultural points on the Silk Route during different eras. It served as the major point of trade in the Persian Empire, and it was the Khorasan province of the Sassanid Empire—the so-called Greater Khorasan. For a long period, it was part of a land where Zoroastrianism was the dominant religion, and this affected the cultural traditions of this region. In the tenth century, it was the center of the Samanid Empire; in 1220, it was besieged by the army of Genghis Khan. It was the capital of the Emirate of Bukhara before the siege by the Russian Red Army in 1920.

It was one of the main trade and cultural centers of the caliphate in the eighth century. Bukhara is frequently called "Holy Buxoro" because it was the center of Islamic culture for many centuries. However, its pre-Islamic history is not well known, because many invasions led to destruction of the city. This city is evidence of many years of hard work by its citizens and their fight for survival in the desert. Located on the banks of the Zerafshan River, Bukhara is famous for its unique architectural monuments, which represent the best surviving examples of the tenth and eleventh centuries and of the fifteenth and sixteenth centuries. A massive fortress, the Ark of Bukhara, was initially built and occupied in the fifth century BCE. This fortress was inhabited by various royal courts. The last Emir of Bukhara, Alim Khan, lived in the Ark until it fell to Russia in 1920 (Fig. 5.37).

This city is also an example of a high and consistent level of urban planning and architecture. A sewerage system existed in the medieval city of Bukhara (Fig. 5.38). Bukhara is famous for its tradition of water saving. Pools in many parts of the city provide water and create higher air humidity, which is crucial in the dry, hot climate (Fig. 5.39). As noted by UNESCO, Bukhara has had a profound influence on the evolution and planning of towns in Central Asia and represents the most complete and unspoiled example of a medieval town in the region.

Samarkand This city has contributed a lot to the world's culture; it is assumed that it was founded in the seventh century BCE. The ruins of Afrosiab are a reminder of the city Marakanda, which was the ancient center of the Sogdiana (Sugd) Kingdom. It was also one of the main centers of the Achaemenid Empire in the sixth century BCE. In Afrosiab, the road system, the sewage and water supply system, and the civil services were well developed. It was an important city at the crossroads between the East and the West, and it

Fig. 5.37 The Ark of Bukhara

Fig. 5.38 The sewerage system in the medieval city of Bukhara was well developed

Fig. 5.39 A water pool in front of the Jome Masjid in Bukhara

served for exchange of goods, people, and ideas. Cyrus the Great expanded his Empire and founded many cities in Central Asia—for example, Cyropolis (the modern Ura-Tyube in Tajikistan) in 544 BCE—to mark the northeastern border of the Achaemenid Empire. Many rulers of Afrosiab contributed to its development.

The story of Samarkand is like a story about a beautiful woman who is a victim of greed and violence. Until the first part of the second century BCE, it was part of the Greco-Bactrian Kingdom and then the Kanguy State. It was then controlled by Hephthalite nomads in the fifth to sixth centuries, a Turk Khanate in the sixth to seventh century, and the Chinese Tan Dynasty in the seventh century. In the ninth to tenth centuries, it became a center of Islamic culture before being finally destroyed by the army of Genghis Khan in the thirteenth century.

In this city, the Turk, Persian, Greek, and Roman cultures met and complemented each other. Little evidence is left of that time. The majority of the architectural monuments were built in the Timurid period in the fourteenth to fifteenth centuries (Fig. 5.40).

The design and construction of the Islamic complexes of this period in Samarkand became the standard of architecture and arts not only in the Central Asian region but also in Persia, India, and Türkiye. In recognition of the architecture, urban structure, and townscape of this city, it is included in the UNESCO World Heritage List.

Fig. 5.40 The Sher Dor Madrassa (Madrassa with Lions), built in 1619–1636, in Samarkand

The modern city of Samarkand was founded to the south of Afrosiab on the floodplain of the Zerafshan River. Today, its expansion is continuing in this direction. The western part of the city was founded by the Russian Empire and consists of a grid of streets oriented from north to south and from east to west, crossing each other. This layout is intrinsic to the building model of the end of the nineteenth century and exists in all cities founded by Russians.

Khiva This city is located in the Lower Amu Darya river area bordering the Kyzylkum and Karakum Deserts. It was an important cultural and trading center of the lost civilization of the Great Khorezm (Khwarezmia) and the Khiva Khanate, and the last resting place for caravans. The eighth-century mosque in Khiva is considered the most ancient in Central Asia (Fig. 5.41). Like many other cities in Uzbekistan, Khiva was destroyed several times by nomadic tribes; the most devastating destruction was caused by the army of Genghis Khan. In 1740, the Iranian ruler Nadirshakh destroyed Khiva once again and ordered that its clay and sand be transported to construct the Nadir Fortress (Nadir-Gala) 100 km to the east of Ashgabat. According to some legends, this was done in retaliation for Nadirshakh's previous captivity in Khiva. Khiva is part of a huge ancient oasis at least two millennia old. The Inner City (Itchan Kala) of Khiva is famous for its outstanding Islamic architecture, urban structure, and design (Fig. 5.42). A sewerage system was developed as well (Fig. 5.43). The style and design of many buildings reflect Turkic traditions. For example, blue (a favorite color of Turkic people) dominates the decoration of many complexes (Fig. 5.44). The nomadic lifestyle is reflected in the composition of the residential complex of palaces (Fig. 5.45), now a museum.

Fig. 5.41 The ancient mosque in Itchan Kala in Khiva

Termez

This city is one of the ancient cities of Central Asia. In 2002, it celebrated its 2500th anniversary. It is in the southernmost part of Uzbekistan near the Hairaton border with Afghanistan and is located on the northern bank of the Amu Darya river.

Its history started in the sixth century BCE when it was a part of the Achaeminid Empire. Alexander the Great captured and annexed this city as part of his empire. It was called Demetris in the time of the Greco-Bactrian Kingdom. In the first and third centuries BCE, the city was known as Ta-li-mi, which became an important center of Buddhism in the Kushan Empire. Different rulers succeeded one another (repeating the history of other cities of Central Asia), such as the Hepthalites and Sassanids in the fifth and sixth centuries CE, and the Gokturks in the seventh century; after conquest by the Arabs in 705, it became one of the centers of Islam during the Abbasid Caliphate and the Sassanid Empire. Its favorable geographic location at the intersection of trade routes from India and China to the West stimulated its prosperity through trade, crafts, culture, and science. In the ninth to twelfth centuries, Termez was a cultural center and a big city of the Khwarazmian Dynasty, or the Khwarazm Shahs. This dynasty of Turkic origin was based on and strongly

Fig. 5.42 Itchan Kala in Khiva

Fig. 5.43 The sewerage system in Itchan Kala in Khiva

influenced by Persian language, culture, literature, and identity. The city was destroyed by the army of Genghis Khan in 1220. It was restored to the east in the second half of the thirteenth century on the western bank of the Surxondaryo river. In 1893, the fortress (Fig. 5.46) and the Amu Darya river port were built by the Russian Empire on the lands of the village of Pattakesar, which was renamed Termez in 1928. It played a very important role as a strategic point at the border between the Russian and British Empires. The officers of the Royal Army represented the majority of its population in the late nineteenth and early twentieth centuries. Several houses remain from that period (Fig. 5.47). Today, it is also an important way point on the highway from Uzbekistan to Afghanistan.

All cities may be defined, depending on their primary functions, as industrial, agrocities, or commercial centers.

5.3.2 Industrial Cities

Many industrial cities were founded along railroads by the Russian Empire and near ore deposits in the Soviet era. At the beginning, they were small villages or incorporated rural neighborhoods. In addition to the main centers—such as Bukhara, Samarkand, and Tashkent—several secondary cities are now important nodes in the whole urban–rural system.

Bekobod is an example of a secondary city; in 2009, its population was 101,200. It was founded as a railroad station in 1889 and was initially named after Michael Hilkov, who was the head of construction of a railroad in Turkestan. As the result of its expansion, it incorporated the rural settlement of Bekov. In 1929, the industrial functions of the city were enlarged by construction of a cement plant there.

Fig. 5.44 The blue color is dominant in Khivan architecture

Fig. 5.45 An element of the nomadic lifestyle: this basement was used for installing a khan's marquee

However, the city grew more rapidly after construction of a large steel mill in 1944, which started operation in 1946. This steel mill is still the only one in Uzbekistan. Before dissolution of the Soviet Union, scrap metal from the whole Central Asian region was delivered there for processing. Construction of the Farkhad Dam and Farkhad Hydroelectric Plant in 1943–1948 near Bekobod added new functionality and importance to it, and it gained city status in 1945. Bekobod is one of the major industrial cities because diverse production businesses are located there, such as a brick factory, a meat-packing plant, and small and medium enterprises providing services.

New cities were founded during the industrialization campaign of the 1930s, and the city of Chirchiq is one of them. Today, this city has become the center of the electrochemical industry (Fig. 5.48). It was established by incorporating the rural settlements of Qirg"izquloq, Qipchoq, and Niyozbek in 1932, and its population has since grown very fast, reaching 15,000 in 1939, 65,000 in 1959, 107,400 in 1970, and 154,600 in 2018.

Fig. 5.46 A fortress built by the Russian Empire in Termez

Fig. 5.47 A house of a high-ranking officer of the Royal Army of the Russian Empire in Termez

Development of the ore-processing industry in the former Soviet Union was a major reason why many cities in Uzbekistan were founded in areas close to mineral and energy resources. Today, they are the industrial centers of the country. The city of Alamalyk (Olmaliq) was founded for extraction of copper, zinc, lead, tungsten, and molybdenum in proximity to a pyrite–polymetallic deposit in the Almalyk (Olmaliq) mining belt of the Kurama Range in the Tashkent region. The city of Angren was founded close to the Angren coalfields. The city of Ahangaran (Ohangaron) (which had about 20,000 inhabitants in 2001) was founded in an area of fields southeast of Tashkent. The name Ohangaron means "smith" in the Tajik language.

Zarafshan (which had 81,400 inhabitants in 2018) and Muruntau were founded in the Muruntau goldfield in the Kyzylkum Desert for mining and processing of gold from the nearby Muruntau open-pit mine. The city of Uchkuduk (which had 37,000 inhabitants in 2018) was founded in 1958 close to deposits of uranium ore. In the late 1960s, this city grew very rapidly, attracting specialists from all over the Soviet Union. Such migration was common in many industrial cities of Uzbekistan.

Navoiy[13] was established in 1958. It is the major industrial city in Uzbekistan, located on the northern bank of the Zarafshan River, close to deposits of natural resources and a railroad station. It incorporates the ancient city Kermine to the north. Enterprises such as Navoyi and Zarafshon Gold Mining and Metallurgic Complexes for extracting gold and Navoiy Azot for producing mineral fertilizers are located in this city. The Navoiy Free Industrial Economic Zone (FIEZ) was recently established there to attract investment for further industrial and technological development. Thirteen joint ventures operate in the Navoiy FIEZ, producing car components, asymmetric digital subscriber line (ADSL) modems,

[13]Data from Official Site of Municipality of the City Navoi (http://navoiy.uz/), accessed on September 18, 2018.

Fig. 5.48 The city of Chirchiq, Tashkent region

light-emitting diode (LED) lamps, cosmetics, diapers, mobile phones, notebook computers, devices for alternative energy sources, and other products.

5.3.3 Agrarian Cities

In Uzbekistan, an agricultural land is 58.14% of land area. Only 9.2% of the land area is arable land (suitable for agriculture). The majority of the cities are located on these lands at oases and in the valleys of major rivers. These lands are a priceless treasure of the Uzbek nation, which has cultivated them for many generations over a period of several thousand years. However, researchers (Hekimoglu 2014) have pointed out the ecological fragility of the Central Asian region as the main challenge that causes difficulties for accommodation of any large-scale commercial agriculture.

In the Soviet era, new cities were founded in areas of cotton and fruit production. Gazalkent in Tashkent province is one such city (Figs. 5.49 and 5.50).

Yangiyul is located 20 km to the south of the capital of Uzbekistan. It was founded as the Kaufmanskaya railroad station in 1899, close to the ruins of ancient Kaunchi-Tepa. Its location on the right bank of the Chirchiq River, in proximity to Tashkent and between the Tashkent–Samarkand railroad and the Great Uzbek Highway, is beneficial from the perspective of access to markets. During World War II, equipment for producing confectionery was relocated from the western part of the Soviet Union to Uzbekistan. At that time, the sugar-processing facility for production sugar was founded in Yangiyul. To supply the sugar beet needed for making confectionery, its cultivation started in this region. Later, it became a leading producer of various sweets, chocolate confections, and candy. Today, several light and food industries are situated in this city. The population is 59,300 (2018).

Guliston (which had a population of 88,200 in 2018) was founded in 1923 and has since changed its administrative status several types with changes in its population. It is located on the banks of the Dostlik Canal 118 km southwest of Tashkent. It originated as a railroad station built in 1895. Via the Tashkent–Havast railroad, this city has access to local and national markets. Its development has been closely related to primary production and processing of cotton, and other industries (Fig. 5.51).

Yangier is founded in 1957 during the project of cultivation the naturally saline virgin lands of Golodnaya Steppe. It was located at the intersection of the railroad and the Great Uzbek Highway on the Southern-Golodnostepskiy canal. The former railroad station built up by Russians in 1880th became the starting point for construction.

5.4 The Fergana Valley

The unique geographic location of the Fergana Valley creates good opportunities for agriculture on this flat plain, which is surrounded on all sides by mountains. It contains rich soil deposits brought by rivers from the surrounding Tien Shan and Gissar–Alay Mountains. This valley has always been praised for the fruit and vegetables it produces. It is not surprising that this valley is frequently called "the pearl of Uzbekistan."

Fig. 5.49 The city of Gazalkent, Tashkent region

Fig. 5.50 Drip irrigation on the outskirts of the city of Gazalkent, Tashkent region

The first written sources mentioning this valley are related to the period of the Greco-Bactrian Kingdom. Over 2300 years ago, the Silk Route passed through this valley. It was colonized by Alexander the Great in 329 BCE. In Chinese chronicles, it was mentioned as a land of the Greeks. Its location on the North Silk Route made this place very attractive for many invaders and conquerors. Many governors ruled the Fergana Valley, and over time it was unified with Transoxiana (a province of the Persian Empire), the Chagatai Khanate in the thirteenth century, and the Kokand Khanate as part of Russian Turkestan in the nineteenth to twentieth centuries.

The ethnicity of the population is a complicated mosaic, making it difficult to define exact borders between the settling of the various ethnicities. Today, this valley is divided between Uzbekistan, Kyrgyzstan, and Tajikistan. Megoran (2006) notes that it is more difficult to outline the nationality than to delineate the territory, and the main problem of the current demarcation is a product not of nations but of nation building.

Fig. 5.51 The first cotton fields in Syr Darya (1923). (Image from the archive of Rabija Axmedova)

In 1924–1925, boundaries between the states were established artificially, and the national delimitations of that time ignored the practice of identification based on tribes, kinship, and clans. This practice is still strong, and it is widespread in many places in the country and in the region. It is also the basis of nationalism and further development of self-identification.

The long history of this valley has made the spatial distribution of its population very uneven. Settlements are located around water resources and have expanded along canals and rivers. Proximity to water resources is the major factor in development of many cities in the Fergana Valley. The population reaches the highest density in mountain foothills, river terraces, and other places close to water resources. The rich natural resources and agricultural productivity have stimulated the growth of the population. Today, it is the most populous part of Central Asia.

In the Fergana Valley, many cities were founded in places with limited land resources for urban sprawl. From the perspective of further development, this part of Uzbekistan is unstable. The rapidly growing young population cannot find jobs in agriculture or industry. The limited land and water resources for agriculture and urban growth make this part of the region vulnerable and problematic in a scenario of poor regional cooperation.

The geographic and administrative boundaries of this valley do not coincide with each other and artificially divide the existing urban–rural systems, creating different enclaves, which are widespread throughout the valley.

The landscape along the rivers of the valley is attractive, represented by melon fields, agricultural villages, apricot orchards, and cotton irrigation canals. Cotton and silk production started in the Fergana Valley.

An oil industry started in the Fergana Valley in 1908 at the Mingbulak oil field. Several cities were founded close to this field; one of them is the city of Oltiariq. An oil spill at Mingbulak in 1992 is considered the largest inland oil spill in history.

The cities of Andijan, Kokand, Namangan, and Fergana are the main centers in the Fergana Valley. Good transportation connects these cities with each other. While they have similar characteristics, their histories and place in the urban–rural system are different.

Andijan

This city is about 2500 years old. Its convenient location on the Silk Route stimulated development of crafts and arts. This city was the birthplace of several rulers and poets, such as Babur (1483–1531) and Nodira Begim (1792–1842). Today, Andijan is a multifunctional city, the administrative center of Andijan region, and one of the important industrial cities. In 2018, its population was 427,400.

The urban sprawl of the city of Andijan is typical of the spatial growth of many cities in Uzbekistan (Fig. 5.52). However, limited spatial opportunities prevent efficient land use without reconsideration of urban and regional policy. This city is an example of the rapid spatial growth that occurred in the Soviet era (1973–1991) and the stagnation that occurred in 1991–2013.

The prospects for further growth in the population of this city are problematic if the current trend of chaotic and unplanned urban sprawl continues in the future (Fig. 5.53).

Andijan represents an example of destruction of strong cultural, economic, and religious ties between cities within the agglomeration. As a result of the division of the Fergana Valley between several states, Andijan was separated from its historical hinterland[14] located to the south of the city. Even in Soviet times, these ties were very strong and were represented by daily commuting for work and education,

[14] The term "hinterland" (borrowed from German by the geographer George Chisholm in the late nineteenth century) refers to a rural area economically tied to an urban catchment area.

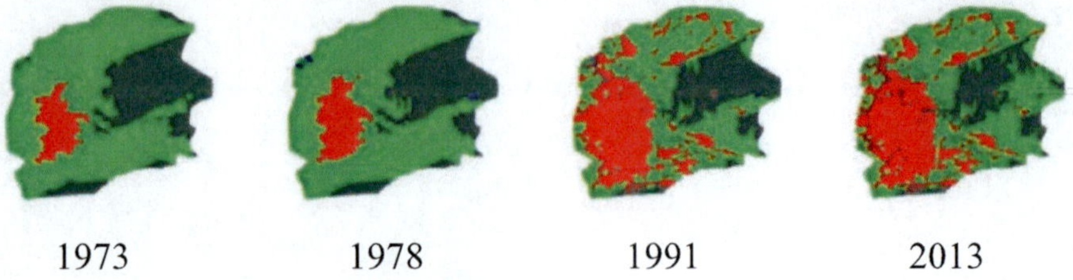

Fig. 5.52 Urban sprawl in the city of Andijan (1973–2013)

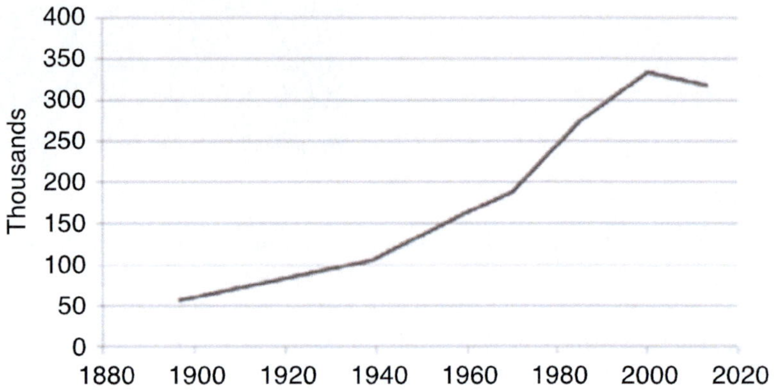

Fig. 5.53 Population growth in the city of Andijan (1880–2020)

established common business entities, and shared services. Actually, Andijan was the center of a metropolitan area with varied opportunities for urban development and well-developed transportation (Fig. 5.54).

Since the fifteenth century, Andijan has been an important craft and trade center on the Silk Route. A railroad was constructed by the Russian Empire in 1889 and stimulated further growth of this city and development of light and heavy industries. Today, chemicals, domestic appliances, electronics, foodstuffs, furniture, plows, pumps, shoes, spare parts for farm machines, and various engineering tools are manufactured by over 50 international companies, five of which produce spare parts for GM Uzbekistan.

Fig. 5.54 A mosque in Andijan. (Source: Carpodacus (CC-BY-SA 4.0), https://creativecommons.org/licenses/by-sa/4.0)) via Wikimedia)

Fergana

This is a relatively new city in the region. It was founded as a garrison town on the border between the Russian and British Empires in 1876. In accordance with the demands of local officials, this city was founded about 20 km southeast of the town of Margilan. In 2009, the population was 197,000. The new city incorporated the settlements of Yarmazar, Sum, and Chirimgan. It was named New Margilan from 1876 to 1907. In 1907, it was renamed in honor of Skobelev, the first Russia military governor of the Fergana Valley. When it was renamed in 1924, it received the name by which it is known today.

Its layout was designed in accordance with a widespread plan of straight streets radiating from a military fortress and regular blocks of residences around it. The building of a former officers' club, which was built especially for Czar Nicholas Romanov (Figs. 5.55 and 5.56), is one of several preserved buildings in Fergana. World War I prevented his arrival in Fergana.

Historically, this city had its foundations in Russian culture and traditions. Before the collapse of the Soviet

Fig. 5.55 The building of a former officers' club in the city of Fergana

Fig. 5.56 The date of construction (1879) is written on the facade of the building

Union, the ethnic population of Fergana was mostly composed of Russians and other European nationalities. Modern Fergana is the center of the Fergana–Margilan metropolitan area. The cities of Fergana and Margilan are connected with each other by good transportation. Today, they constitute a single urban area with strong daily commuting, cultural ties, and business ties. Actually, the border between the two cities is a mere technicality and does not exist in real life. In 2018, the population of Fergana was 278,500 and that of the entire metropolitan area was 506,800.

Today, it is the center of Fergana region and is a well-developed industrial city with many enterprises and businesses.

Kokand

Like many other cities in the Fergana Valley, Kokand was conquered by the Han Dynasty in the first century BCE. Later, it was ruled by Arabs and destroyed by the army of Genghis Khan. This city was the center of the Kokand Khanate (Fig. 5.57) and an important trade point on the Silk Route. In 2018, its population was 245,500. Kokand's name is derived

Fig. 5.57 The Palace of Khudoyorkhan. (Source: User Doron under CC-BY-SA 3.0 via Wikimedia)

from the well-known "Kokan" tribal family group, who belong to the Kongrat tribe of Uzbeks.

In 1883, Kokand became part of the Russian Empire and the capital of the short-lived anti-Bolshevik Provisional Government of Autonomous Turkestan.

Modern Kokand is located at the crossroads of the two main ancient trade routes into the Fergana Valley, one leading northwest over the mountains to Tashkent and the other leading west through Tajikistan. This location gives Kokand an opportunity to serve as the main transportation junction in the Fergana Valley.

Kokand has been an important cultural and educational center in the eastern part of Uzbekistan. The first publishing house in Uzbekistan was founded in Kokand. The city is also a big industrial center. Fertilizers, chemicals, machinery, cotton, and food products are manufactured there.

Namangan

This city was in a different location before an earthquake occurred there in 1610. Namangan has survived several devastating earthquakes. It has been rebuilt several times—for example, after earthquakes in 1620 and 1926. It also survived destruction by the Mongols in 1219. The ruins of Akhsikath close to Namangan are a reminder of this destruction. This ancient settlement was founded in the third to the second centuries BCE (Ribakov 1985) The design of the city was well planned. Even in the second century BCE, Akhsikath had a bathhouse. Today, it is designated as a UNESCO World Heritage Site. Its location on the Silk Route between Sogdiana (Samarkand) and China, and near ore deposits, made this city a rapidly growing trading and craft center. Akhsikath was the only place in Central Asia that produced high-quality steel. Knives produced in the Namangan region are considered the best.

Table 5.9 Population growth in the city of Namangan in 1897–2018

Year	Population (thousands)	Growth (%)
1897	62	–
1926	74	+19.4 since 1897
1939	77	+04.0 since 1926
1959	123	+59.8 since 1939
1979	227	+84.6 since 1959
1989	306	+34.8 since 1979
1990	306	–
2000	386	+26.1 since 1990
2010	441	+14.3 since 2000
2018	600.2	+36.1 since 2010

Sources: Soliev et al. (2010), p. 136; [CC-BY-4.0] public domain via State Committee on Statistics of the Republic of Uzbekistan, open data (2019)

In the fifth and sixth centuries, migration from Sogdiana to the Fergana Valley, and especially to Namangan, resulted in intensification of trade between China and the West. Namangan was the city of the dominant Sogdians, who were praised for their proficiency in trade. Today, the Persian-speaking Tajik population is still dominant in some parts of Namangan region (Chust, Kosonsoy, and Pop), while the Kyrgyz population is dominant in the Yangiqo'rg'on and Uchqo'rg'on districts. Because of an alliance between the Sogdians and the Uyghurs, the eastern Fergana Valley was part of the Uyghur Empire of the Karakhanid State between the eighth and ninth centuries. That alliance benefited both sides and the intermediary cities.

Namangan was originally populated by Sogdian people, but gradual Turkification of the city started after an influx of Turkic people into the region in medieval times. Persian was replaced by the Turkish language. This process lasted for several centuries until the nineteenth century, when Namangan still had a Persian-speaking Tajik majority. This city is an example of the ethnolinguistic complexity of its population. The mosaic of ethnic groups in the population is inherited from ancient times. As researchers (Hughes and Sasse 2014) have pointed out, state building is challenging in ethnically and regionally divided societies.

Namangan became an important craft and trade center in the Fergana Valley in the seventeenth century. Before the invasion by the Russian Empire in 1867, Namangan was a part of the Kokand Khanate.

A railroad, constructed in 1912, connects Namangan to other cities in Uzbekistan. Its economic development is related to periods of industrialization in the 1930s and 1970s.

Today, Namangan is a rapidly growing center for light industry, mainly in the food industry. It is the second-largest city, by population, in Uzbekistan. Its population was only 62,000 in 1897 but doubled to 123,000 by 1959 and reached 600,200 in 2018 (Table 5.9).

References

Alimova D, Filanovich M (2007) The history of Tashkent from ancient times until today (in Russian: Д.Алимова. М.Филанович История Ташкента. От древнейших времен до наших дней- "Правда Востока"– 26 мая 2007). Pravda Vostoka, 26 May 2007

Center for Economic Research (2013) Urbanization in Central Asia: challenges, issues and prospects. Analytical report 2013/03. United Nations Economic and Social Commission for Asia and the Pacific. Accessed on September 15, 2018 from https://www.unescap.org/sites/default/files/Urbanization-in-CA-ENG.pdf

Angel S et al (eds) (2016) Atlas of urban expansion—2016 edition. Volume 1: areas and densities. Institute of Land Policy, New York. http://www.atlasofurbanexpansion.org/cities/view/Tashkent. Accessed 18 Mar 2018

Bennigsen A, Wimbush SE (1985) Mystics and commissars: Sufism in the Soviet Union. Hurst, London

Chey YS (1994). Koreans. In: Encyclopedia of World Cultures Vol 6: Russia and Eurasia, pp 204–207. Accessible from http://1.droppdf.com/files/BuapW/encyclopedia-of-world-cultures-volume-6-russia-and-eurasia-china-1994-.pdf

Crowe D, Demeter NG, Lemon A (1994) Gypsies. In: Encyclopedia of world cultures. Vol 6, Russia and Eurasia. Macmillan, Boston, pp 144–148

Derevyanko AP et al (2013) Results of Kulbulak (Uzbekistan) site research in 2013. In: Problems of archeology, ethnography, anthropology of Siberia and adjacent territories: materials of the Annual Session of the Institute of Archeology and Ethnography of the Siberian Branch of the Russian Academy of Sciences, Novosibirsk, 2013 (in Russian: Деревянко А.П., Исламов УИ., Павленок К.К., Шнайдер С.В., Рыбин Е.П., Павленок Г. Д., Раджабов А., Лазарев С.Ю., Мухтаров Г.А. Результатыисследованиястоянки Кульбулак (Узбекистан) в 2013 году // Проблемыархеологии, этнографии, антропологииСибири и сопредельныхтерритор ий:материалыгодовойсессииИн-таархеологии и этнографии СО РАН 2013 г. – Новосибирск, 2013). https://cyberleninka.ru/article/n/zubchatoe-mustie-stoyanki-kulbulak-novye-dannye-i-interpretatsii. Accessed 20 May 2018

Hekimoglu L (2014) The heartland fallacy: Central Asia, geography and globalization. In: Megoran N, Sharapova S (eds) Central Asia in international relations: the legacies of Halford Mackinder. Hurst, London, pp 275–277. https://doi.org/10.1093/acprof:oso/9780199327973.003.0011

Hughes J, Sasse G (eds) (2014) Ethnicity and territory in the former Soviet Union regions in conflict. Routledge, London. https://doi.org/10.4324/9780203045732

Hwang LB, Lee BS (2008) Cultural relations of Korea and Uzbekistan. Uzbekistan National Encylopedia, Tashkent (in Russian: Л.Б. Хван; Б.С. Ли/ КультурноесотрудничествоКореи и Узбекистана. -Т.: УзбекистанМиллийЭнциклопедияси, 146 стр

Irwin JA, Ikramov A, Saunier J et al (2010) The mtDNA composition of Uzbekistan: a microcosm of central Asian patterns. International Journal of Legal Medicine 124(3):195–204. https://doi.org/10.1007/s00414-009-0406-z

Kotler I (1994) Bukharan Jews. In: Encyclopedia of world cultures. Vol 6, Russia and Eurasia. Macmillan, Boston, pp 111–114

Levinson D (1994) Koreans In: Encyclopedia of world cultures Vol 6 Russia and Eurasia Macmillan, Boston, pp 204–207

Marushiakova E, Popov V (2016) Gypsies in Central Asia and the Caucasus. Palgrave Macmillan, Cham. https://doi.org/10.1007/978-3-319-41057-9_2

Megoran N (2006) For ethnography in political geography: experiencing and re-imagining Ferghana Valley boundary closures. Polit Geogr 25(6):622–640. https://doi.org/10.1016/j.polgeo.2006.05.005

Patnaik A (2016) Central Asia: geopolitics, security and stability. Routledge, Abingdon

Pohl JO (1999) Ethnic cleansing in the USSR, 1937–1949. Greenwood, Westport

Ribakov BA (1985) The ancient cities of Caucasus and Central Asia. "Science" Publishing House, Moscow (in Russian: Археология СССР. Древнейшие города Кавказа и Центральной Азиие Под ред. Б.А.Рыбакова М.:"Наука» 1985)

Soliev A et al (2010) The social and economic development of regions of Uzbekistan. "Mumtoz so`z" Publishing House, Tashkent (in Uzbek: Солиев А., Назаров М., Курбонов Ш. Ўзбекистон худудлари ижтимоий–иқтисодий ривожланиши. Монография. Т.: MUMTOZ SO'Z, 2010. 348 б)

State Committee on Statistics of the Republic of Uzbekistan (2017) Analytical materials: demographic analysis. https://stat.uz/en/435-analiticheskie-materialy-en1/2066-foreign-trade-statistics. Accessed 6 Dec 2017

State Committee on Statistics of the Republic of Uzbekistan (2018) Open Data. Demography. Accessible from https://stat.uz/en/official-statistics/demography

UNDP (2018) Human development indices and indicators. 2018 statistical update. Retrieved on July 16, 2022 from https://hdr.undp.org/system/files/documents//2018humandevelopmentstatisticalupdatepdf.pdf.

United Nations Department of Economic and Social Affairs (2014) World Urbanization Prospects. Retrieved from https://esa.un.org/unpd/wup/publications/files/wup2014-highlights.Pdf

United Nations Department of Economic and Social Affairs (2017) Population Division database. http://www.un.org/en/development/desa/population/ http://www.worldometers.info/world-population/uzbekistan-population/

World Bank Data. Available from 27 April, 2017 http://data.worldbank.org/indicator/SP.URB.TOTL?locations=UZ&page=3

6 Environmental Geography: Nature and Human Interaction

Abstract

In Uzbekistan, differences in human occupation are dependent on access to the main resources needed to survive in an arid climate. Water is unevenly distributed across the country, resulting in significant contrasts in conditions for life between eastern and western Uzbekistan, and between the plains and the mountains. The natural conditions determine the variety of the histories, economic life, and social life of the people. Human activity and global geological processes have significantly changed the landscape within the country.

Keywords

Nature · Uzbekistan · Subregions · Physical geography

6.1 The Geography of the Regions

6.1.1 Physical Geography

Significant contrasts in the conditions for life between eastern and western Uzbekistan, and between the plains and the mountains, have resulted in uneven human occupation across the country. Survival in the continental arid climate depends on access to water, which is unevenly distributed over the whole of Central Asia. This determines differences in the histories, economic life, and social life of the people. The landscape has been changed significantly by global geological processes and human activity all over the country.

The Great Silk Route passed through oases on plains and in intermountain depressions. The caravan routes lay close to sources of fresh water in the semidesert and the desert. The locations of the major cities of the Great Silk Route on the rivers (the Amu Darya, Syr Darya, and Zerafshan River) played important roles in their development. These cities became the urban centers of the civilization of the whole Central Asian region and accumulated cultural and social treasure from all Central Asian nations.

The physical geography of Uzbekistan is a combination of the natural conditions, which have many similarities to those of all neighboring states. Uzbekistan shares high mountains with Tajikistan and Kyrgyzstan, the Kyzylkum Desert with Turkmenistan, the great Turan Plain with Kazakhstan, and the Amu Darya, Syr Darya, and Zerafshan River with all of them.

Tectonics

In connection with a planar fracture in a volume of rock across which there has been significant displacement, and the thickness and structure of the surface of the Earth, the territory of Uzbekistan can be divided into two tectonic subregions: (1) an orogen (mountainous) region and (2) a plain platform.

The orogeny of Uzbekistan in the south, east, and central mountainous parts consists of several tectonic structures. The main ones are the ranges of the Chatkal–Kurama, Nurata–Alai, and Gissar–Zerafshan Mountains; the northern–western part of the Afghan–Tajik fold and the residual mountains in the Kyzylkum Desert and Tashkent; and the Zerafshan and Fergana intermountain depressions within the alpine epi-platform fold zone.

The Turan Plain platform is located in the western part of this country and includes several syneclises (i.e., large, relatively shallow depressions formed in a continental platform setting by slow and steady subsidence). They are the southern part of the Syr Darya syneclise, the northern–eastern part of the Amu Darya hemisyneclise, and the syneclise of southern Ustyurt and the hills of central Ustyurt. All of them are part of the Turan platform. The Paleozoic basement of the Turan platform lies 1–6 km beneath the Earth's surface. This basement is divided by several faults, and the biggest of them are the Gissar–Mangishlak fault (which is oriented from east to west) and the Ural–Omon fault (which is oriented from north to south). The basement is divided into four blocks by

L. Gulyamova, *The Geography of Uzbekistan*, World Regional Geography Book Series,
https://doi.org/10.1007/978-3-031-07873-6_6

these faults. The main part of the country is in the northern–eastern block, and just the southern part is in the southern–eastern one. The northern–eastern block's genesis is associated with the Meso-Cenozoic Era, and the basement is no deeper than 1 km; it is only in some deep folds that it may be more than 2 km deep. The southern–eastern block's rise happened in the Mesozoic Era; after that, in the Cenozoic Era, it was cutting down, which is why the basement is 8–10 km deep there.

Geomorphology

Orogens and plain platforms are formed during different periods of the formation of mountains and volcanic activities, resulting in creation of reserves of many ores and rare and precious metals. The composition of the mountain ranges inclined to the west and the depressions between them are unique features of the relief of eastern Uzbekistan. The characteristics of the relief of the Tien Shan Mountains can also be applied to many other mountains. These are the combination of mountain ranges and intervening valleys, with basins trending generally from east to west (Allan et al. 2011).

Two major topographic regions can be identified in the mountains: (1) the mountain systems and (2) the intermountain depressions (Khasanov and Gulomov 2006). The mountain systems are represented by three ones: (1) the Chatkal–Kurama, (2) the Nurota–Turkestan, and (3) the Gissar–Zerafshan. They differ from each other in terms of their absolute elevations and average heights, the erosion of their slopes, and their steepness.

The form of the relief of the mountains tells the history of their geology, modern tectonic and erosion–accumulative processes, and formation of rocks. In the alpine zone of the mountains, the relief is formed by tectonic, glacial, and erosion processes. Deep valleys with extremely steep slopes and rocky peaks and upper parts are dominant in these mountains. They are glacial moraine forms (with accumulation of rocks and debris carried and deposited) and other glacial forms. These relief forms are present in the ranges of the Piskom, Chatkal, Gissar, and Zerafshan Mountains.

In the low mountains—such as Nurota, Kuhitang, Kurama, Mo'g'iltog', Morguzar, Qoratepa, and Bobotag—these relief forms are not formed.

Many forms are developed under erosion that is still leveling the surface. The erosion forms are frequently located in the median mountains, and they are represented by steep slopes, deep valleys, and vertical disruption from 400 to 1000 m. In the low mountains, the erosion relief forms are mainly on their outer sides. The slopes have a steepness of about 10–15° and are part of the valley. In many places, the former terraces exist. The upper parts of the mountains are flat and wide. The low hills are integrated into the accumulative plains in the foothills (Fig. 6.1).

The intermountain depressions (Figs. 6.2 and 6.3) are of accumulative origin and were formed under wind and water erosion. Oblique plains are primarily located on the borders of depressions and filled with sediments brought from the mountains. All of these processes are still continuing in the mountains. Rocks divided by erosion are brought to the valleys by flooding, snow melting, and under gravitation forces, causing damage to nature and to people.

Under water and wind erosion, a lot of different sediments from mountains have been brought and spread over the plain part of Uzbekistan.

The Relief Formation The relief formation is related to different tectonic processes and fluctuations of the ocean over long geological time periods. Several times, the ocean came to these places when the surface was higher and was formed by Carboniferous rocks (the Ustyurt Plateau). The influence of a denudation was not strong after the shrinking of the Neogene Ocean. The main depressions were formed during that very period. Such depressions are formed under tectonic, suffusion, and aeolian processes. The Asakaovdan and Barsa-Kelmes depressions are examples of these. Some

Fig. 6.1 Low mountains, Nurata Ridge, Jizzakh region

Fig. 6.2 Intermountain depressions, Karjantau and Ugam Ridges, Tashkent region

Fig. 6.3 Intermountain depressions, Chatkal and Ugam Ridges, Tashkent region

stretched depressions (such as the Janadaryo in the Lower Amu Darya) are formed under wind erosion in the setting of a dried meander or tectonic depression. They are filled with salty rocks or salt.

The orogeny process of the mountain building in Uzbekistan includes folding, faulting, volcanic activity, igneous intrusion, and metamorphism. From the Paleozoic Era to the Neogene Period, it was the Tethys Ocean. In the Upper Cretaceous Period, as a result of some closure of this ocean, several big and small islands were formed in what is now the Kyzylkum Desert. The strong seabed had folded in the zone along a continental margin in places where the process of evolution of mountains and mountain systems was continuing. In the Early Silurian Period of the Caledonian orogeny (550–400 million years ago (Mya)), no folding process occurred in Uzbekistan. At the end of the Paleozoic Era after the Hercynian (back-arc basin) orogeny (350–200 Mya), new lands appeared instead of ocean water. It is assumed (Khasanov and Gulomov 2006) that at the beginning of the Hercynian orogeny in central Kyzylkum, the rising of the Western and the Southern Tien Shan mountain systems occurred during the Carboniferous Period. The flat mountainous plains appeared at the end of the Hercynian orogeny during the Permian Period and the beginning of undermining mountains until the end of the Paleozoic Era and the beginning of the Mesozoic Era.

In the Cretaceous Period and the Paleocene Epoch, the ocean came back and again covered plains and folds and reached the Karatag and Alay mountain systems. The

Fergana and Tajikistan seas were linked to each other. The ranges of the Alai, Turkestan, Fergana, Chatkal, and Nurota Mountains were like islands in the middle of an ocean. Mechanical scraping of rock surfaces by friction between rocks and moving particles during the sea transgression and settling of gravels, sand gravels, and clay on the bottom of the sea brought the formation of abrasion plains. The thickness of these sediments is up to 2500 m in central Fergana. By the end of the Paleocene Epoch, the sea withdrew from Uzbekistan's territory, and formation of red beds or sedimentary rocks consisting of sandstone and siltstone with a predominant presence of ferric oxides had started.

During the Neogene Period, the alpine orogeny began, and it resulted in a new rising of the Hercynian orogeny mountains. According to research (Seltman et al. 2011), the Hercynian Tien Shan orogen formed during Late Paleozoic collision between the Karakum–Tarim and Kazakhstan paleocontinents. They are the base of the modern relief forms in Uzbekistan.

The mountain building continued in the Quaternary Period, which resulted in the rising of the Western Tien Shan in combination with stronger fluvial and glacial processes due to the warming of the climate and the melting of ice and snow in the mountains. Thus, the plain was covered by sand and clay alluvial sediments. Wind erosion helped to distribute these sediments and to build the modern relief forms in the deserts and steppes.

In the Quaternary Period, not just mountains but also low-lying sunken rocks between the mountains began their rise. As a result of this process and water erosion during the Neogene Period, terraces formed in valleys. The modern state of Uzbekistan's natural environment was formed mainly in the Quaternary Period.

The history of the relief formation tells stories about many abrupt changes that occurred after different tectonic processes during a long geological period. The whole topographic picture is as much of a mosaic as the geology itself. In Uzbekistan, four major geomorphologic groups have been identified:

(1) Multiple cliff erosion reliefs in the foothills
(2) Sculpture reliefs in places with soft deposits
(3) Nival reliefs in the high mountains
(4) Accumulative intermountain depressions

In Uzbekistan, a widespread form of relief is hills on plains. This formation occurred in the Kyzylkum Desert during the Quaternary Period, and sandy hills of different forms can be found everywhere there. They originated from the spreading of alluvium rocks by the wind. In the mountains, one can find the history of geological, modern tectonic, and erosion–accumulative processes.

6.1.2 Geographic Subregions

The complexity of the orogenesis, the climate differences between the plains and mountains, and soil and water distribution create difficulties in determination of borders between subregions. The landscapes are mosaics with very complicated relations between them. Several approaches exist to determine their properties. In accordance with research by L. N. Babushkin and N. A. Kogay, Uzbekistan (except for Ustyurt) is included in the Turan natural geographic province, which is characterized by a dry continental climate and a homogeneous plain surface.

The Turan Plain in the eastern and northern–eastern part of the country constitutes 78.7% of its area; the Tien Shan and Alai (Alay) Mountains systems (including mountain ranges, foothills, and intermountain plains) constitute only 21.3% of the whole area. The lowest point in the country is the Mingbulak depression in Kyzylkum (12 m below sea level), and the highest peak is Hazrati Sultan (4643 m high) in the Gissar Mountains system. The plains tend to lower from the east and the southeast to the northwest; the main rivers flow in the same direction.

The lowest parts of this country are at altitudes of 30–100 m above sea level in northern–western Kyzylkum, the Amu Darya river delta, and the dried Aral Sea seabed. Several parts (Sultan Uvays, Bukantag, Kuljuktag, and Tamditag) of the Turan Plain are as high as 900 m.

The Turan Plain was under the Turkestan Ocean for a long geological period. As a result, it is covered by thick Mesozoic–Cenozoic rocks. The mountains are primarily composed of Paleozoic rocks. Their rising happened in the Hercynian Epoch and under the Himalayan orogeny.

According to Babushkin and Kogay's (1975) classification, Uzbekistan is divided into nine physical–geographic subregions. Their borders do not coincide with the administrative borders (Fig. 6.4).

The Delta of the Amu Darya River Subregion

This is a part of Uzbekistan where significant and even irreversible landscape changes have occurred. The environment has been affected by the geological evolution of the region and human activity in combination with global climate change. The shrinking of the Aral Sea has created many challenges for human beings and nature. The ecology is worsening for inhabitants, animals, and plants, reaching critical levels in most of this subregion. This subregion includes the former delta of the Amu Darya river, which was one of the biggest in the world, at 50,000 km^2 in size (Fig. 6.5). It stretched from the point where the modern Tuymuyun reservoir has been constructed toward the Aral Sea. Today, this delta starts near the city of Tahiatash and does not reach the Aral Sea. It has lost about half of its area. These lands were the lands of the Great Kwharezmiya (Xorazm). Water and

Fig. 6.4 Physical-geographic regions

good soils were the biggest treasures that enabled them to flourish. Historically, these lands were suitable for irrigated agriculture.

Three alluvial plains of different ages—the Sarygamysh, the Aqchadarya, and the modern delta—constitute the Lower Amu Darya. The Aral–Sarygamysh depression was formed on the Pliocene plain that comprises the main part of the Turan lowland. The Aralkum plain is a new relief in the dried bed of the Aral Sea. This saline desert covers an area of about 30,000 km^2.

One study (Le Callonnec et al. 2005) estimated that the total mass of salt contained in the Aral Sea is about 69,109 t, which would cover the entire Aralkum in a salt layer at least 0.5 m thick. Today, the local people are trying to cover some parts of this area with salt-tolerant shrubs and trees. Their plan is to cover the entire former seabed with a forest of saxaul, in the hope that the trees will stop the wind picking up contaminated sand from the dried seabed and spreading it around.

Two different parts can be identified on the site of the former Amu Darya river delta: (1) a sandy plain 10–20 km wide and (2) salt plains.

The delta of the Amu Darya river was formed over 7000 years ago. However, scientists (Skvortsov 1959; Tolstov and Kes 1956) have suggested different scenarios regarding the genesis and age of this plain. Geological evolution caused cyclic fluctuations of the Aral Sea and changes of flow in the Amu Darya and Syr Darya rivers. Long-term natural fluctuations in the river runoff contributed significantly to the formation of this delta.

The base of the delta was formed by the most ancient Cretaceous rocks, covered with Paleocene, Neogene, and Anthropogenic rocks. The Quaternary rock layer reaches a thickness of 100 m in the old delta (in the modern Xorazm province) and is from 12 m to 60 m thick in the modern delta. This delta is represented by sandy and clay deserts (Zonn et al. 2009), whereas the upper part is surrounded by the Kyzylkum and Karakum Deserts.

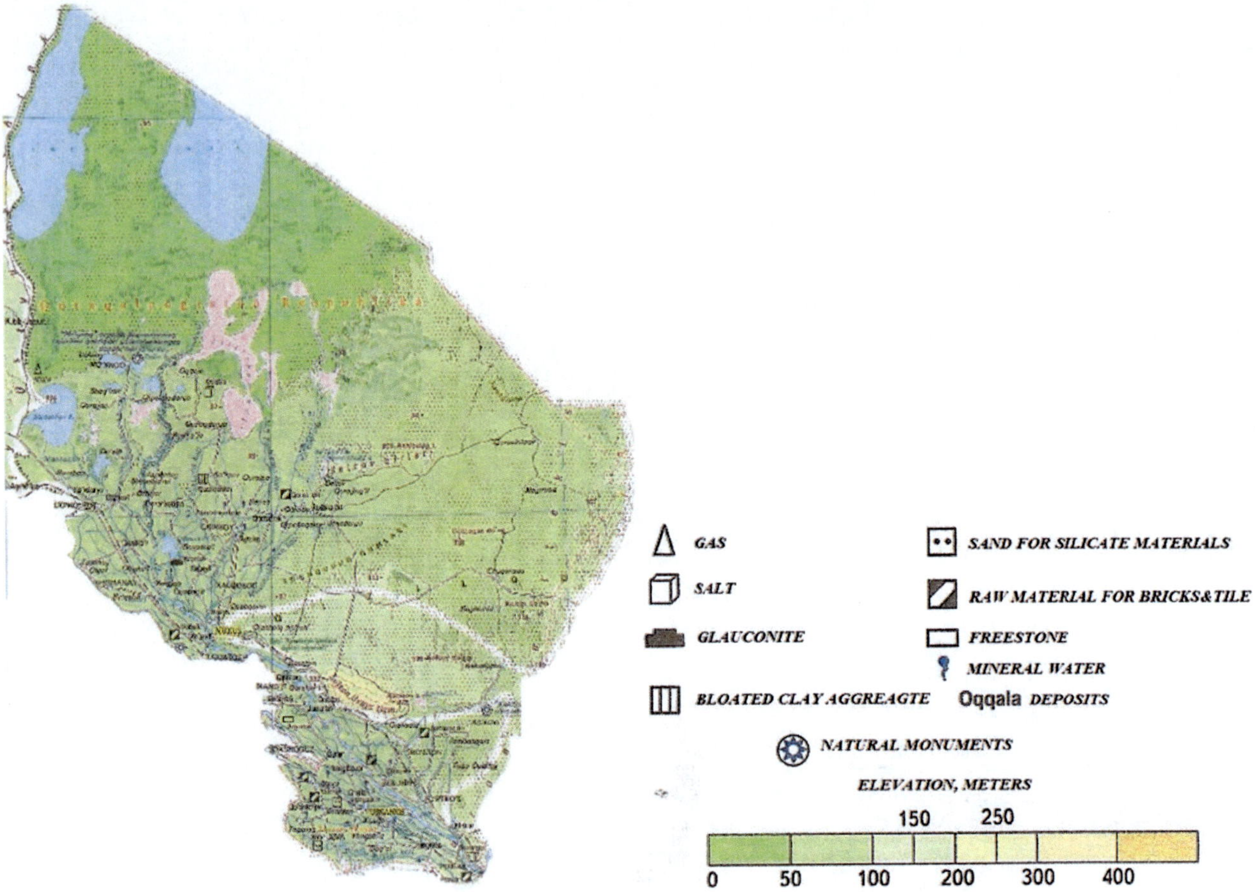

Fig. 6.5 The Amu Darya river delta. (Source: Atlas c, courtesy of the State Committee for Land Resources, Geodesy, Cartography and State Cadastre of the Republic of Uzbekistan)

The flat plain is gently inclined to the Aral Sea and the Sarygamysh depression. The highest southern point is at 114 m, and the northernmost point is at 35 m on the coast of the Aral Sea. Some hills (Beltag, Kushkantag, and Kiziljar) differentiate the monotonous landscape. This type of surface causes formation of many small deltas and slows the flow of water. The delta is divided into several parts by meanders, oxbow lakes, sandy hills, canals, and collectors.

This subregion is situated in the northern part of the Turan geographic province. The Arctic and Siberian anticyclones play significant roles in forming its climate. According to Khasanov and Gulomov (2006), the average annual temperature is 10–12 °C. In January, the average temperature is between −7.6 °C and −4.5 C, reaching a minimum of −32 °C or even −40 °C. In summer, the average temperature is between +27 °C and +29 °C, but sometimes it reaches +41 °C or +42 °C. This subregion is the driest in Central Asia, with annual precipitation of 79–108 mm, which occurs in winter and spring (Fig. 6.6).

Khasanov and Gulomov (2006) argue that climate change and the shrinking of the Aral Sea have led to an increase in the average summer temperature of at least 2–2.5 °C and a decrease in the winter temperature of 1–2 °C, making the winter colder and the summer hotter, eliminating precipitation, shortening the warm period, and causing dust storms. The consequences include drops in the groundwater level from 0.8 to 3.6 m in irrigated areas, from 10 to 15 m in nonirrigated areas, and from 20 to 30 m in the desert, along with an increase in the mineralization level to 3–5 g/l (Khasanov and Gulomov 2006). The salinity of the soil is increasing.

The tugay (tugai) was another treasure of the Lower Amu Darya subregion. However, the change in the water regime has caused the area of tugay to shrink by at least at 700,000 ha (Khasanov and Gulomov 2006), and the drought has afflicted many species. The vegetation in the tugay consists of two levels: trees (tol, jiyda, and turong'il) in the upper level and bushes (yulg'un, ko'g'alar, ruvak, and qamish) in the lower level. Many of them are under threat of extinction (Figs. 6.7 and 6.8).

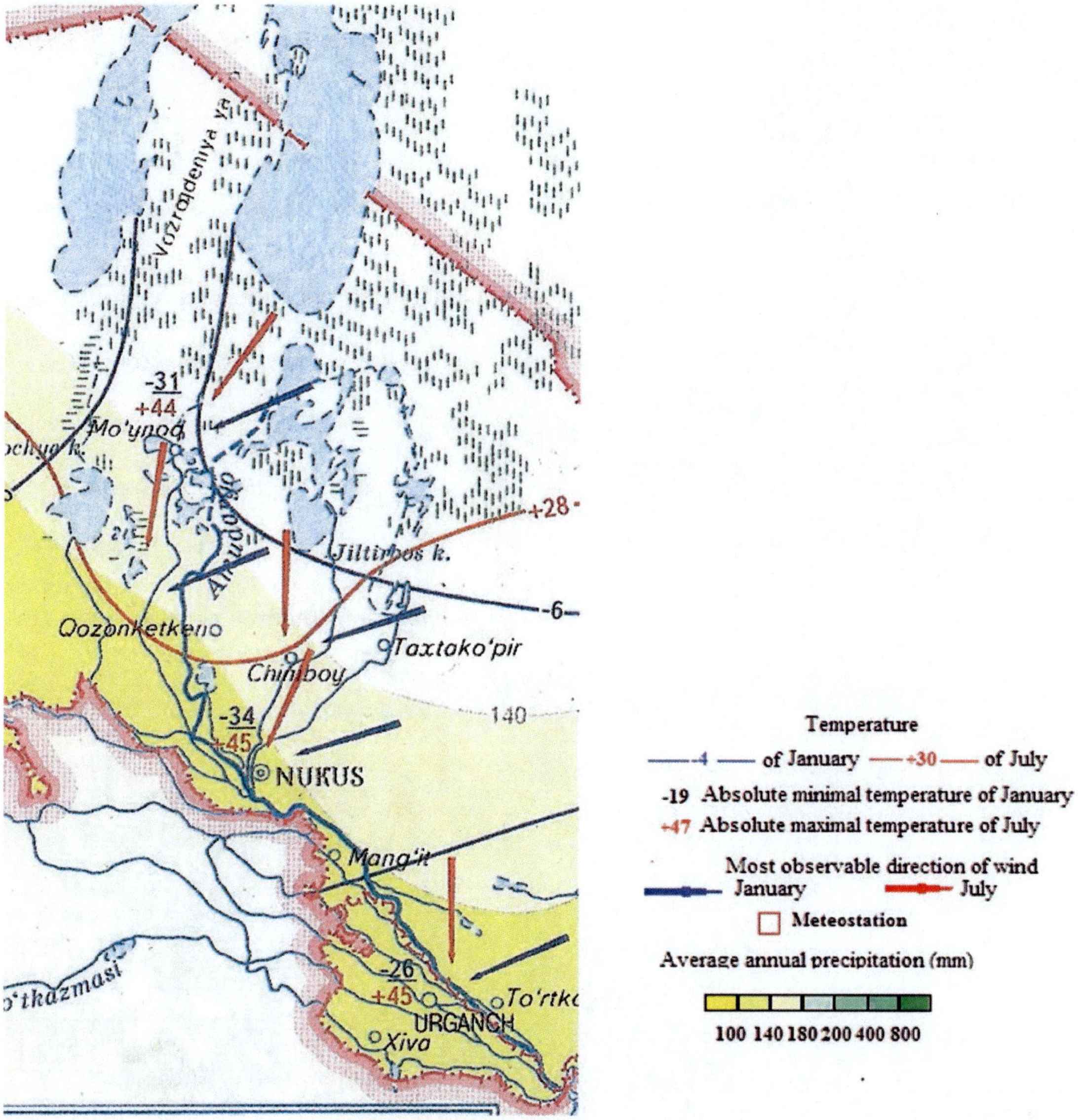

Fig. 6.6 The climate zone of the Amu Darya river delta. (Source: Atlas 2017, courtesy of the State Committee for Land Resources, Geodesy, Cartography and State Cadastre of the Republic of Uzbekistan)

Fig. 6.7 The dried bed of the Amu Darya river, Khorezm region

Fig. 6.8 Halophyte species have replaced previous tugay vegetation in Khorezm region

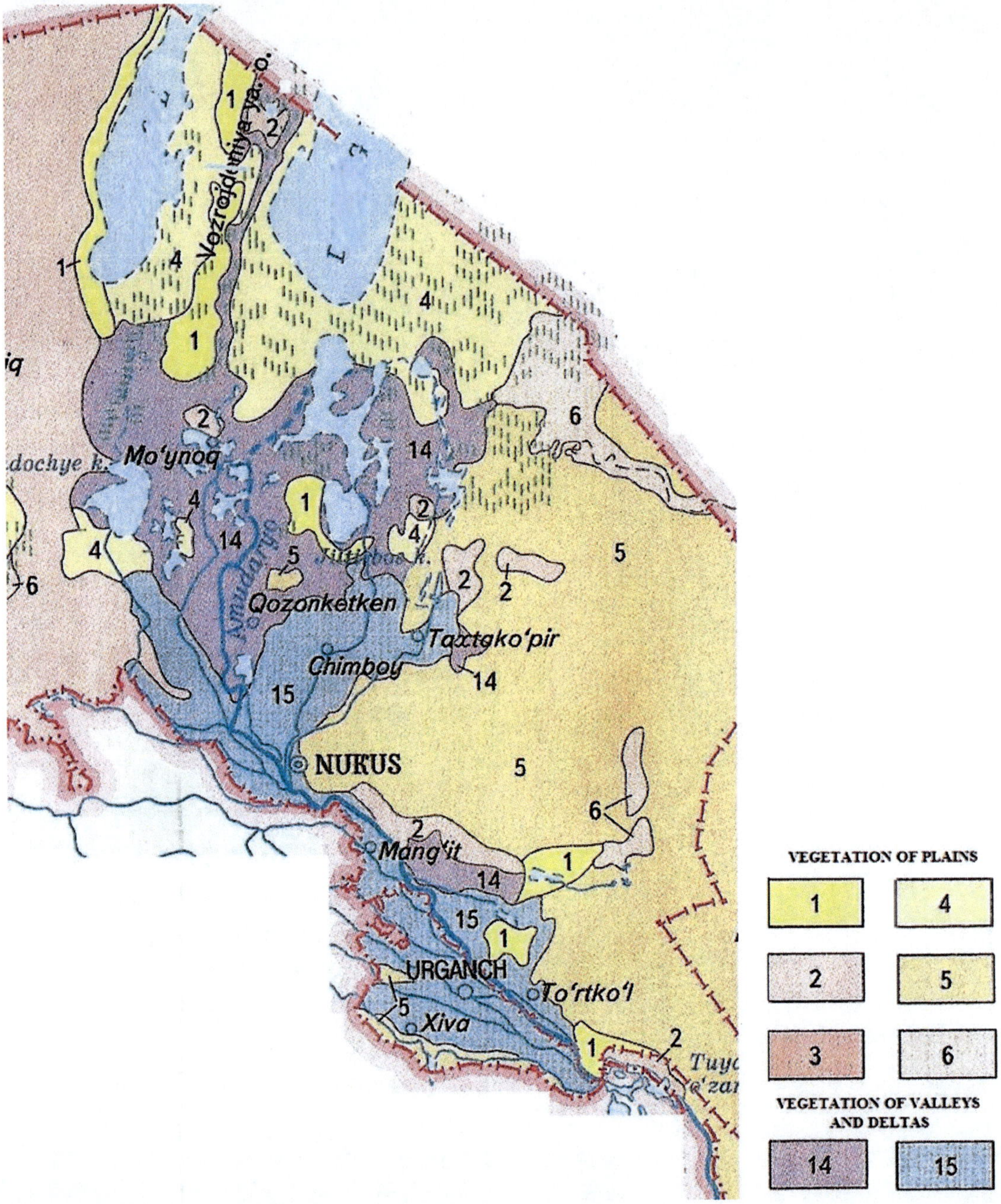

Fig 6.9 Vegetation of the Amu Darya river delta. *1* indicates sparse desert psammophyte shrub vegetation; *2* indicates desert wormwood and Biyurgun (a saltwater barnacle); *3* indicates saline black saxaul desert; *4* indicates sparse saline desert vegetation; *5* indicates white saxaul, sagebrush, white saxaul, and black saxaul–white saxaul deserts; *6* indicates salt desert black saxaul; *14* indicates tugai vegetation; and *15* indicates cultural vegetation. (Source: Atlas 2017, courtesy of the State Committee for Land Resources, Geodesy, Cartography and State Cadastre of the Republic of Uzbekistan)

Halophyte[1] species (sarsazan, qorabarak, yulg'un, shohiloq, it siygak) have replaced the previous tugay vegetation in places with high soil salinity and in the southern part of the Sudochye Lake. Psammophyte[2] species (selin, juzg'un, oqsaksovul, kuyonsuyak, qizilcha, sho'ra) have spread in the sandy parts of the delta and in the neighboring Karakum and Kyzylkum Deserts (Figs. 6.9 and 6.10).

[1] A halophyte is a plant adapted to growing in saline conditions.

[2] A psammophyte is a plant that thrives on or requires sandy soil.

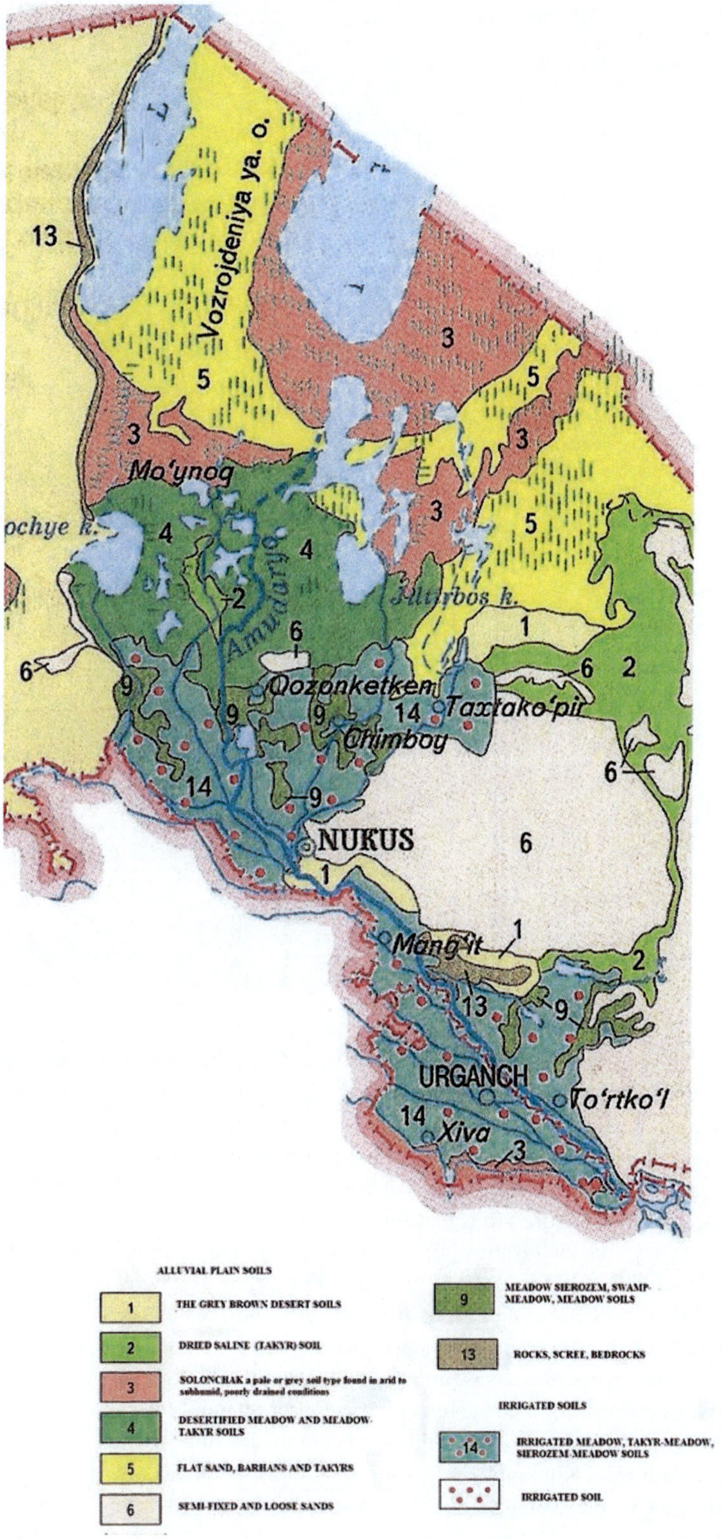

Fig. 6.10 Soil of the Amu Darya river delta. (Source: Atlas 2017, courtesy of the State Committee for Land Resources, Geodesy, Cartography and State Cadastre of the Republic of Uzbekistan)

Fig. 6.11 Sandy forms (0.2 m high) in the dried part of the delta in Khorezm region

Fig. 6.12 A salty plain on the dried bed of the Aral Sea

The modern relief of the delta differs from that which existed in the period of its formation. Signs of the former delta may be found in its western and central parts, inside the border between Kiziljar, Kushkantag, and Beltag (Khasanov and Gulomov 2006). The relief in this part is represented by meanders and depressions that formed before 1960.

Today, the wind is the main driving force of modern relief formation in the dried part of the delta (Fig. 6.11). The wind spreads sand from the bed of the Aral Sea and creates the new relief, consisting of sandy plains, sandy hills and barchan[3] dunes. Barchans are not huge; often their height is only a few meters. The movement of these sandy relief forms fits the wind blowing from the northeast to the southwest. According to Khasanov and Gulomov (2006), barchans are located in the Muynak and Ajiboy dried plains, close to chinks. The area of the barchans is up to 4 km^2, and their height is about 1–2 m; in places close to the chinks of the Ustyurt Plateau, they are no higher than 0.5 m. Wind erosion leads to leveling of hills, filling of depressions with sediments, and prevalence of sandy forms. Small deltas are filled with sandy alluvial sediments; the thickness of the alluvial rocks is 20–25 m and sometimes exceeds 30 m. Human activity (i.e., irrigation system construction) transformed the relief of the Sarygamysh delta. Today, it is a combination of deep canals, depressions, and hills between them. Drainage water has created many small lakes, swamps, and salty swamps (solonchaks) in the lowest parts of the delta.

The salty plains are primarily flat; their surface lowers toward the western and central parts of the bed of the dried Aral Sea. The salinity type of the surface is dependent on the proximity to the sea body of the Aral Sea. The high salinity of the soil prevents growth of plants in the drying parts. Now these places are lifeless except for rare halophytes (Fig. 6.12).

Exploration for mineral resources (oil and gas) has taken place on the dried bed of the Aral Sea and the northern part of the delta. Mining is replacing fishing and agriculture.

The Kyzylkum Subregion

The red color of the sand in many parts of this desert is the reason why it is called Kyzylkum (which means "red sand"

[3]Barchans are traveling sand dunes; they have an isolated crescentic form with arms that stretch downwind.

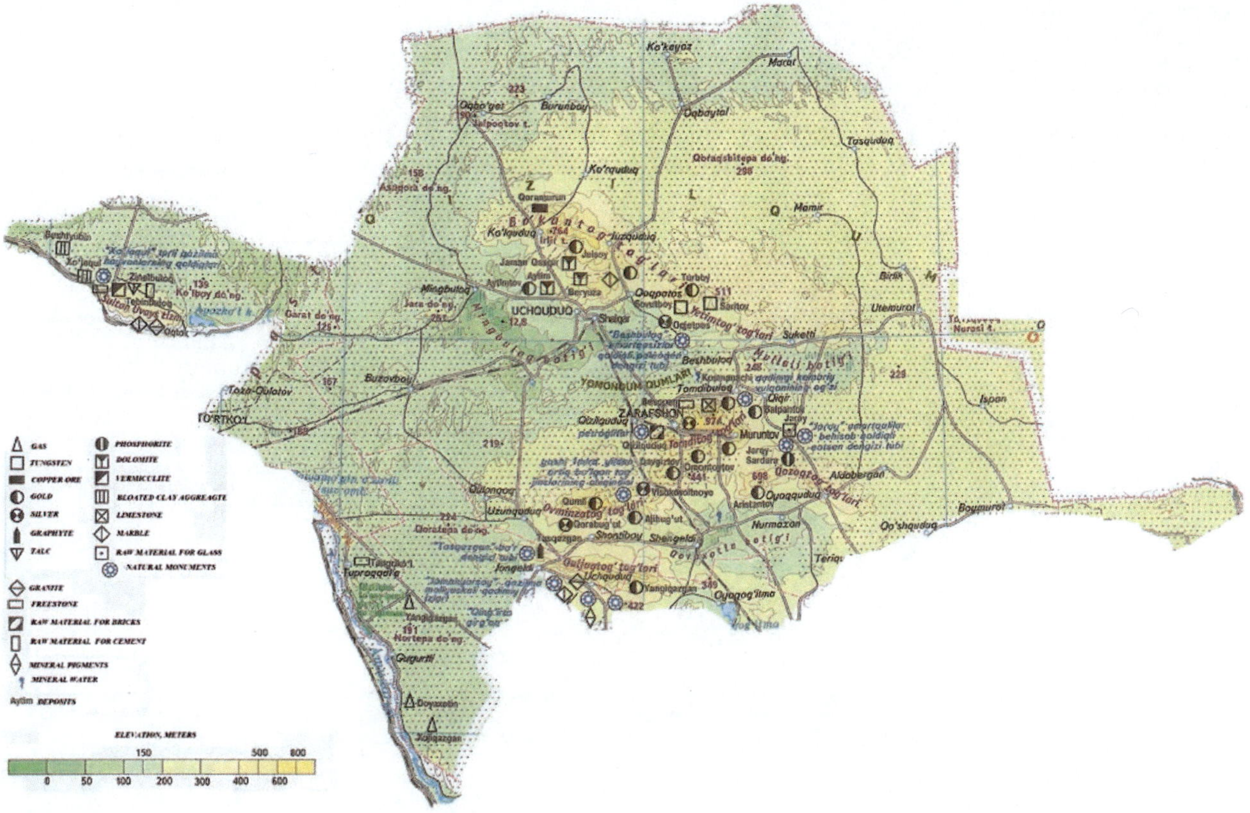

Fig. 6.13 The Turan Plain. (Source: Atlas 2017, courtesy of the State Committee for Land Resources, Geodesy, Cartography and State Cadastre of the Republic of Uzbekistan)

in Uzbek). However, the red rocks are not spread over the whole area. The Zerafshan River brings sediments of a gray color to the southern part of this subregion. The base of this plain is formed by Neogene and Anthropogenic rocks, and is covered by a sand layer 10–12 m thick (Fig. 6.13).

The Kyzylkum Desert is located between the "three sisters"—the Amu Darya, the Syr Darya, and the Zerafshan River. Its northern side borders the Aral Sea (Fig. 6.14). The main relief forms are the lowlands and flat and low mountains. The depressions are widely spread between the low mountains. The biggest of them are Mingbulak, Ayakatigma, Karakata, Kulkuduk, Akjetpas, and Davudbay. The salted swamps (*takir*) are mainly spread in depressions between the clay–sandy plains and the linear sandy hills in southern Kyzylkum.

Many hundreds of millions of years ago, the high mountains in Kyzylkum were part of the chain consisting of the Ural and Tien Shan Mountains. Their raising occurred in the Hercynian orogeny, and ever since the Mesozoic Era, they have become lower and flatter under erosion. They were formed by the rocks of the Paleozoic Era (from 541 to 252.17 Mya). Today, the rocky outcrops are reminders of the former mountainous landscape. The height of the modern relief does not exceed several hundred meters. For example, Kuljuktag in the center of Kyzylkum is 784 m high, Tamdytag is 974 m high, and Sultan Uvays is 448 m high. Many deposits of mineral reserves (gold, metallic ores, tungsten (wolfram), and turquoise) are located in this part of Kyzylkum. The deep bedding of these rocks is frequently covered by rocks from the Mesozoic Era (between 252 million and 66 Mya) and the Cenozoic Era (from 66 Mya to the present day) in the plain part of the Turan plate and between the mountains. The red rocks were formed in the Mesozoic Era and at the beginning of the Quaternary Period when a subtropical climate existed there. The red rocks were spread in the foothills of those mountains. Paleocene rocks in the form of sandstone (arenite) composed of quartz, clay, and aleurolite are spread on the intermountain plateaus. Deposits of oil, gas, uranium, phosphorite, and sulfur have been explored and exploited on these plains.

Kyzylkum is inclined from the southeast to the northwest. It is relatively flat, with an average elevation of between 200–300 m and 350–400 m in the eastern–southern part and 90–100 m in the northern–western part. The topography is represented by different sandy forms, low plateaus, sinks, and beds of dried rivers. Sandy forms are stretched along the

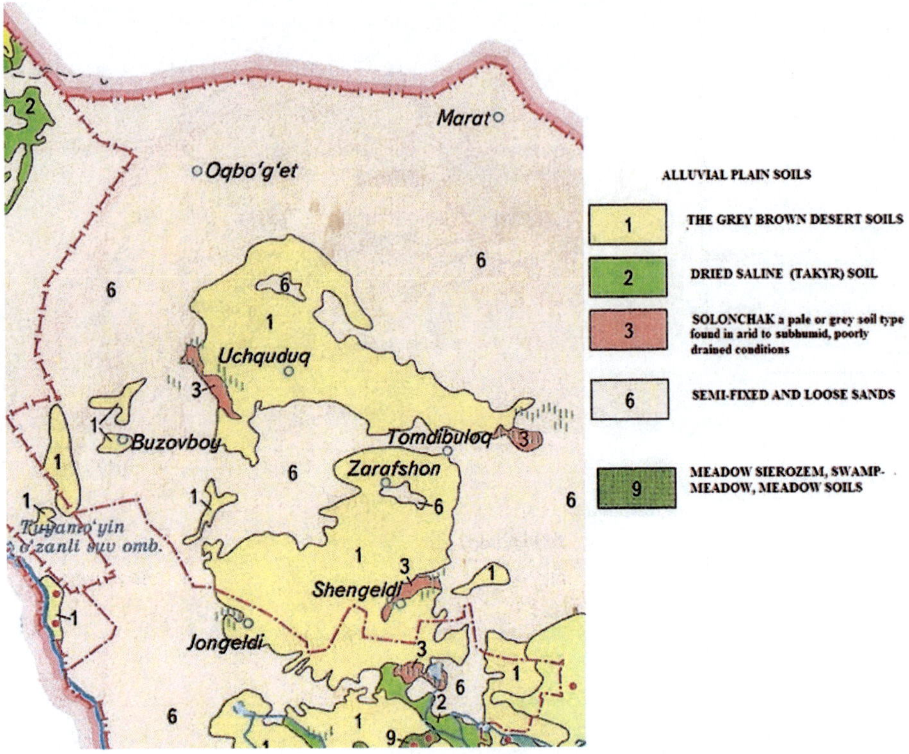

Fig. 6.14 Soils of the Turan Plain subregion. (Source: Atlas 2017, courtesy of the State Committee for Land Resources, Geodesy, Cartography and State Cadastre of the Republic of Uzbekistan)

meridian; their height is about 4–15 m, and at some sites, they may reach 50–70 m. These sandy forms are covered with psammophyte vegetation (Figs. 6.14 and 6.15). Other sandy forms—barchans—are not reinforced by any vegetation, and they move in accordance with the wind direction. Their primary location is in western Kyzylkum, near settlements and wells, where vegetation is destroyed by the feeding of the cattle.

In Kyzylkum, beds of ancient dried rivers are the widespread forms of relief. They are relics of the former flow of the Syr Darya and Amu Darya rivers. The name of the Syr Darya river translates as the "secret river" since it has changed its flow many times. Janadarya is one of the former beds of the Syr Darya; it dried up around 1810. The ruins of settlements on the banks of this river are good evidence of its fluctuations. The Akchadarya river is a relic of the Amu Darya river.

The extremely dry climate prevents formation of water resources. However, groundwater is spread in many places, although some deposits are unsuitable for drinking because of high levels of mineralization. Only the groundwater 100 m deep under the barchans has a mineralization level of less than 1 g/l. These deposits are mainly located in the foothills of the mountains in central Kyzylkum.

The climate is more extreme than those of the neighboring subregions. The average temperature in January is between −4.1 °C and −7.8 °C; the lowest is between −31 °C and −34 °C. The average temperature in July is 30 °C; the highest is 46 °C. The annual precipitation is 70–108 mm (Fig. 6.16). This amount is enough to grow only ephemeral, xerophyte, and psammophyte species. Ephemeral species have a very short period of vegetation in spring and dry out in summer. Xerophyte and psammophyte vegetation continues to grow even in the hot and dry days. This determines the low humus level in the soil, which is less than 0.4–0.5%.

The rare settlements in this area were situated near freshwater sources; they served as the main points on the caravan routes in the desert. Sheep and cattle farming and hunting were the primary occupations of the people who temporarily lived in the sandy environment. Today, cities built by the mining industry (Uchkuduk, Zerafshan) are found in central Kyzylkum.

Industrial development of the desert and overgrazing in the pasture zone negatively impact the natural desert environment. Scientists (Azimov et al. 2016) have pointed out that some endemic plants and animals (such as the Severtsev ram (Kyzylkum ram), golden eagle, bearded vulture, and black griffon vulture) are now threatened with extinction; they are listed in the *Red Data Book of Uzbekistan*.

The Zerafshan Subregion

This subregion consists of the mostly flat Lower Zerafshan plain and the Zerafshan intermountain depression (Fig. 6.17). These are stretched along the Zerafshan River valley. Their natural environments are semiarid and arid. The major cities on the Silk Route—Bukhara (Buxoro in Uzbek) and

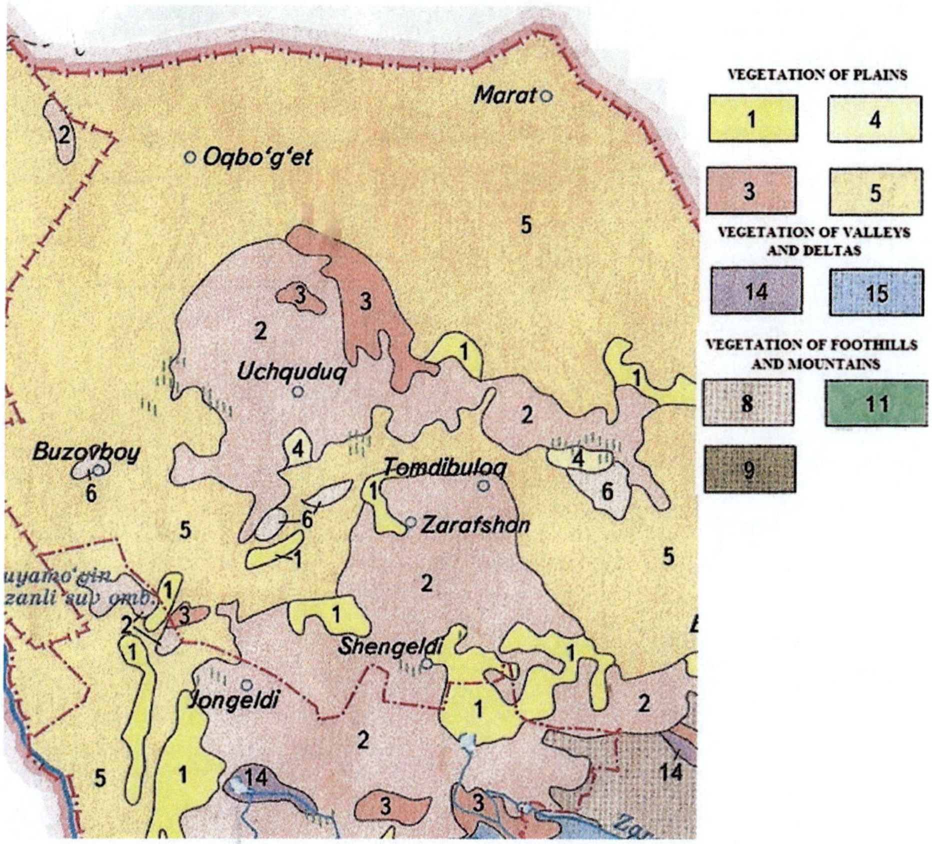

Fig. 6.15 Vegetation of the Turan Plain subregion. 1 indicates sparse desert psammophytes shrub vegetation; 3 indicates saline black saxaul desert; 4 indicates sparse saline desert vegetation; 5 indicates the white-saxaul, sagebrush, white-saxaul, black-saxaul-white saxaul deserts; 8 indicates the semidesert foothill and low-mountain grass and wormwood; 9 indicates the mountain dry grass steppes; 11 indicates sparse archa trees on the background of dry steppe herbs; 14 indicates tugai vegetation; 15 indicates cultural vegetation. (Source: Atlas 2017, courtesy of the State Committee for Land Resources, Geodesy, Cartography and State Cadastre of the Republic of Uzbekistan)

Samarkand (Samarqand in Uzbek)—are located in this subregion. Long-term human activities have impacted the natural environment here, leading to significant landscape change. Irrigated lands have replaced natural landscapes in all reaches of the Zerafshan River.

Lower Zerafshan includes the Bukhara and Karakul oases, the plains around the river, and the Dengizkul Plateau. The base is formed by Paleozoic rocks covered by Mesozoic and Cenozoic rocks. The Zerafshan River transfers Quaternary sediments (gravel, sand, and clay), enriching the soil. The major gas and gas condensate deposits of Uzbekistan have been explored and exploited in this subregion.

The surface of the deltas of the Zerafshan and Karakul Rivers is flat. The elevation of the Zerafshan delta is about 200–250 m above sea level. The Karakul delta is slightly lower. Its elevation changes from the northeast (200 m) to the southwest (178 m). These deltas are gently inclined to the southwest toward the connection with the Amu Darya river valley.

The climate is extremely arid, with a short winter (lasting 1–2 months) and a very hot summer (Fig. 6.17). The average temperature in January is between −2 °C and 1 °C; the absolute minimum is −26 °C. The average temperature in July is between 29.5 °C and 36 °C; the absolute maximum is 46 °C. The annual precipitation is 95–125 mm. This subregion is the richest in solar resources (4500–5000 hours of sunshine per year).

The Amu Darya and the Zerafshan River are the main water sources for irrigation in this area. The water brought by the Zerafshan River is not enough to irrigate the cotton fields. The 55-km Amu–Karakul canal and the 268-km Amu–Bukhara canal were constructed to fill the gap in the water supply. A chain of pumping stations (at Alat (the southernmost point), Karakul, and Hamza) has been constructed to pump water from the Amu Darya up to 66 m. The Zerafshan River is also divided into 50 canals. Khasanov and Gulomov (2006) note that the annual consumption of water is 4.3–4.5 km^3; 20% of this water is sent from irrigated fields to outside sinks. More than ten lakes have now been formed, with a total volume of about 1.0 km^3. Only some of this water is returned to the Amu Darya river.

Agriculture, not only cotton farming, requires irrigation in a dry climate. Irrigated agriculture at oases during many centuries led to leveling of the terraces of the Zerafshan River, changing the natural landscape significantly. The ecological system is at high risk. The depth and quality of the groundwater are dependent on the thickness of the underlying Neogene and Quaternary rocks. It is only in the eastern part of the subregion that the groundwater at a depth of 10 m is of good

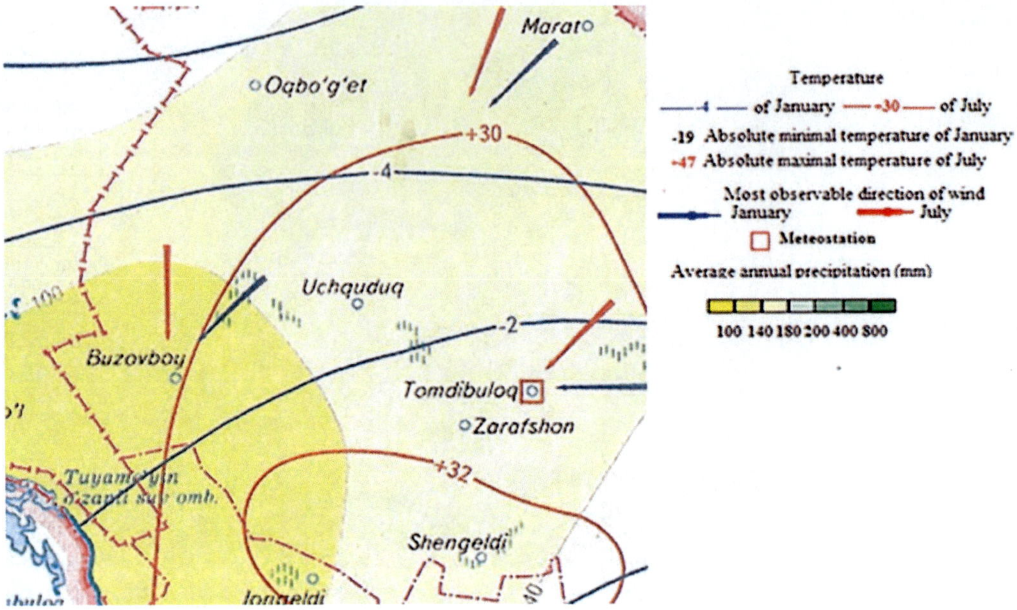

Fig. 6.16 Climate of the Turan Plain subregion. (Source: Atlas 2017, courtesy of the State Committee for Land Resources, Geodesy, Cartography and State Cadastre of the Republic of Uzbekistan)

Fig. 6.17 The Zerafshan subregion. (Source: Atlas 2017, courtesy of the State Committee for Land Resources, Geodesy, Cartography and State Cadastre of the Republic of Uzbekistan)

quality with a low level of mineralization and does not cause soil salinity. In other parts of the delta, the mineralization of the water may reach 10–15 g/l. Intensive irrigation also results in a rise of the level of highly mineralized groundwater, which can destroy the whole ecological system. According to estimates (Khasanov and Gulomov 2006), soil with high salinity occurs in 95% of the total area of Bukhara province. Desertification is another challenge, caused by wind erosion, air pollution, and overgrazing in the pasture zone.

The soil quality is better in some central parts of oases, where the humus level is between 1.5% and 6.0%. In other places, the soil is primarily of poor quality, with a humus level of 0.5%. This determines the prevalence of psammophytes in the sandy soil outside oases (Figs. 6.18 and 6.19).

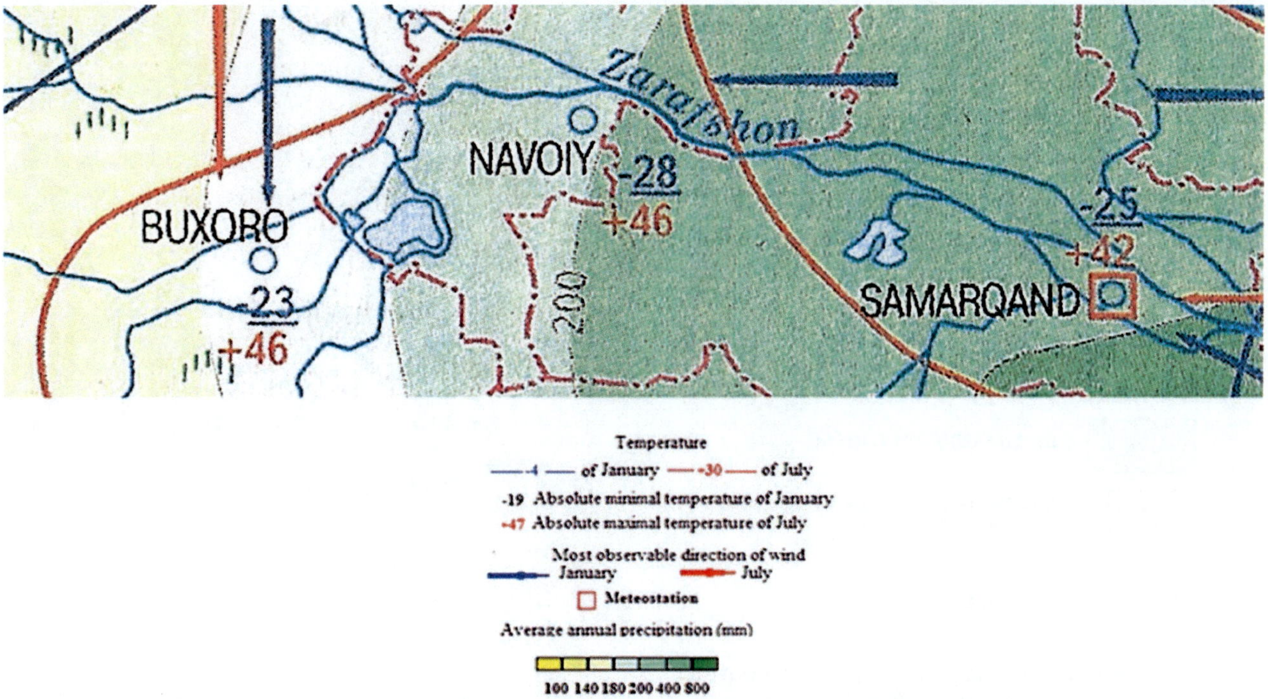

Fig. 6.18 Climate of the Zerafshan subregion. (Source: Atlas 2017, courtesy of the State Committee for Land Resources, Geodesy, Cartography and State Cadastre of the Republic of Uzbekistan)

The Zerafshan depression is partly located in the eastern part of the subregion. It is surrounded by the Nurota and Sangzor Mountains to the north; the Karatepa, Ziyoviddin, and Zirabulak Mountains to the south; and the Chumkartag Mountain to the east. The western part of the Nurota–Turkestan Mountains system is relatively lower than their eastern part in Tajikistan. The highest point of the Nurota Ridge is Hayotboshi Mountain (2169 m high). The Turkestan, Zerafshan, Nurota, Ziyoviddin, and Zirabulak Mountains contain aleurolite, mudstone, quartzite, and limestone from the Ordovician period (485.4–443.8 Mya).

At the start of the Sangzor River, the northern side of the Turkestan Mountains is divided into two smaller ranges. One of them is Morguzar, stretching to the northwest, and Chumqor is to the west. These mountains are not as eroded as the Western Tien Shan. The Morguzar Mountains surround the Sangzor River Valley. Erosion is detected mainly on the northern side of the ridge. The composition of the depressions and highlands creates a wave-shaped landscape, which is divided by many small and dried valleys. Rocky sediments fill these valleys.

The climate differs from that of Lower Zerafshan. The summer is slightly cooler, with an average July temperature of 25–28 °C; the absolute maximum is 42 °C (Fig. 6.19). The difference is greater in the eastern foothills and mountains. The annual precipitation is 2–3 times as great, reaching 300–400 mm on the plains and in the foothills. The solar resources are lesser—between 4200 and 4600 hours of sunshine per year (Khasanov and Gulomov 2006). The Zerafshan and Sangzor Rivers are the main water sources for irrigated agriculture. Their total flow module is about 14 l/s/km^2. This amount of water is supplemented by water from about 50 creeks in the foothills of the surrounding mountains. The water is fully distributed through a system of canals.

The landscape varies in the intermountain depression and the foothills, with differences in soils and vegetation (Figs. 6.19 and 6.20). The natural environment is of the desert, semidesert, and dry steppe types in the intermountain depressions whereas it is of the steppe type in the low mountains. Forests and alpine vegetation are not widespread. Hydromorphic soils occur in the valleys, and brown soils are found in the foothills. Tugay vegetation can be found along the rivers.

This part of the subregion is industrially well developed. Such cities as Navoiy, Samarkand, and Katta Kurgan are the centers of many industries. Human activity has a negative impact on the environment. Overuse of chemicals and pollution of air and water lead to salinity of the soil. Soil degradation is intensifying in the western parts close to the center of the chemical industries in the city of Navoiy.

The Kashkadarya Subregion

The surface of this subregion is significantly inclined to the west. It includes the Karnabchul and Karshi (Qarshi) plains in the west and the high Gissar Mountains in the east. The elevation of the Gissar Mountains ranges from 2000 to 4152 m at the highest point (Gova Peak). The Zerafshan Mountains surround this region on the northern–eastern side, with the highest points being Chaqilqalon (2197 m high) and Qoratepa (2616 m high) (Fig. 6.21).

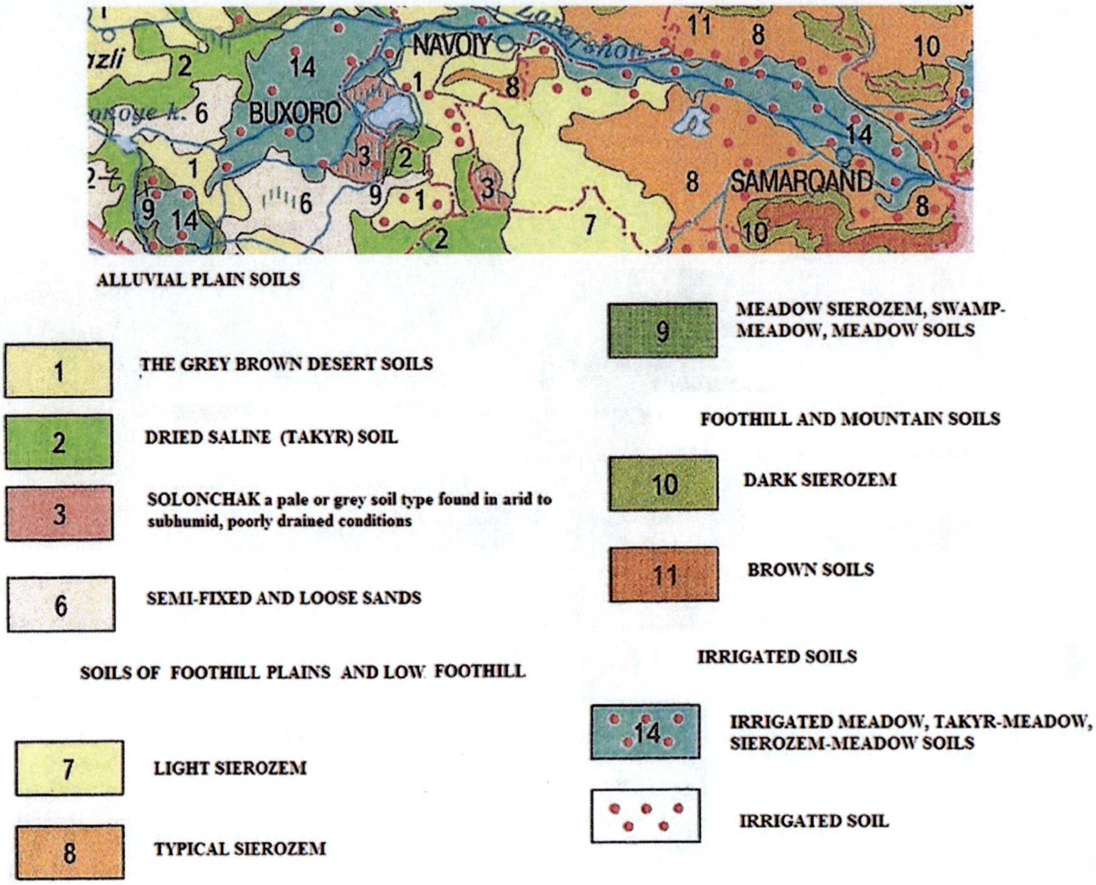

Fig. 6.19 Soils of the Zerafshan subregion. (Source: Atlas 2017, courtesy of the State Committee for Land Resources, Geodesy, Cartography and State Cadastre of the Republic of Uzbekistan)

The Gissar Mountains were formed during the Hercynian Epoch and contain Cretaceous, Paleocene, and Neogene rocks. The thickness of the Paleocene rocks may reach 160 m in a form of sandstone (arenite) composed of quartz, clay, and aleurolite combined with dolomite in the southern–western parts of the Gissar Mountains. These rocks have been pushed to the surface, resulting in steep slopes. The Gissar and Zerafshan Mountains constitute the main tectonic structures in Uzbekistan. In the alpine zone of the mountains, the relief is formed by tectonic, glacial, and erosion processes. The dominant relief forms are deep valleys with extremely steep slopes, rocky peaks, and upper parts.

The plains in the western part of this subregion are rich in hydrocarbons, which have been intensively exploited. Exploration for various metallic ores has taken place in the mountains.

The rivers have transferred sediments from the Zerafshan, Yakkabag, and Gissar Mountains to the Kashkadarya, Kitab–Shahrisabz, Miraki, Yakkabag, and Guzar intermountain depressions. These depressions are composed of plains of different heights (*adirs*).

There is a significant difference in the average annual precipitation between the plains (190–200 mm) and the mountains (550 mm) (Fig. 6.22). The climate is relatively mild in comparison with that of the Lower Zerafshan subregion. The winter is warm, and it is only on mountains higher than 1500–1600 m that the snow does not melt. The maximum runoff occurs mainly in the winter and spring. The runoff and the flow module are dependent on the orientation of the slopes and their elevation.

The Kashkadarya river is a major water source, supplemented by water from the Amu Darya river. This subregion suffers water shortages in summer. The resources of the basin of the Kashkadarya river (where the flow module is 6.2 $l/s/km^2$) are comparatively less than in the basin of the Zerafshan River (10.7 $l/s/km^2$) or in the mountains of Surxondaryo (18.2 $l/s/km^2$). Several reservoirs (at Chimkurgan, Kamashi, Pachkamar, and Talimarjan) have been constructed to collect water.

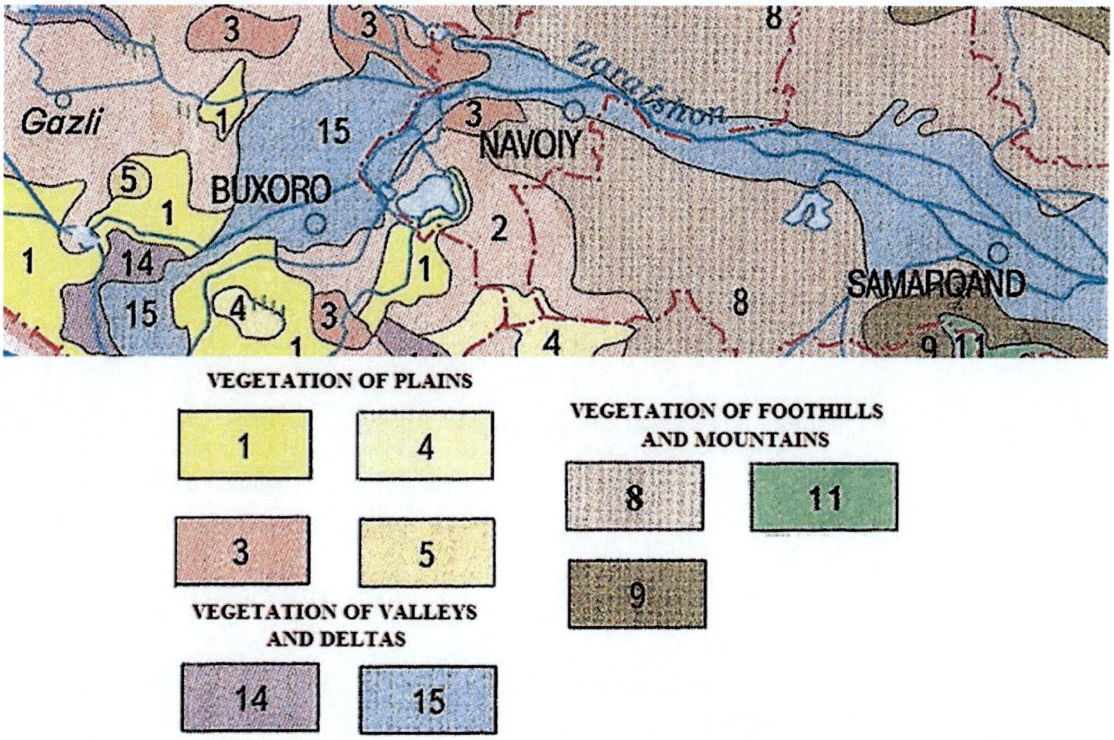

Fig. 6.20 Vegetation of the Zerafshan subregion. 1 indicates sparse desert psamophyte shrub vegetation; 2 indicates the desert wormwood and biyurgun (saltwater barnacle); 7 indicates ephemerae semi-desert vegetation; 8 indicates the semi-deser foothill and low-mountain grass and wormwood; 10 indicates the mountain broad-leaf forests, sparse trees, shrubs; 11 indicates the sparse archa trees on the background of dry steppe herbs; 13 indicates subalpine meadow steppes and alpine meadows; 15 indicates cultural vegetation. (Source: Atlas 2017, courtesy of the State Committee for Land Resources, Geodesy, Cartography and State Cadastre of the Republic of Uzbekistan)

The nature of the plain (soil, vegetation, and animals) changes from the desert type in the west to the semidesert type in the east (Figs. 6.23 and 6.24).

Intensive cotton cultivation has caused degradation and high salinity of the soils on the plains.

The Surkhandarya Subregion

This subregion is the southernmost one in Uzbekistan. The highest point of this country is in the Gissar Mountains, which surround the Surkhanndarya Valley to the north. This valley is surrounded by the Bobotag Mountains to the east, with the Baysun and Kuhitang Mountains to the west (Figs. 6.25 and 6.26).

This intermountain depression is filled with thick sediments from the Neogene and Anthropogenic Periods. The tectonic process is still causing lowering of the Surkhandarya Valley, which is a part of the synclinal depression. Karst is widespread. The "Dark Star" karst cave is in the Baysun Mountains (Synnott 2017).

The Surkhandarya and Sherabad Rivers are tributaries of the Amu Darya from the north. The valleys of these rivers are filled with loess-like rocks, which are valuable for agriculture. The surface of the depression is inclined to the south, toward the Amu Darya river, with the elevation changing from 700 m in the northern–eastern part to 300 m at the conjunction.

The Gissar–Zerafshan mountain system is the most eroded one, with extremely asymmetric slopes. The Gissar Ridge is the highest one in this system, which includes Uzbekistan's highest peak, Hazreti Sultan (4643 m high). The southern side of the Gissar mountain range is wide, with many plateaus and plains in the foothills. Several mountain ranges—Boysuntog, Yakkabog, Guzor, Ko'xitangtog, Surhontog, and Babaytag—stretch to the southwest.

Different kinds of mineral resources—such as oil and gas (the Kokaydi, Lalmikor, Uchkizil, and Hovdogdon deposits), coal (the Shargunsoy deposit), and rare metallic ores—have been explored and exploited in Surxondaryo.

This subregion is located at a subtropical latitude and is isolated from cold air masses from the north. The geographic location and topography are natural advantages of this subregion. The climate is milder, with the average temperature in January being 2.5–2.8 °C on the plains and in the foothills. The vegetation period is longer. The average temperatures of July are between 29 and 32 °C, and the maximum are

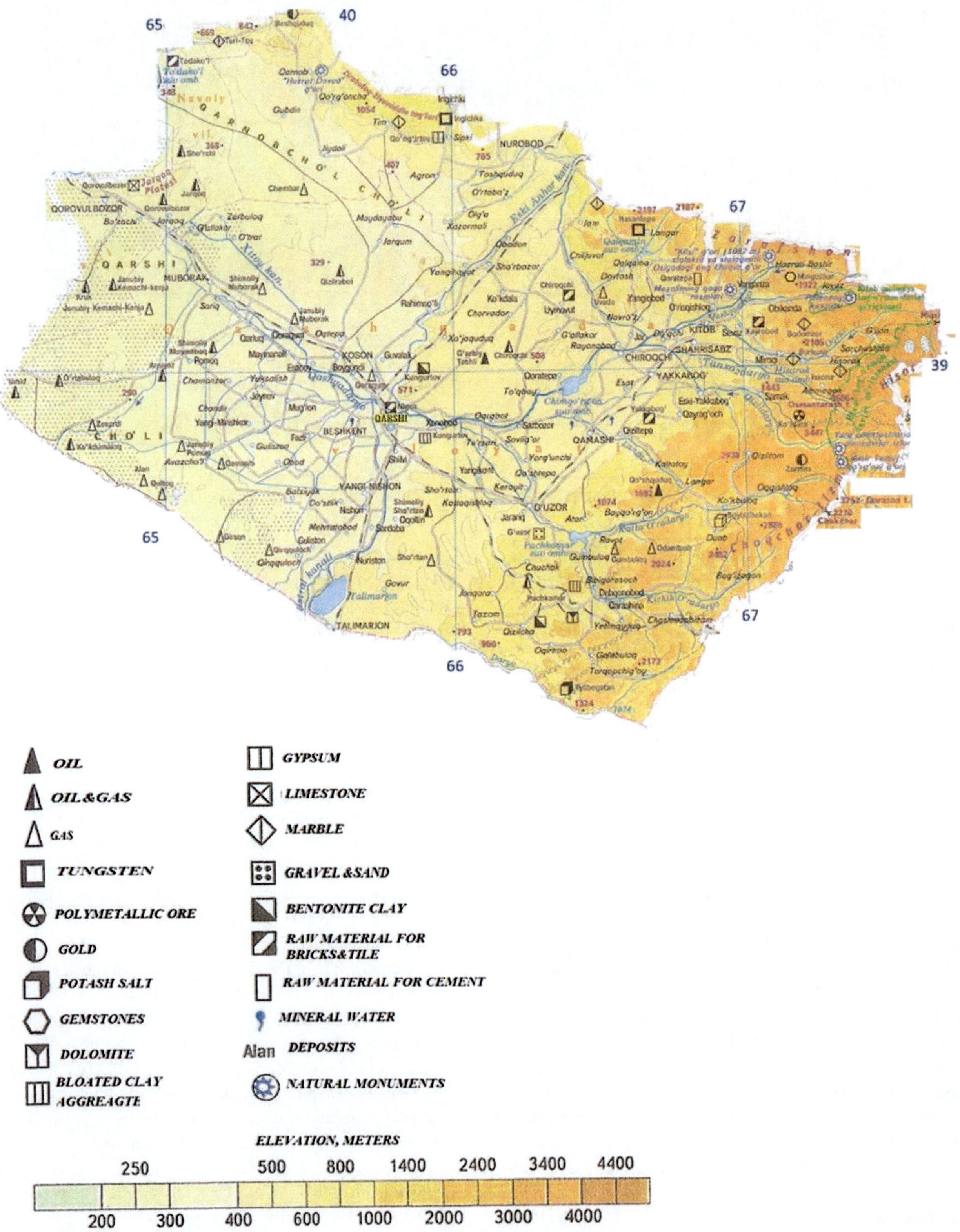

Fig. 6.21 The Kashkadarya subregion. (Source: Atlas 2017, courtesy of the State Committee for Land Resources, Geodesy, Cartography and State Cadastre of the Republic of Uzbekistan)

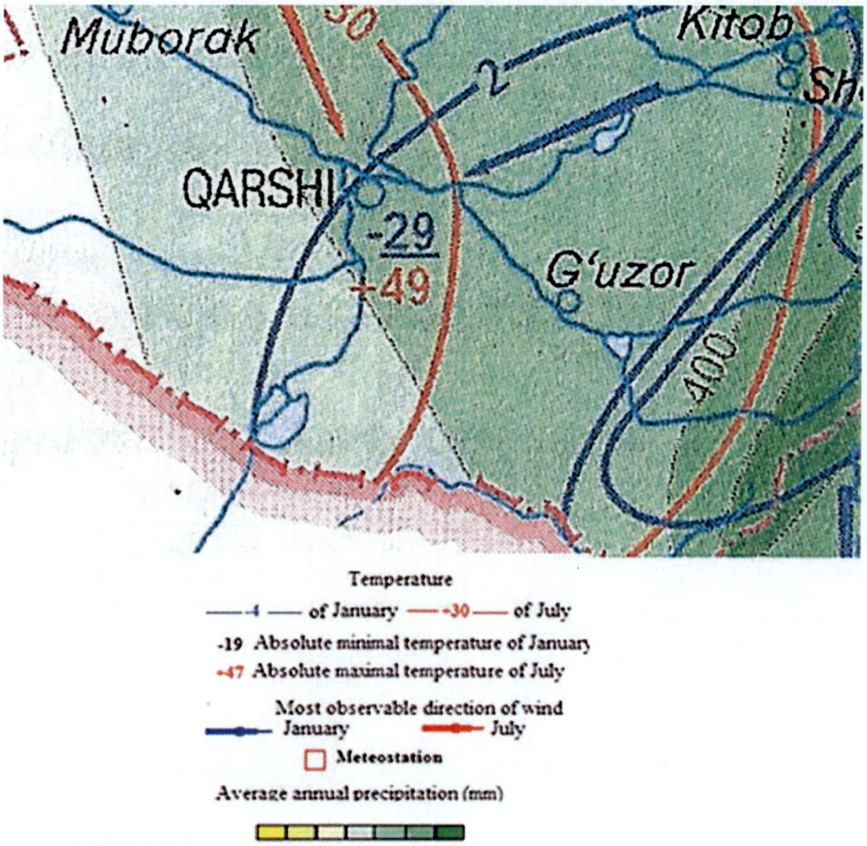

Fig. 6.22 Climate of the Kashkadarya subregion. (Source: Atlas 2017, courtesy of the State Committee for LandResources, Geodesy, Cartography and State Climate)

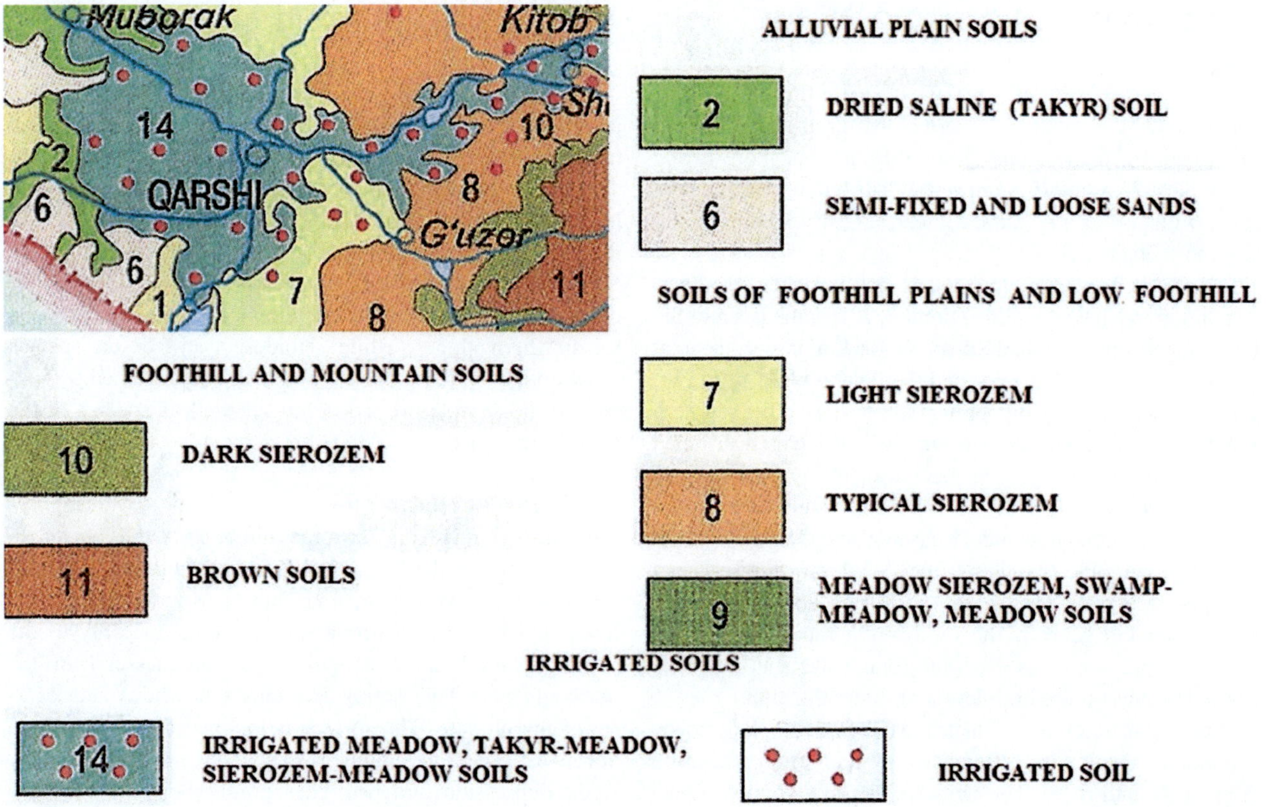

Fig. 6.23 Soils of the Kashkadarya subregion. (Source: Atlas 2017, courtesy of the State Committee for Land Resources, Geodesy, Cartography and State Cadastre of the Republic of Uzbekistan)

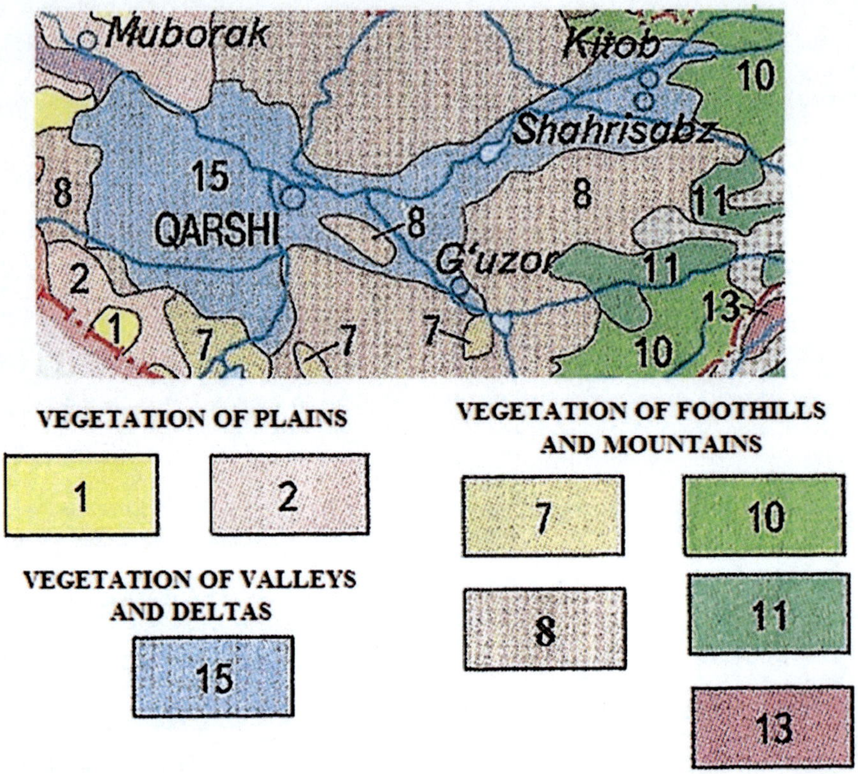

Fig. 6.24 Vegetation of the Kashkadarya subregion. (Source: Atlas 2017, courtesy of the State Committee for Land Resources, Geodesy, Cartography and State Cadastre of the Republic of Uzbekistan) 1 indicates sparse desert psamophyteshrub vegetation; 3 indicates saline black saxaul desert; 4 indicates sparse saline desert vegetation; 5 indicates the white saxaul, sagebrush, white-saxaul, black-saxaul-white saxaul deserts; 8 indicates the semi-desert foothill and low-mountain grass and wormwood; 9 indicates the mountain dry grass steppes; 11 indicates the sparse archa trees on the background of dry steppe herbs; 14 indicates Tugai vegetation; 15 indicates cultural vegetation

between 42 °C and 48 °C. There is less precipitation in the intermountain depressions (133–700 mm) but more in the mountains. The runoff is unevenly distributed over the seasons, with wet springs and winters, and dry summers and autumns (Fig. 6.27).

The Surkhanndarya subregion is rich in water resources. According to Khasanov and Gulomov (2006), the basin of the Surxondaryo and Sherabad Rivers is 8406 km^2 in size, the total water resources are 153 m^3/s, and the flow module is 18.2 l/s. These values are much higher than those in the other southern subregions. However, the water supply is unevenly distributed over the seasons. Fifty-eight percent of the water comes in March–June, with only 25.7% in summer (July–September) and 18.3% in autumn–winter (Khasanov and Gulomov 2006). Several reservoirs (including the Southern Surkhan and Uchqizil reservoirs) have been constructed for water collection and land irrigation in the Surkhan–Sherabad Desert.

This subregion is also rich in groundwater, but its use is limited because it has high levels of mineralization.

The contrast of the climates on the plains and in the mountains results in differences in soil and vegetation (Figs. 6.28 and 6.29). The humus content is about 1.5% in the gray soil of the plains but reaches 2.5% in the foothills at an elevation of 500–600 m above sea level. The salinity and humidity of the soil are higher in the southern–western part. The vegetation changes depending on the topography and location. Ephemeral and psammophyte species are spread in the mostly hot and dry southern parts and deserts of the Surxondaryo subregion. The vegetation of the mountains is dependent on the orientation of their slopes. Forests grow on the northern slopes, while semidesert and desert species grow on the southern ones. Animals are represented by species of the semidesert and desert, as well as species of the subtropical zone.

The Mirzachul Subregion

This subregion is situated in the midreach of the Syr Darya river. It includes the Mirzachul plain; the northern slopes of the Turkestan, Morguzar, and Nurota Mountains; and the Sangzor River. It is surrounded by mountains only to the south and southeast. It is open to cold air masses from the north and west. This setting determines the aridity and severity of the climate. The surface is inclined to the north and northwest. The lowest point (230 m above sea level) is at the Aydar depression, and the highest point (about 4000 m high) is on the Turkestan Ridge (Fig. 6.30).

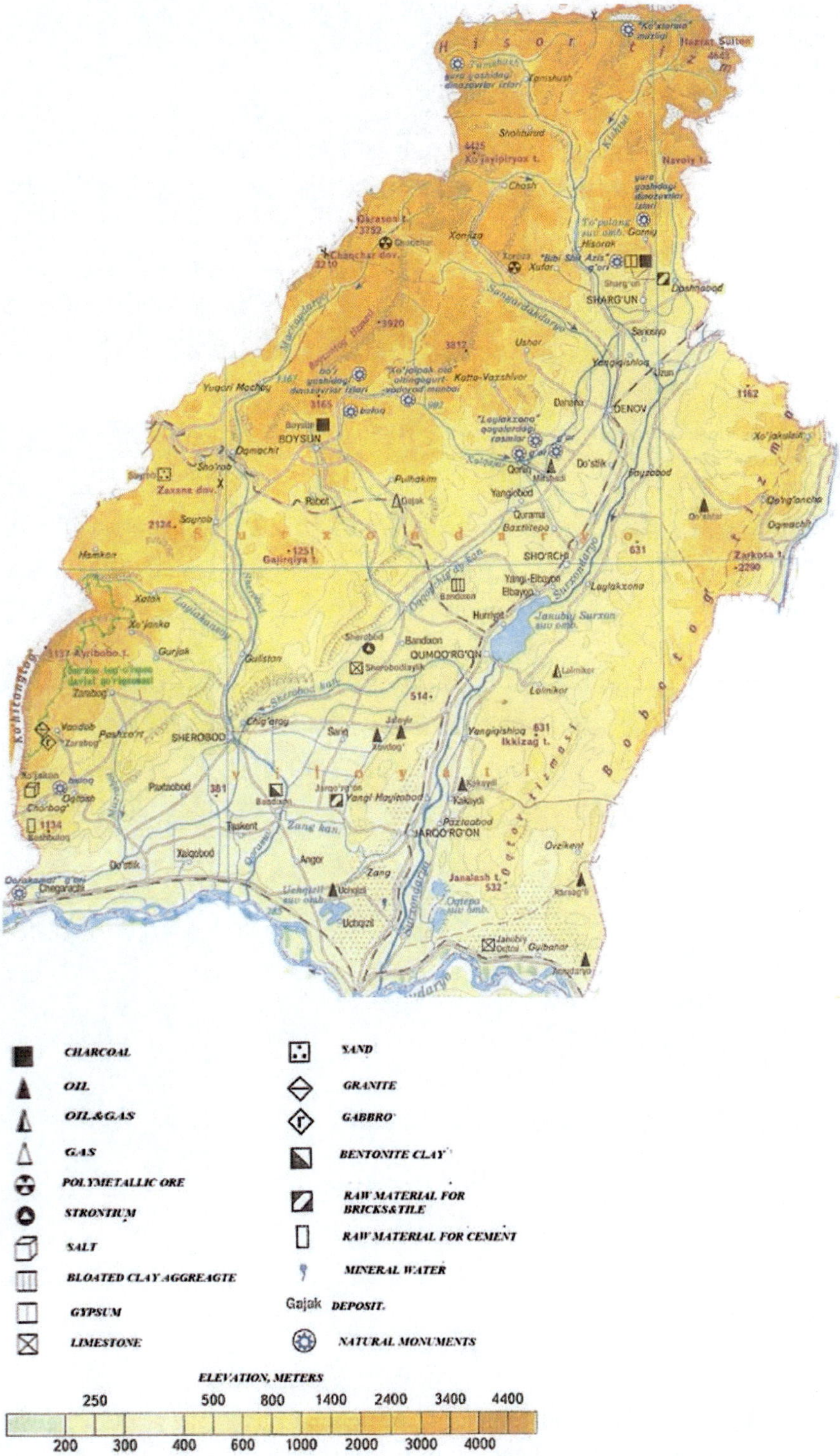

Fig. 6.25 The Surkhandarya subregion. (Source: Atlas 2017, courtesy of the State Committee for Land Resources, Geodesy, Cartography and State Cadastre of the Republic of Uzbekistan)

Fig. 6.26 The Gissar Ridge, Surkhanndarya. (Source: Farsizabon Creative Commons Attribution-Share Alike 4.0 International license via Wikimedia)

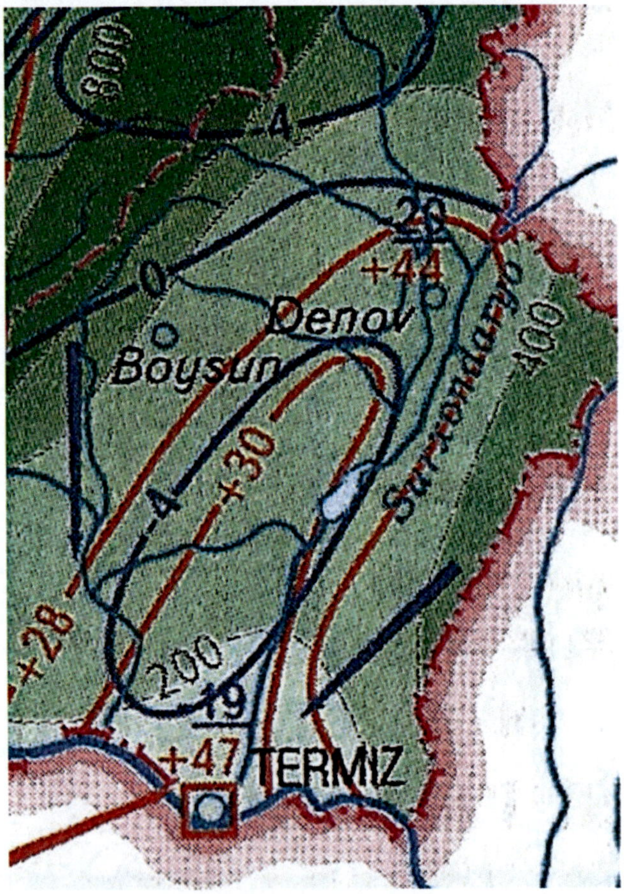

Fig. 6.27 Climate of the Surkhanndarya subregion. (Source: Atlas 2017, courtesy of the State Committee for Land Resources, Geodesy, Cartography and State Cadastre of the Republic of Uzbekistan)

The Turkestan and Nurota Mountains are formed from Paleozoic gneiss, slate, and dolomite (Fig. 6.31). The northern slopes of the high Turkestan Mountains are extremely eroded by water. The Morguzar Mountains are lower, with the highest point being at 2620 m (Fig. 6.32). The Turkestan Mountains surround the Sangzor sink, where the Sangzor River starts its flow. This sink is filled with loess and loess-like rocks, which the Sangzor River transfers to the foothills and plains. The Sangzor River created a narrow pass between the Nurota and Turkestan Mountains, which is called the Gates of Tamerlang (Fig. 6.33). The Nurota Mountains (Fig. 6.34) stretch to the northwest, and the Morguzar Mountains stretch to the southeast.

The plain and foothills are the dominant relief forms. They replace the former bed of the Syr Darya river, which moved to the northeast. The former bed is filled with loess-like

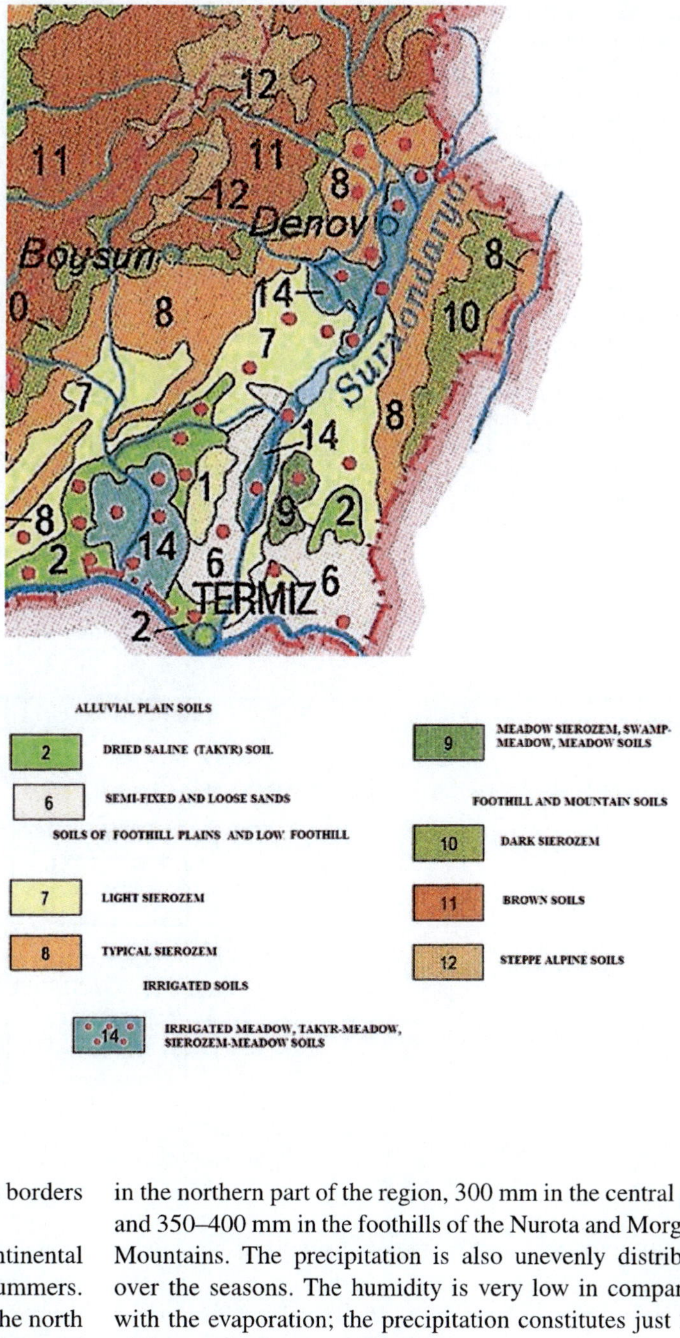

Fig. 6.28 Soils of the Surkhandarya subregion. (Source: Atlas 2017, courtesy of the State Committee for Land Resources, Geodesy, Cartography and State Cadastre of the Republic of Uzbekistan)

sand and clay, with clay in the sinks. A sandy desert borders Kyzylkum in the west.

The topography and location determine the continental subtropical climate, with cold winters and dry, hot summers. Cold and wet air masses easily enter the plain from the north and the west. The microclimate depends on the existence of water, proximity to the mountains, and the orientation of the plains. The annual average temperature ranges from 12.5 °C in the northern part to 15.1 °C in the southern part. This means that the difference in the summer temperature between the north and the south is about 5 °C, with a difference of 15 °C between the temperatures on the plains and those in the mountains. At the same time, the Turkestan Mountains serve as a natural barrier and prevent movement of cold air masses further to the south. This leads to more contrasts in the weather. The annual precipitation rates are only 200 mm in the northern part of the region, 300 mm in the central part, and 350–400 mm in the foothills of the Nurota and Morguzar Mountains. The precipitation is also unevenly distributed over the seasons. The humidity is very low in comparison with the evaporation; the precipitation constitutes just 20% of the required amount (Fig. 6.35).

The scarcity of water resources is exacerbated by the fact that there are few water source formations within this subregion. The Syr Darya river supplies water only to selected lands, since it is the transit and transboundary river. The banks of the Syr Darya river are formed of loess. Its width is about 15 km in the period of maximum flow in May–July. The river freezes in winter. A dense network of artificial canals divides the subregion (Fig. 6.36). Water from the Sangzor and Zaminsuv Rivers is intensively used for irrigation. Several artificial lakes

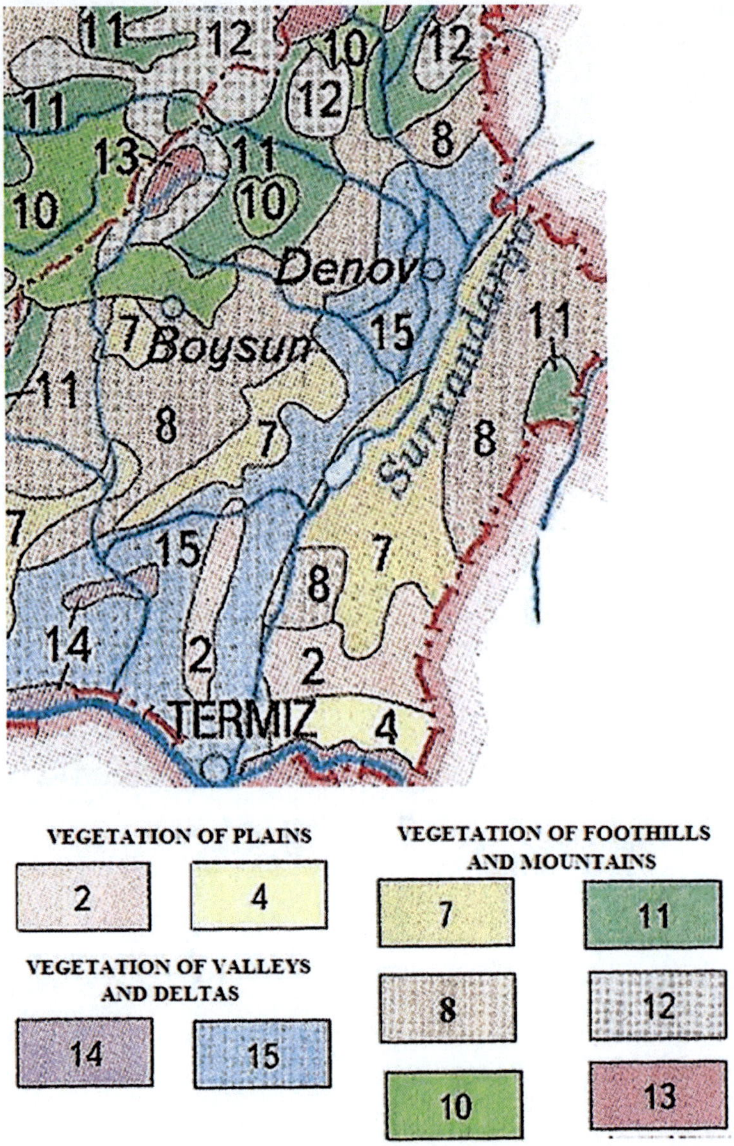

Fig. 6.29 Vegetation of the Surkhandarya subregion. 2 indicates the desert wormwood and the biyurgun (saltwater barnacle); 4 indicates sparse saline desert vegetation; 7 indicates the ephemeral semi-desert vegetation; 8 indicates the semi-desert foothill and lowmountain grass and wormwood; 10 indicates the mountain broad-leaf forests, sparse trees, shrubs; 11 indicates the sparse arxha trees on the background of dry steppe herbs; 12 indicates the high mountains steppe; 13 indicates subalpine meadow steppes and alpine meadows; 14 indicates tugai vegetation; 15 indicates cultural vegetation. (Source: Atlas 2017, courtesy of the State Committee for Land Resources, Geodesy, Cartography and State Cadastre of the Republic of Uzbekistan)

(including Tuzkon and Aydarkul) have been created in depressions, and their water is of high salinity, with a magnesium sulfate level of 10 g/l. The Jizzakh reservoir helps irrigate 5000 ha of virgin lands.

Groundwater is formed by runoff and water from rivers and canals. The flat surface of the main part of the subregion is the reason why the groundwater here flows slowly and rises up. This causes high salinity of the soil. Khasanov and Gulomov (2006) note that the groundwater rises up annually by 1–2 m. The rising of the groundwater increased after the Chardara reservoir was constructed and the Arnasay sink was filled with water. Today, the depth of the groundwater is 1–10 m, and the mineralization level is high.

On the plains, the soil is of the desert and semidesert types; in the foothills and mountains, it is of the gray and brown types (Fig. 6.37). The vegetation is also very different, ranging from psammophyte and ephemeral species on the plains to forests and subalpine species in the mountains (Fig. 6.38). Irrigated lands have replaced the natural landscape in most of this subregion (Fig. 6.39).

The Chirchiq–Ahangaran Subregion

This subregion is separated from Kazakhstan by the Karjantau and Ugam Mountains, and by the Kalas Valley. The border with Kyrgyzstan passes through the Talas Valley and the Piskom and Chatkal Mountains. The Kurama Mountains separate the Chirchiq–Ahangaran subregion and the Fergana Valley. All of these mountains are western branches of the Tien Shan. Their glaciers are the sources of the Chirchiq and Ahangaran Rivers.

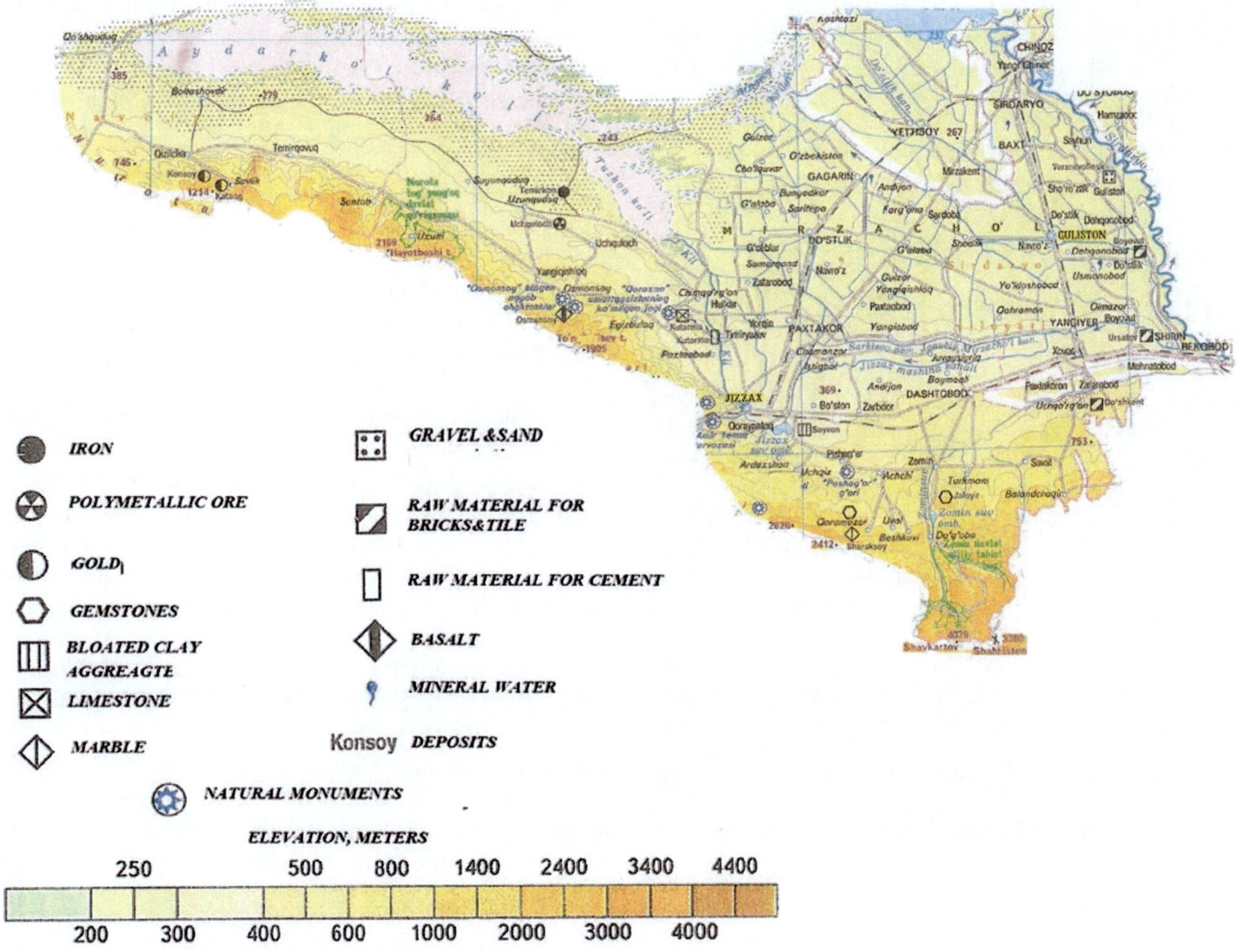

Fig. 6.30 The Mirzachul subregion. (Source: Atlas 2017, courtesy of the State Committee for Land Resources, Geodesy, Cartography and State Cadastre of the Republic of Uzbekistan)

Their formation began in the Caledonian and Hercynian Epochs. A variety of rocks from different epochs have been identified in the mountains. Rocks from the Carboniferous Period (limestone of greater thickness containing a thin layer and lenses of bauxite, dolomite, marl, silicon, and slate in a mixture with other rocks) have been explored in the Chatkal and Kurama Mountains. Paleozoic, Mesozoic, and Cenozoic rocks are represented in the forms of granite, Paleozoic limestone, sandstone, and shale in the mountains, and in the form of Paleocene, Neogene, and Anthropogenic gravel, sand, and clay in the foothills and valleys. The valleys of the Chirchiq and Ahangaran Rivers were covered by the water of the Turkestan Ocean during the Mesozoic Era. The alpine orogeny was affected by strong vertical motion of the surface and the closing of the ocean. Today, this tectonic activity is continuing, making this subregion seismically active. According to estimates (Khasanov and Gulomov 2006), the land under the city of Tashkent is still rising by 1–2 mm/year.

The surface is a composition of high mountains and valleys in the northern–eastern part of the subregion, with foothills in its southern–western part (Fig. 6.40).

The Piskom Mountains (Fig. 6.41) are the highest in this subregion, with Adelung Peak reaching 4301 m above sea level.

The Chatkal Mountain range (the Western Tian Shan) is located in the territory of the Ugam-Chatkal National Park (Fig. 6.42).

The Chatkal Mountains are also high, with the highest point being Bobotag (3555 m) (Fig. 6.43).

Fig. 6.31 The Koyatsh Ridge, Nurota Mountains, Mirzachul subregion, Jizzakh region

The Koksuv Ridge separates Uzbekistan and Kyrgyzstan. The shape of these mountains tells of the volcanic activity that raised the surface (Fig. 6.44).

The Korjantau is lower; its highest point is Mingbulak, at an elevation of 2823 m (Fig. 6.45).

The valleys of the Chirchiq and Ahangaran Rivers are inclined to the southwest toward the Syr Darya river. The intermountain depressions, plateaus, and plains have been intensively changed by geological processes (erosion and denudation) and by human activity (Fig. 6.46).

Greater Chimgan is a laccolith composed of Paleozoic rocks (Fig. 6.47). It is also a resort (Figs. 6.48, 6.49, and 6.50).

The Chirchiq River transfers loess from the mountains to the piedmont plains and further to the Syr Darya river. The Ahangaran and Chirchiq Rivers have similar sources of water in the high mountains and similar water regimes.

The Tashkent foothill depression (Fig. 6.51) is located between the Chatkal, Kurama, Turkestan, and Nurota Mountains, neighboring the Kyzylkum Desert in the west. *Adirs* are sparsely located on the southern and eastern sides of this depression. The surface is a wave-shaped plane in combination with terraces of the Syr Darya, Chirchiq, and Ahangaran Rivers. The height of these terraces is 1–7 m; in some places, they are about 40 m higher than the river body. In general, this depression is mainly flat but slightly inclined to the northwest. In the south, this depression borders the Mirzachul oblique plain in the foothills of the Turkestan Mountains.

Fig. 6.32 The northern slopes of the Morguzar Mountains, Mirzachul subregion, Jizzakh region

Fig. 6.33 The Sangzor River created this narrow mountain pass in the Mirzachul subregion, Jizzakh region. (Image courtesy of Magfirat Muminova)

Fig. 6.34 The Nurota Mountains, Mirzachul subregion, Jizzakh region

The mountains are rich with various mineral resources (nonferrous metals, molybdenum, tungsten, gold, silver, and silicon). The raw material for aluminum (alunite) has been explored in the Chatkal, Kurama, and Korjantau Mountains. There are coal deposits in the Ahangaran Valley.

The climate is formed by cold north and wet west air masses. The average temperatures in July are 26–27 °C on the plains and 20–24 °C in the foothills, the maximum is +42 °C, +44 °C. The avergae temperatures in January are −0.7 °C on the plains and −5 °C in the foothills; the minimum temperatures are −29 °C, −32 °C respectively. The precipitation is unevenly distributed over the subregion, ranging from 400–500 mm on the plains to 800–900 mm in the mountains. More runoff occurs on mountain slopes oriented to the west and southwest slopes; less occurs in the foothills and on the plains. The winter season lasts for 65 days on the plains and 129 days in the mountains (Fig. 6.52).

The basin of the Chirchiq River is 13,240 km^2 in size, and its flow module is 21 $l/s/km^2$. It is one of a few big rivers that start inside Uzbekistan. The watershed is in the glaciers of the Chatkal Mountains, which contain a total of 222 glaciers in an area of 173 km^2). The water of this river is used for irrigation. Only 31% of its hydropower energy is used. A cascade of 22 hydroelectric power plants (HPPs) have been constructed on the Chirchiq River. The Chatkal River is the main tributary of the Chirchiq, contributing 55% of its water.

This subregion is self-sufficient in water resources; the Chirchiq River has an annual balance of 224 m^3/s, and that of the Ahangaran River is 43 m^3/s. The Charvak reservoir (with a capacity of 2.006 km^3) on the Chirchiq River and the Tuyabuguz reservoir (with a capacity of 0.25 km^3) on the Ahangaran River are used to collect water (Fig. 6.53).

Vertical belts are a peculiar feature of the landscape of both valleys, ranging from semidesert and dry steppe types in the foothills (at altitudes of 400–500–1200 m above sea level) to forests in the middle mountains (at altitudes of 1200–1500 m) and alpine and nival types in the high mountains (at altitudes of 2500–4000 m), where there are glaciers (Figs. 6.54 and 6.55).

The landscapes of all plains and foothills have been changed. Irrigated soil and crops at oases have replaced the natural soil and vegetation everywhere except in the high mountains. This subregion is one of the most developed ones in this country. The comfortable natural conditions make it attractive for different kinds of activities, including industry, agriculture, and services. The capital of Uzbekistan is located in this subregion. The Tashkent agglomeration includes almost all cities in the Tashkent region. At the same time, such intensive human activity impacts the environment.

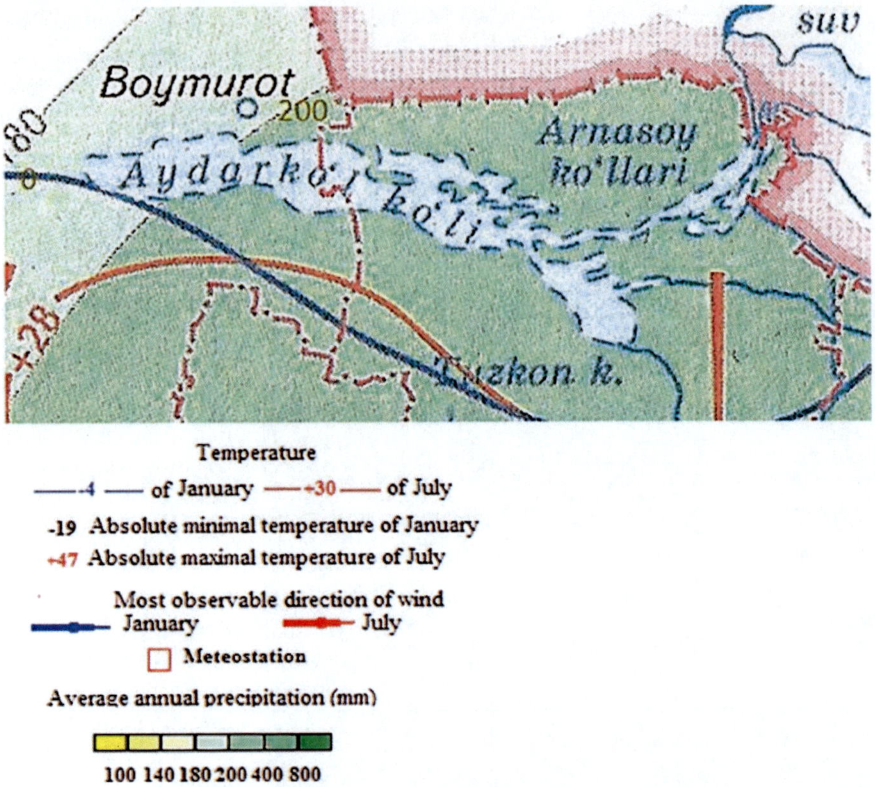

Fig. 6.35 Climate map. (Source: Atlas 2017, courtesy of the State Committee for Land Resources, Geodesy, Cartography and State Cadastre of the Republic of Uzbekistan)

Fig. 6.36 Irrigation system of artificially constructed concrete trays in the Mirzachul subregion, Syrdarya region

The Fergana Subregion

Intermountain depressions such as the Fergana Valley represent a group of topographic features of relief, consisting of plains in the foothills and inclined terraces in the valleys of the rivers. The majority of the population centers have been created in such depressions that have filled with deposits of silt and mud.

The Fergana Valley is considered the gem of Uzbekistan. All routes from China went through the cities of this valley. It has a very convenient geographic location (Fig. 6.56), surrounded by the Mo'g'iltog' Mountains to the west, the Kurama and Chatkal to the northwest, the Oto'ynoq to the east, and the Turkestan and Alay ranges to the south. The Dalvarzin and Mirzachul plains are connected to each other,

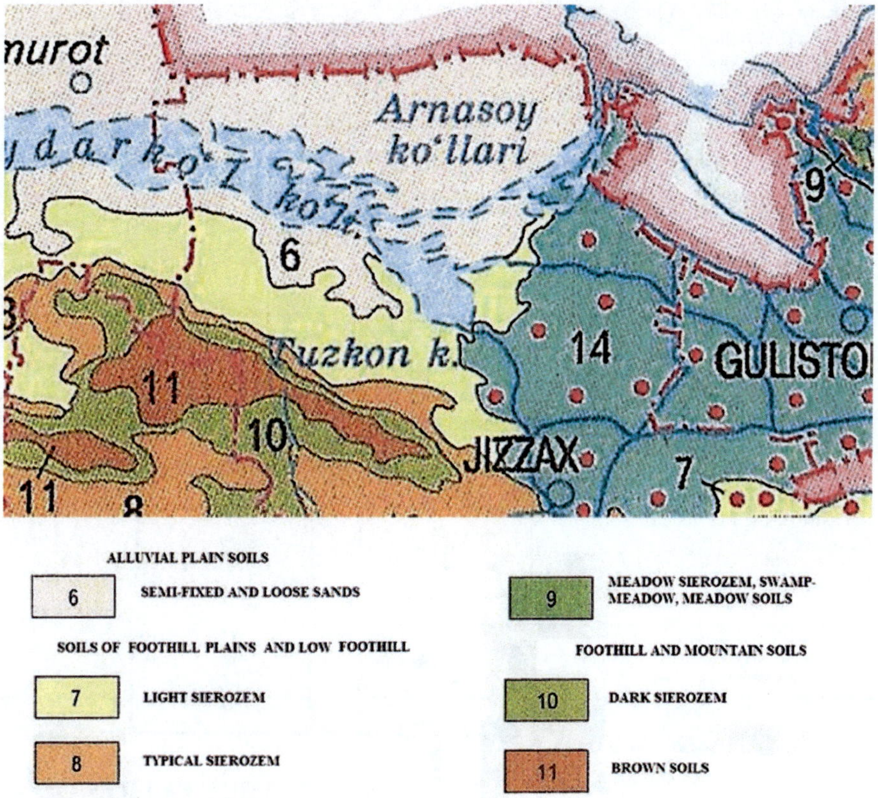

Fig. 6.37 Soils of the Mirzachul subregion. (Source: Atlas 2017, courtesy of the State Committee for Land Resources, Geodesy, Cartography and State Cadastre of the Republic of Uzbekistan)

and this place, 9–40 km wide, opens the western side of the Fergana Valley. The valley is a flat, homogeneous plain inclined from east to west, repeating the general slant of the country's surface. In its eastern part in the Kampirovat, the altitude is 812 m above sea level, and it is 320 m in the Ho'jand in the west. The altitude of the surrounding hills is between 600 and 1200 m above sea level.

The Syr Darya river divides the valley into two parts that are unequal in size. The southern part consists of wide foothills and the third terrace of the Syr Darya river. The northern part consists of a narrow plain in the foothills without any terrace. The valley is surrounded in the south by plains (*adirs*) formed by Neogene and Quaternary rocks (conglomerates, sandy rocks, and marl) and covered by loess and loess–sandy rocks. These plains are eroded in their upper parts. The depressions between the *adirs* and the mountains create wave-shaped topography.

The climate here is different from those of the neighboring Tashkent and Mirzachul valleys. Cold mountain air masses may freeze the center of the valley in winter. About 30% of the runoff occurs in the spring, which is very short with unstable weather. The summer is hot with the an average temperature in July of 26–27 °C and a maximum temperature of 40–42 °C. The average annual precipitation ranges from 100 mm in the western part to 230 mm in the eastern part, and 10–16% of the runoff occurs in summer, 36% in winter, and 48% in the spring and autumn (Fig. 6.57).

The Syr Darya is the transit river in the Fergana Valley, with annual flow of 568 m^3/s. Water from it and from the rivers that flow from the surrounding mountains is used for irrigation. A dense network of canals (the Great Fergana, Great Andijon, Northern Fergana, Great Namangan, and Southern Fergana, named after Ahunbabaev) and reservoirs (Qayroqqum, Uchqurgan, Andijon, Bozorqorgan, and Karkidon) have been constructed in the valley. The valley is the most populous place, with the highest density of population in Central Asia. Cotton cultivation started in this valley. Many generations of local people have been involved in irrigated agriculture. As a result of this human activity, the landscape has been changed everywhere within the subregion, except in the high mountains (Figs. 6.58, 6.59, 6.60, 6.61, and 6.62).

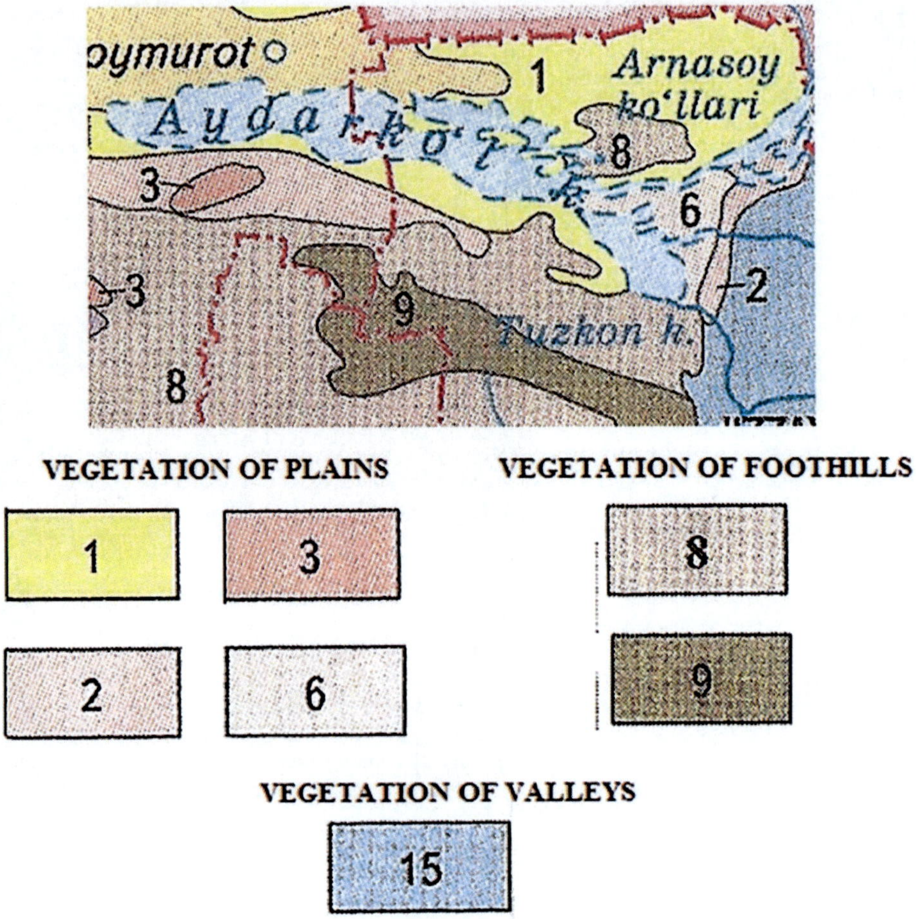

Fig. 6.38 Vegetation of the Mirzachul subregion. 1 indicates sparse desert psamophyte shrub vegetation; 2 indicates the desert wormwood and the biyurgun (saltwater barnacle); 3 indicates saline black saxaul desert; 6 indicates the salt-desert black saxaul; 8 indicates the semi-desert foothill and low-mountain grassand wormwood; 9 indicates the mountain dry grass steppes; 15 indicates cultural vegetation. (Source: Atlas 2017, courtesy of the State Committee for Land Resources, Geodesy, Cartography and State Cadastre of the Republic of Uzbekistan)

Fig. 6.39 Agricultural crops have replaced natural vegetation in the Mirzachul subregion

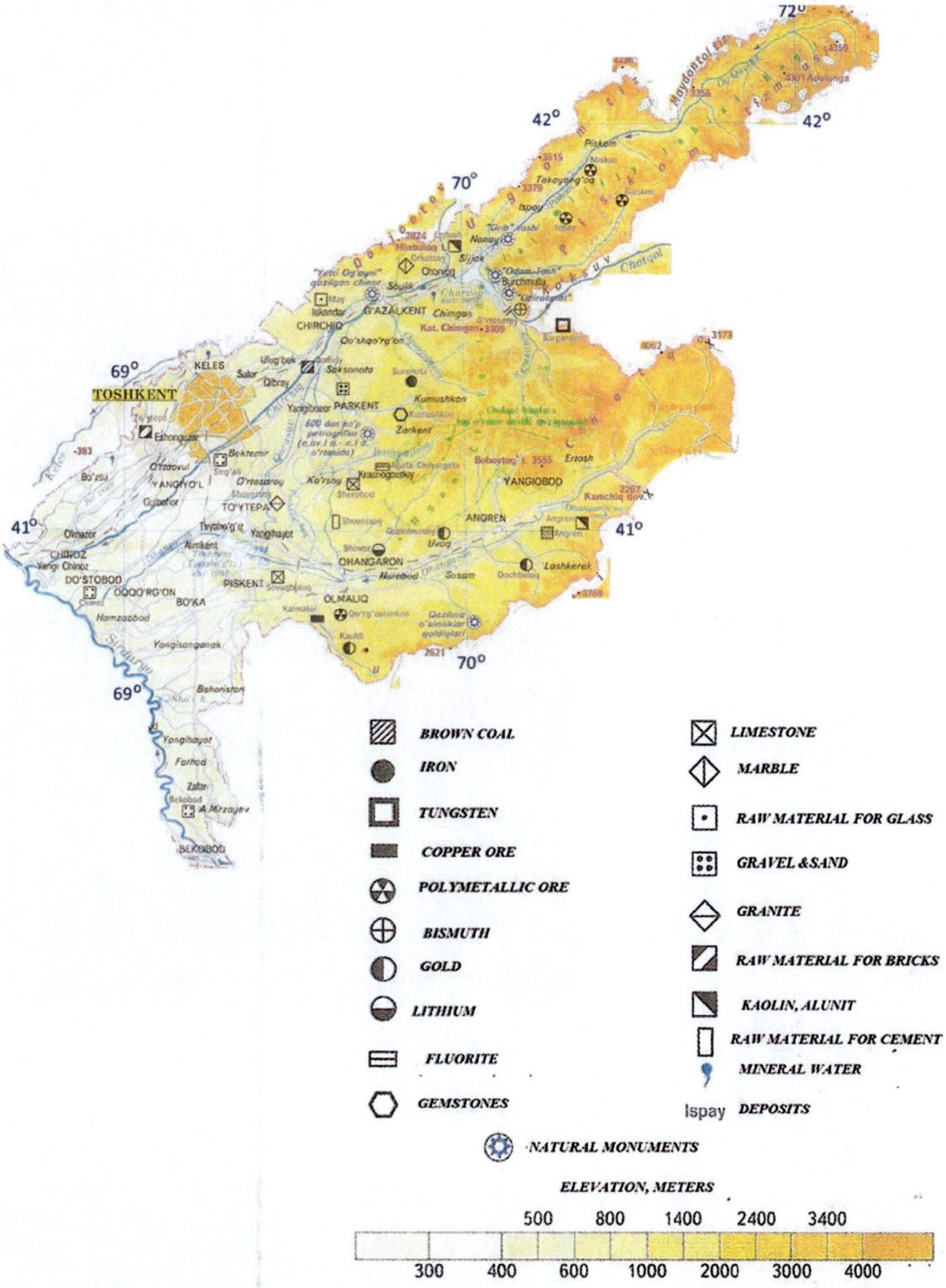

Fig. 6.40 The Chirchiq–Ahangaran subregion. (Source: Atlas 2017, courtesy of the State Committee for Land Resources, Geodesy, Cartography and State Cadastre of the Republic of Uzbekistan)

Fig. 6.41 The Piskom Ridge, Chirchiq–Ahangaran subregion, Tashkent region

Fig. 6.42 The Chatkal Ridge, Western Tien Shan, Tashkent region

Fig. 6.43 Bobotag, Chatkal Ridge, Western Tien Shan, Tashkent region

Fig. 6.44 The Koksuv Ridge, Chirchiq–Ahangaran subregion, Tashkent region

Fig. 6.45 The Korjantau Mountains (which separate Uzbekistan and Kazakhstan), Chirchiq–Ahangaran subregion, Tashkent region

The Ustyurt Subregion

The nature and location of this subregion between the Caspian and Aral Seas makes it different from the others. According to Babushkin and Kogay (1975), this subregion is included in the Central Kazakhstan geographic province. The total size of this plateau is 160,000 km^2. Only one-quarter (40,000 km^2) of the Ustyurt Plateau is included in Uzbekistan. The height of this flat arid denudation plateau is 50–80 m; in some places, it reaches 200–300 m. The highest point of the plateau is at Karabavur, 292 m high (Fig. 6.63).

The waveform surface is a combination of the flat surface and various depressions (Asakaovdan, Barsa-Kelmes). The relief forms repeat the tectonic features; the hills are within the anticline, and there are sinks in the depressions. These features make the Ustyurt Plateau an attractive sightseeing place in the western surroundings of the Aral Sea.

This plateau consists of Sarmatian limestone, gypsum, Neogene marl, and Paleocene clay. These rocks are covered by pebbles, sand, and crushed stone 1–1.5 m thick. Horizontally oriented dolomite, gypsum, and Neogene marl compose this plateau, and dissolution of these soluble rocks has resulted in formation of karst with underground caves. The longest of these caves is 190 m long and 120 m deep, with a lake at the bottom (Potapova 1971).

The borders of this plateau are steep, and they are known as "chinks" (Fig. 6.64). Chinks are denoting steep, hard-to-reach ledges (cliffs) up to 350 m high. Their formation is the result of a combination of tectonic, abrasion, and erosion processes.

The climate is extremely continental, with a low temperature in winter (averaging −6 °C to −8.6 °C), a long period of cold days (4 months), a very hot summer (with an average temperature of 26.8–28.5 °C) and little precipitation (annual average 100–122 mm) (Fig. 6.65). The dry climate and absence of any rivers result in very poor soil and vegetation (Fig. 6.67). It is only in spring that ephemeral species cover the plain. Many animals are under threat of extinction, and they are listed in the *Red Data Book of Uzbekistan*.

The current development of Ustyurt is related to the gas and oil industry.

Fig. 6.46 Erosion in the Chatkal–Ugam intermountain depression

Fig. 6.47 Greater Chimgan (3309 m high), Chatkal Mountain range, Chirchiq–Ahangaran subregion

Fig. 6.48 The Greater Chimgan resort is famous for its opportunities for skiing

Fig. 6.49 Kumishkan, Chatkal Ridge, Chirchiq–Ahangaran subregion. (Image courtesy of Magfirat Muminova)

Fig. 6.50 The landscape of the intermountain depression, Chatkal Ridge, Chirchiq–Ahangaran subregion

Fig. 6.51 The Tashkent foothill intermountain depression, Tashkent region

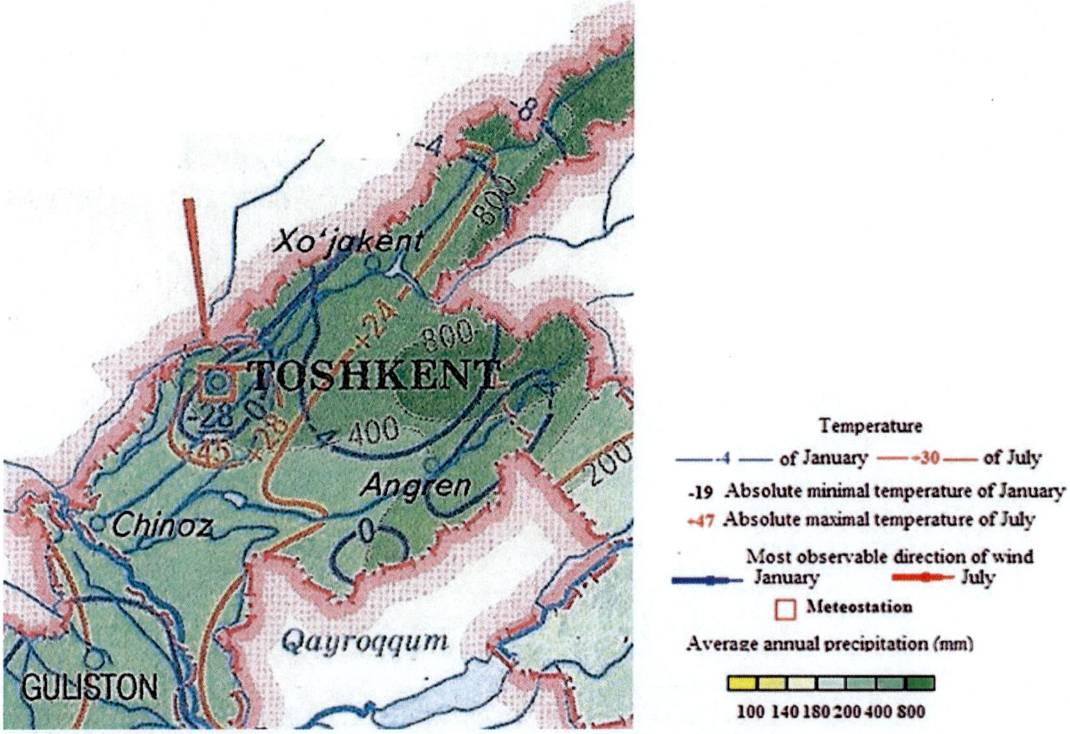

Fig. 6.52 Climate of the Chirchiq–Ohangoron subregion. (Source: Atlas 2017, courtesy of the State Committee for Land Resources, Geodesy, Cartography and State Cadastre of the Republic of Uzbekistan)

Fig. 6.53 The Charvak water reservoir

Fig. 6.54 Soils of the Chirchiq–Ahangaran subzone. (Source: Atlas 2017, courtesy of the State Committee for Land Resources, Geodesy, Cartography and State Cadastre of the Republic of Uzbekistan)

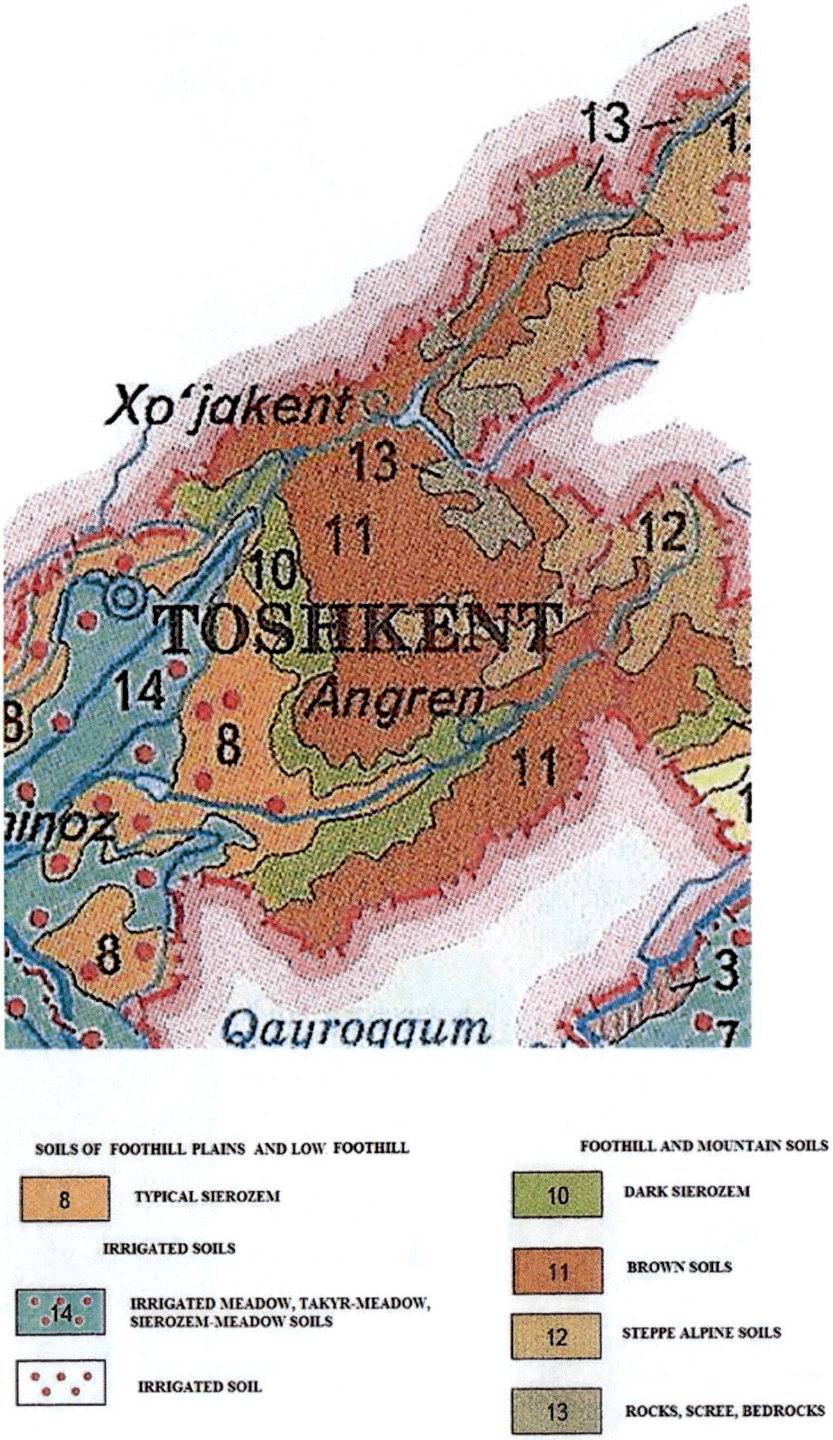

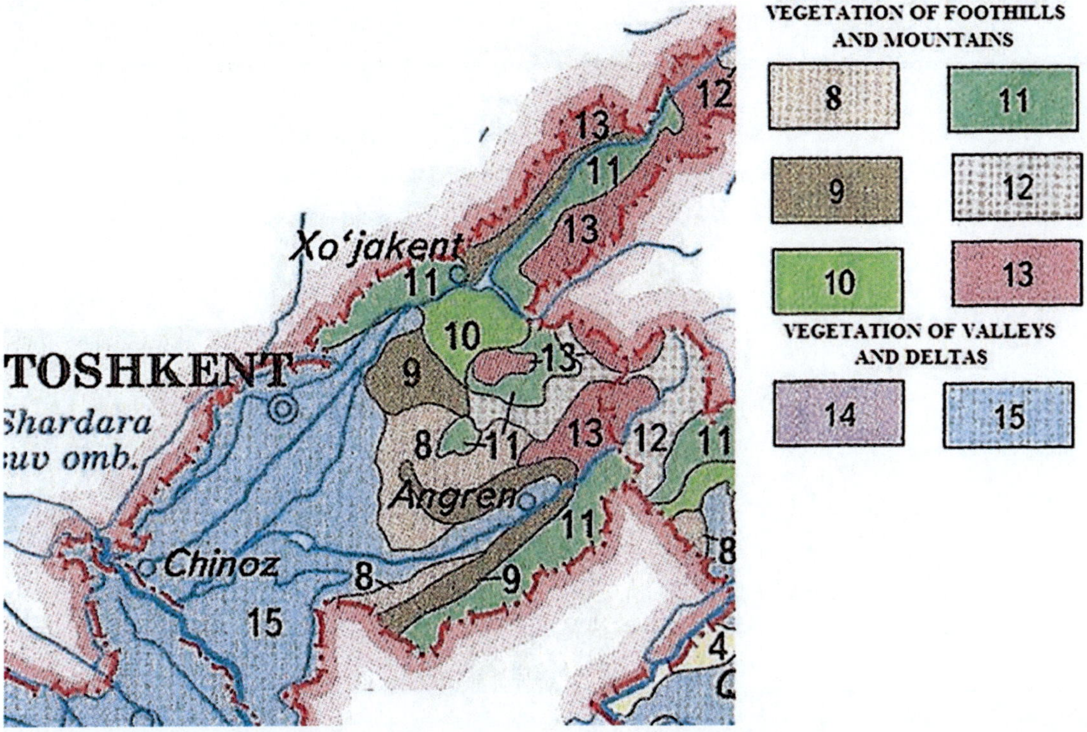

Fig. 6.55 Vegetation of the Chirchiq–Ahangaran subzone. 4 indicates sparse saline desert vegetation; 8 indicates the semi-desert foothill and low-mountain grass and wormwood; 9 indicates the mountain dry grass steppes; 10 indicates the mountain broad-leaf forests; 11 indicates the sparse archa trees on the background of dry steppe herbs; 12 indicates the high mountain steppe; 13 indicates subalpine meadow steppe and alpine meadows; 15 indicates cultural vegetation. (Source: Atlas 2017, courtesy of the State Committee for Land Resources, Geodesy, Cartography and State Cadastre of the Republic of Uzbekistan)

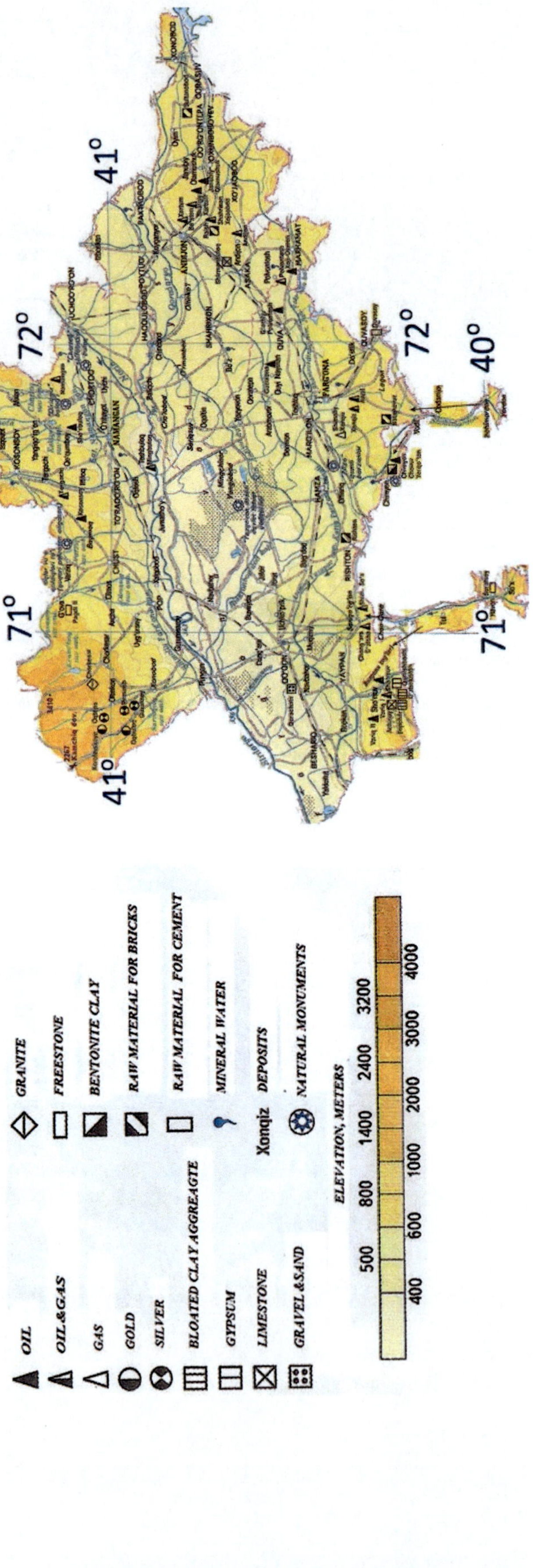

Fig. 6.56 The Fergana Valley. (Source: Atlas 2017, courtesy of the State Committee for Land Resources, Geodesy, Cartography and State Cadastre of the Republic of Uzbekistan)

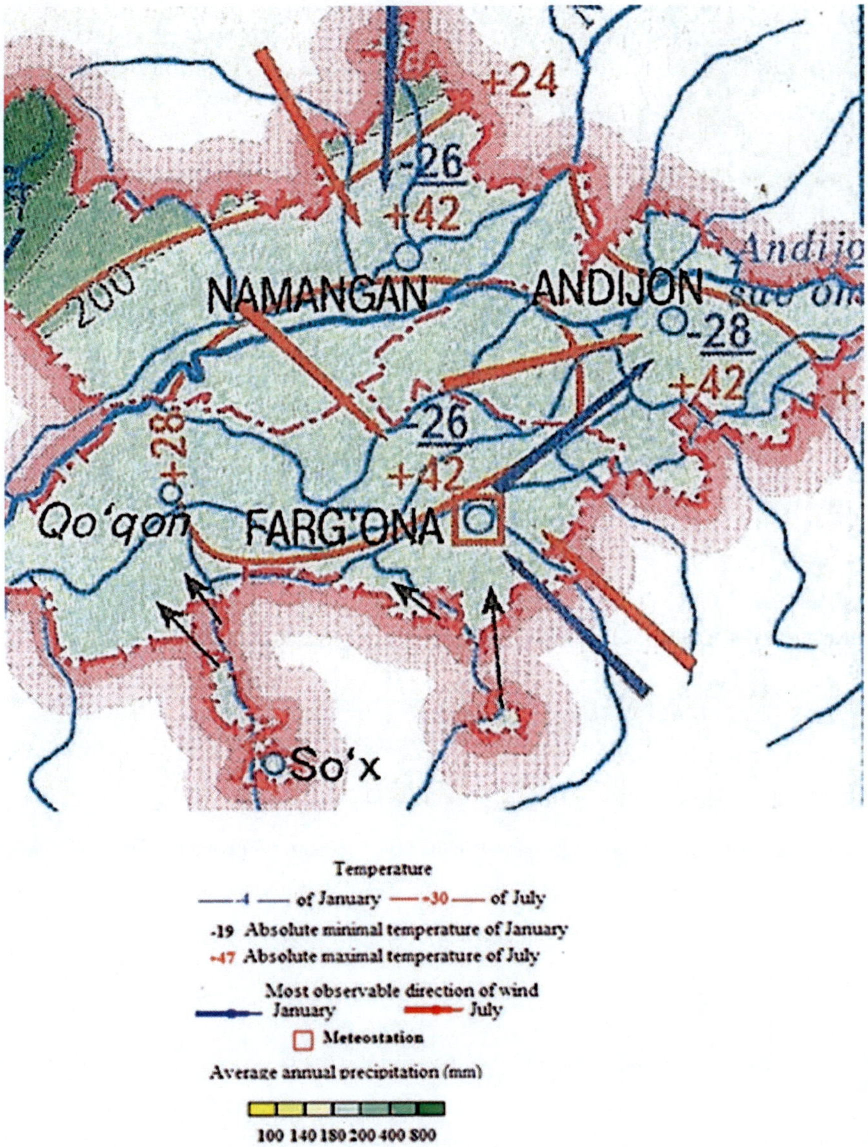

Fig. 6.57 Climate of the Fergana Valley. (Source: Atlas 2017, courtesy of the State Committee for Land Resources, Geodesy, Cartography and State Cadastre of the Republic of Uzbekistan)

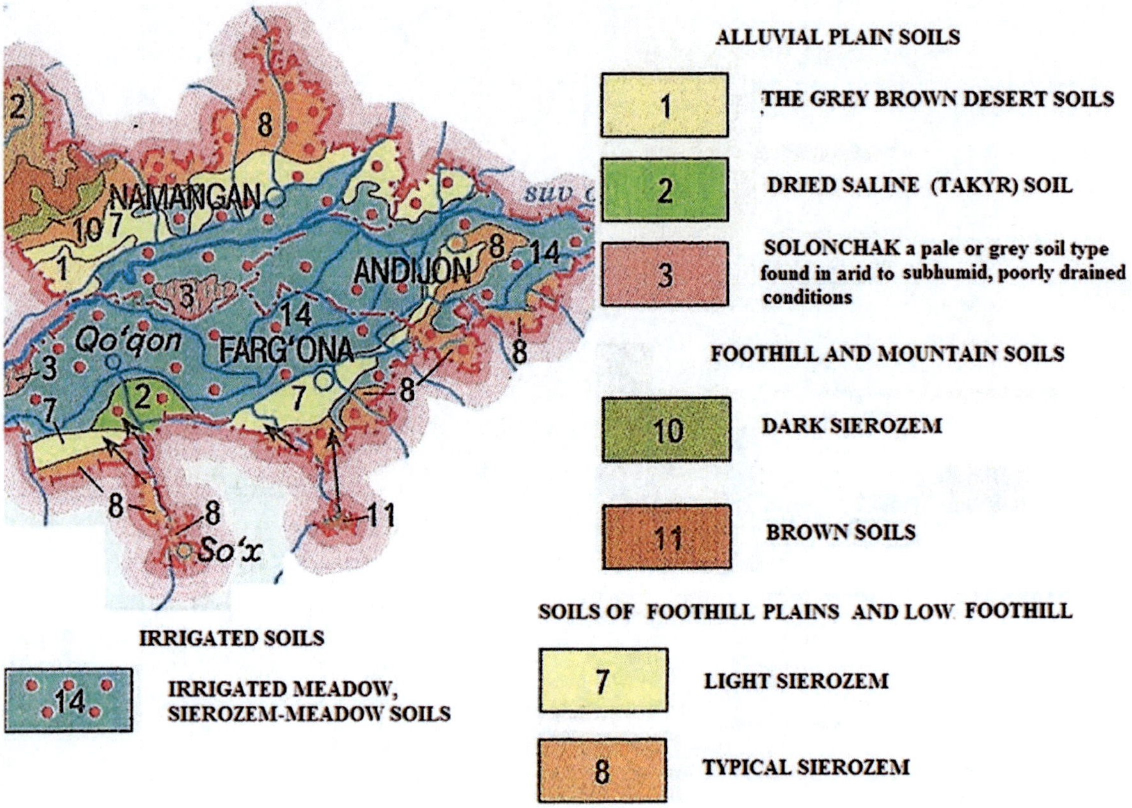

Fig. 6.58 Soils of the Fergana Valley. (Source: Atlas 2017, courtesy of the State Committee for Land Resources, Geodesy, Cartography and State Cadastre of the Republic of Uzbekistan)

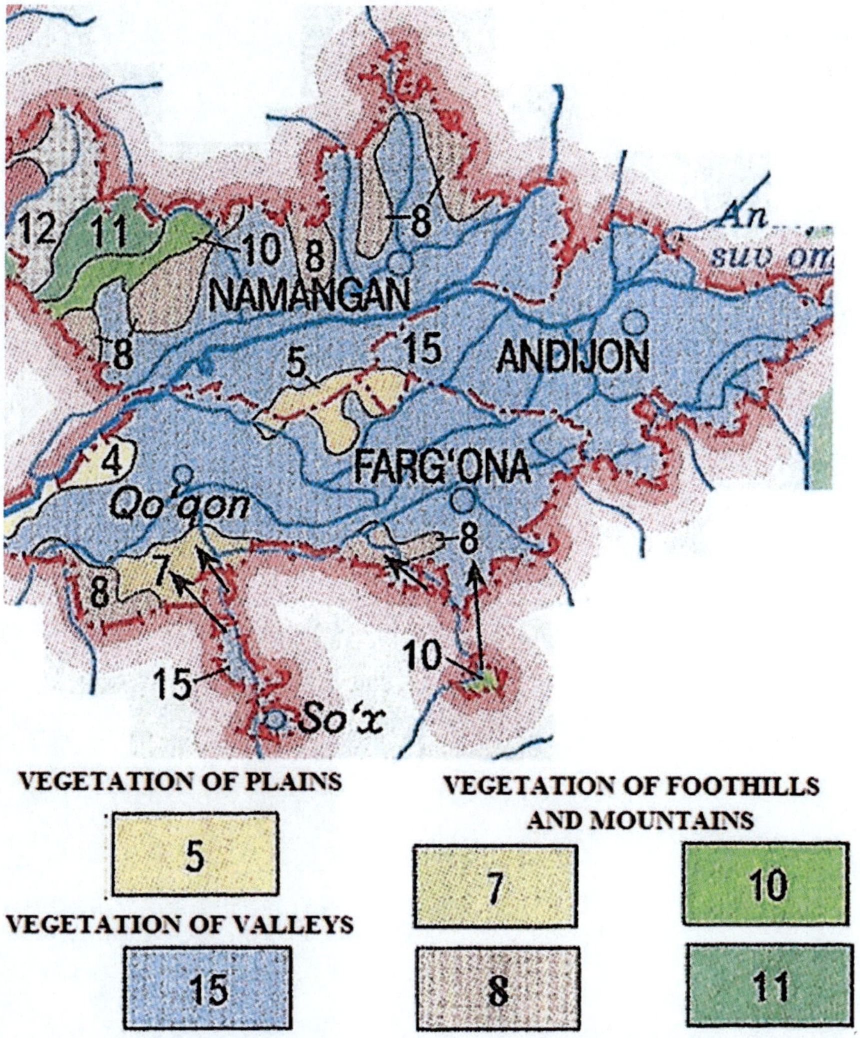

Fig. 6.59 Vegetation of the Fergana Valley. 5 indicates the white saxaul, sagebrush, white-saxaul, black-saxaul-white saxaul deserts; 7 indicates the ephemeral semi-desert vegetation; 8 indicates the semi-desert foothill and low-mountain grass and wormwood; 10 indicates the mountain broad-leaf forests, sparse trees, shrubs; 11 indicates the sparse archa trees on the background of dry steppe herbs;12 indicates the high mountain steppe; 15 indicates cultural vegetation. (Source: Atlas 2017, courtesy of the State Committee for Land Resources, Geodesy, Cartography and State Cadastre of the Republic of Uzbekistan)

Fig. 6.60 The modern landscape of the Fergana Valley

Fig. 6.61 Fergana Valley Cotton fields have replaced natural vegetation

Fig. 6.62 Fergana Valley Gardens have replaced the natural landscape

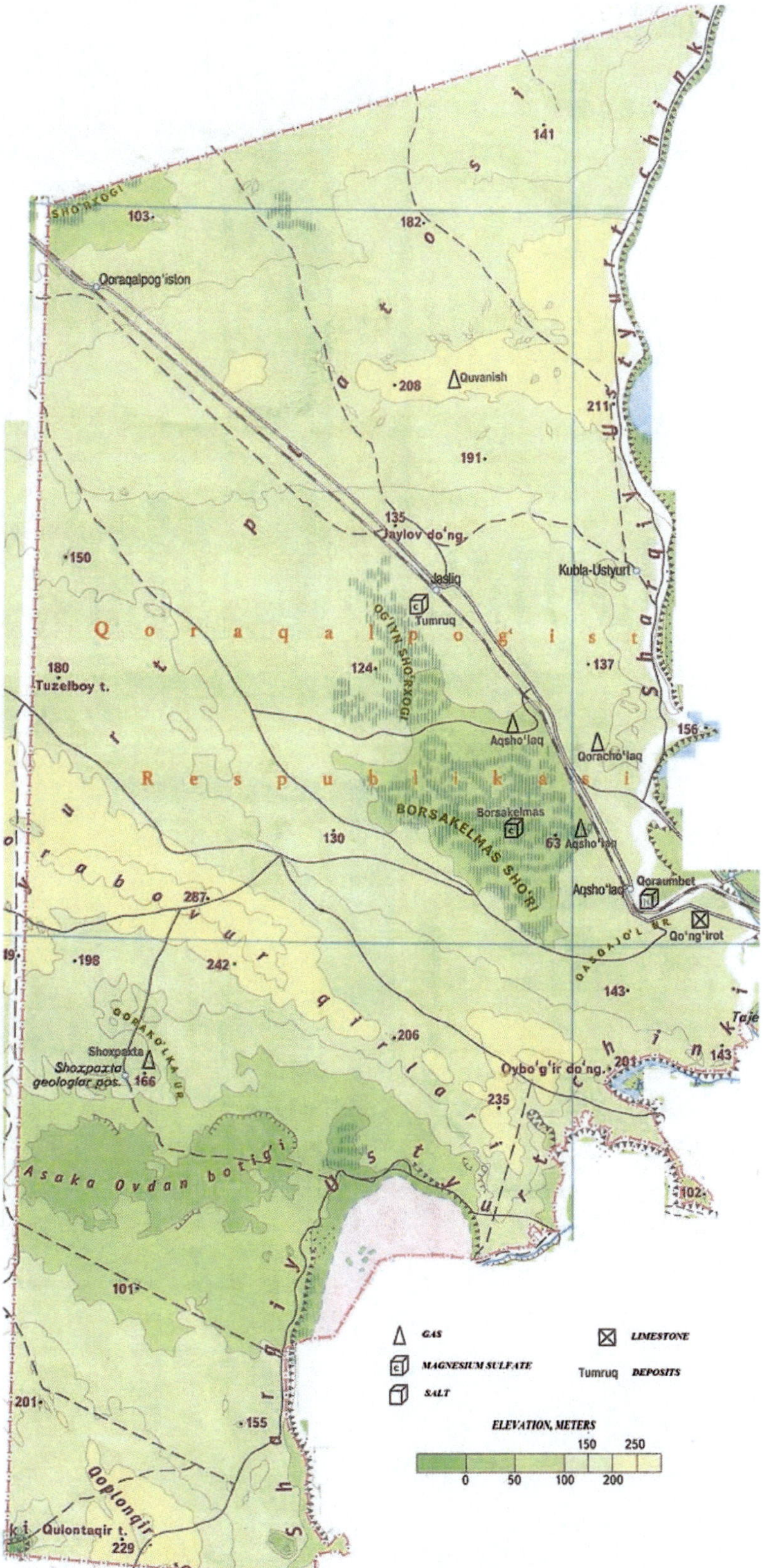

Fig. 6.63 The Ustyurt Plateau. (Source: Atlas 2017, courtesy of the State Committee for Land Resources, Geodesy, Cartography and State Cadastre of the Republic of Uzbekistan)

Fig. 6.64 Chinks in the western surroundings of the Aral Sea. (Image courtesy of RIAN, Lev Ustinov/Sputnik)

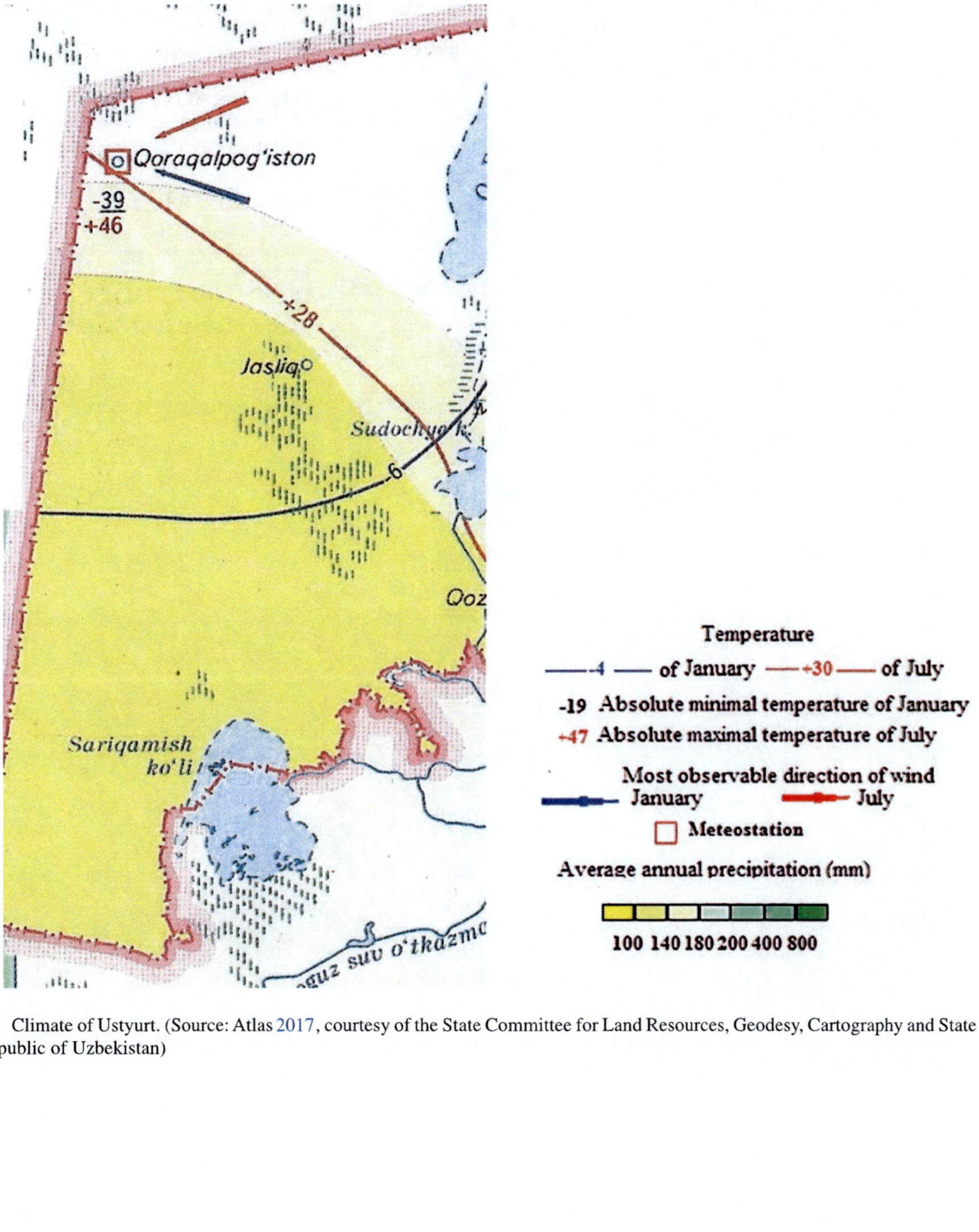

Fig. 6.65 Climate of Ustyurt. (Source: Atlas 2017, courtesy of the State Committee for Land Resources, Geodesy, Cartography and State Cadastre of the Republic of Uzbekistan)

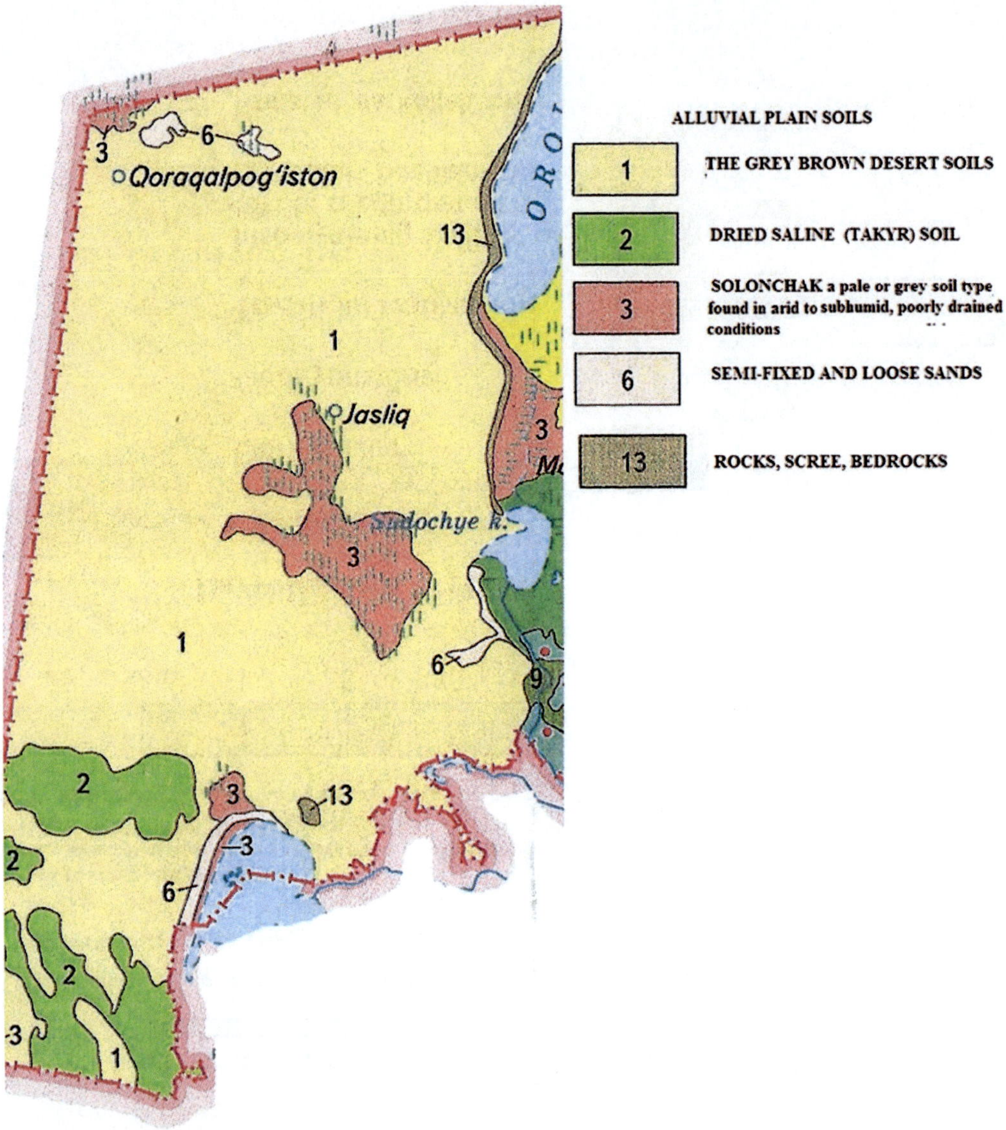

Fig. 6.66 Soils of Ustyurt. (Source: Atlas 2017, courtesy of the State Committee for Land Resources, Geodesy, Cartography and State Cadastre of the Republic of Uzbekistan)

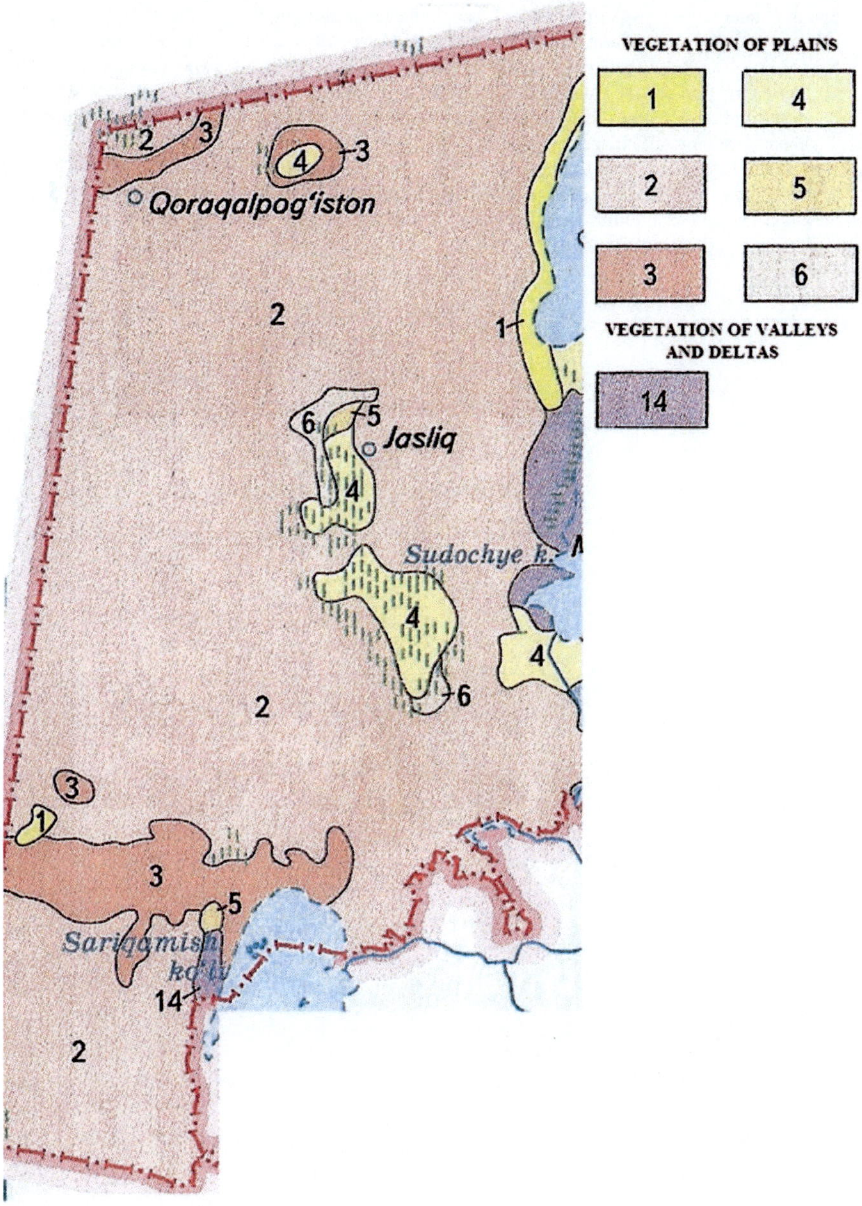

Fig. 6.67 Vegetation of Ustyurt. 1 indicates sparse desert psamophyte shrub vegetation; 2 indicates the desert wormwood and the biyurgun (saltwater barnacle); 3 indicates saline black saxaul desert; 4 indicates sparse saline desert vegetation; 5 indicates the white saxaul, sagebrush, white-saxaul, black-saxaul-white saxaul deserts; 6 indicates the salt-desert black saxaul; 14 indicates tugai vegetation. (Source: Atlas 2017, courtesy of the State Committee for Land Resources, Geodesy, Cartography and State Cadastre of the Republic of Uzbekistan)

References

Allan NJR, Bruk SI, Rantsman YY (2011) Tien Shan. Mountains in Asia. Accessed on August 17, 2018. https://www.britannica.com/place/Tien-Shan

Atlas (2017) Central Asia and Uzbekistan (in Uzbek: Atlas. O'rtaOsiyo va O'zbekiston, 2017 Toshkent). Tashkent: State Cadastre of the Republic of Uzbekistan

Babushkin LN, Kogay NA (1975) The natural territorial complexes of south-eastern Central Asia (in Russian: Л. Н. Бабушкин, Н. А. Когай Природные территориальные комплексы Юго-Запада Средней Азии Ташкент: Фан, 1975). Tashkent: Fan

Khasanov I, Gulomov PN (2006) The physical geography of Uzbekistan (in Uzbek), Tashkent

Le Callonnec L, Person A, Renard M, Letolle R et al (2005) Preliminary data on chemical changes in the Aral Sea during low-level periods from the last 9,000 years. Comptes Rendus Geosci 12:1035–1044

Potapova GM (1971) The caves of southern Mangyshlak and southern-western Ustyurt (in Russian). In J Caves 14:12–13. Perm State University Perm

Seltman R et al (2011) Hercynian post-collisional magmatism in the context of the Paleozoic magmatic evolution of the Tien Shan orogenic belt. J Asian Earth Sci 42(5):821–838. https://doi.org/10.1016/j.jseaes.2010.08.016

Skvortsov Yu A (1959) To the question of studying the geomorphology of Quaternary deposits (in Russian: Скворцов Ю.А. К вопросу об изучении геоморфологии четвертичных отложений // Материалы по производительным силам Узбекистана. – Вып. 10. Ташкент: Изд-во АН УзССР, 1959. – С. 53–80). In: Materials of the Productive Forces of Uzbekistan 10: 53–80. Tashkent: Tashkent Academy of Sciences

Synnott M (2017) Is this underground Everest? National Geographic. https://www.nationalgeographic.com/magazine/2017/03/dark-star-deepest-cave-climbing-uzbekistan/

Tolstov SP, Kes AS (1956) The history of primitive settlements on the channels of the ancient deltas of the Amu Darya and Syr Darya (in Russian: Толстов С.П., Кесь А.С. История первобытных поселений на протоках древних дельт Амударьи и Сырдарьи // Вопросы географии. Сб. докл. 18 Международного географического конгресса. М.-Л.: Изд-во АН СССР, 1956.-С. 327–336). Moscow: Academy of Sciences of the USSR

Zonn IS, Glantz M, Kosarev AN, Kostianoy AG (eds) (2009) The Aral Sea Encyclopedia. Springer. https://doi.org/10.1007/978-3-540-85008-5

Index

L. Gulyamova, *The Geography of Uzbekistan*, World Regional Geography Book Series,
https://doi.org/10.1007/978-3-031-07873-6